Advanced
Microprocessors

Advanced Microprocessors

Daniel Tabak

*Electrical and Computer
Engineering Department
George Mason University
Fairfax, Virginia*

Second Edition

McGraw-Hill, Inc.

New York   San Francisco   Washington, D.C.   Auckland   Bogotá
Caracas   Lisbon   London   Madrid   Mexico City   Milan
Montreal   New Delhi   San Juan   Singapore
Sydney   Tokyo   Toronto

Library of Congress Cataloging-in-Publication Data

Tabak, Daniel,
    Advanced microprocessors / Daniel Tabak.—2nd ed.
        p.   cm.
    Includes bibliographical references and index.
    ISBN 0-07-062843-2
    1. Microprocessors.   I. Title.
QA76.5.T28   1995
004.16—dc20   94-27979
                                                    CIP

1 2 3 4 5 6 7 8 9 0   DOC/DOC   9 0 9 8 7 6 5 4

ISBN 0-07-062843-2

*The sponsoring editor for this book was Stephen S. Chapman, the
editing supervisor was Joseph Bertuna, and the production supervisor
was Suzanne W. Babeuf. It was set in Century Schoolbook by McGraw-
Hill's Professional Book Group composition unit.*

Printed and bound by R. R. Donnelley & Sons Company.

To the countries of my heritage and education
Poland, Russia, Israel, United States
And to my universities
Technion—Israel Institute of Technology, Haifa, Israel;
University of Illinois, Urbana, Illinois

Trademarks

Contents

Part 2   The Intel x86 Family

Part 3   The Motorola M68000 Family

Part 4   Advanced RISC Microprocessors

Part 5   System Development and Comparison

Preface to the Second Edition

The development and appearance of new advanced microprocessors advances at a tremendous pace. Only 2 years after the appearance of the first edition of this book in 1991, there is a whole generation of new, far more advanced and higher-performing systems. The older systems have new generations of more powerful microprocessors. There are also some new "kids on the block," coming from manufacturers that can be considered as "old timers" in the computer industry. A new edition, describing the new systems, is certainly called for.

The general structure of the first edition has been maintained: the first part dealing with general principles of computer architecture involved with advanced microprocessors, followed by parts dealing with major microprocessor complex instruction set computer (CISC) families, a separate part dealing with reduced instruction set computer (RISC) microprocessors, concluding with the last part which deals with system development and comparison.

The detailed structure of each part has been considerably changed however. The main changes are discussed in the following paragraphs.

1. Architecture description. In the first edition the discussion of each system was preceded by a detailed description of the particular major system implementation (i486, MC68040, for instance), followed by the detailed presentation of the system architecture. The system architecture is generally designed to survive (more or less intact) several generations of specific microprocessor implementations. For this reason, in the second edition, the detailed presentation of the system architecture, without relation to its specific implementations, is placed in the beginning, followed by a short description of its various implementations. As was the policy in the first edition, the top-level implementation is presented in greater detail. This is the feature which makes this book different from many others in this area. Many

other texts dedicate most of the space to the lower-level models (8086, MC68000) and relatively little space to the top-level models.

2. General principles.   This part was considerably expanded. The first edition was used as a text in a George Mason University (GMU) graduate course entitled ECE 516 Advanced Microprocessor. While students who graduated from GMU had a solid undergraduate preparation in basic computer design (partially using the Hennessy and Patterson text), most of the graduate students enrolled in ECE 516 have graduated elsewhere and did not have such a basis. Some didn't even know how the cache is managed or the details of a pipelined operation. For this reason it was decided to broaden the scope of Part 1 to include some basic, brief, preliminary reading for students and readers who do not possess the necessary background to fully appreciate the technical features of modern microprocessors. There were three chapters in Part 1 in the first edition. The first two chapters have been completely rewritten to reflect the new changes in the modern microprocessor environment. Chap. 3 on microprocessor architecture was only slightly modified. Chap. 4 on memory hierarchy and Chap. 5 on pipelining are completely new. Some new practices, which became common in advanced microprocessors, such as dual and secondary caches, instruction level parallelism (ILP), and others, are covered in these chapters. Chap. 6 on RISC (formerly Chap. 13 in the first edition) was brought forward into Part 1, as part of the general principles discussion. It should be noted that there is no intention to compete with regular texts on computer architecture and design. The purpose of Part 1 is to give the unexperienced reader the basics of architectural features implemented in modern advanced microprocessors and to direct them for more detailed study of the topics covered in Part 1, in other texts and papers, cited in the list of references at the end of the book.

3. Choice of material.   Because of reduced worldwide applications and reluctance to exceed a certain page limit, some systems described in the first edition were dropped from the second edition. On the other hand, some notable new systems with a great future applications potential (Intel Pentium, Digital Equipment Corporation (DEC) Alpha AXP, PowerPC, SuperSPARC, Microprocessor without Interlocked Pipeline Stages (MIPS) R4400, MC88110, Hewlett-Packard (HP) PA-RISC, and others) have been added.

As was the case in the first edition, this book is intended primarily as a reference text on advanced, top-level microprocessors, rather than as a classroom textbook. However, it may also be used in a course on advanced microprocessors, as it was done by the author for the past 3 years. For this reason, a set of problems was added at the end of Parts 1, 2, and 3. The book is intended for students and professionals in electrical and computer engineering, computer science, and other engineering disciplines. It requires as a background some basic knowledge and understanding of logic circuitry and basic computer organization. Readers well versed in computer architecture and organization may skip Part 1 and start directly with Part 2.

As was the case with the first edition, the author obtained valuable comments from outstanding professionals directly involved with the development of the advanced microprocessors discussed in the text. The author would like to express his thanks for valuable comments and information to Prof. Alan J. Smith (University of California, Berkeley), Dr. Richard Sites (DEC), Charles Moore (International Business Machines [IBM]), Steve Krueger (Texas Instruments [(TI]), Dr. Arie Harsat (Intel—Israel), Roger Golliver (Intel), Keith Diefendorff (Motorola), Dennis Brzezinski (HP), and Martin Whittacker (HP). The initial part of the book ws prepared during the summer of 1993 while the author was a Visiting Professor at the Basser Department of Computer Science, University of Sydney, Sydney, Australia. The help of Professor John Rosenberg and Dr. Frans Heskens of the University of Sydney is highly appreciated.

Last but not least, the author would like to thank his wife Pnina for her everlasting patience, understanding, and moral support.

Daniel Tabak

Preface to the First Edition

Since the appearance of microprocessors in the early 1970s, a vast number of books and manuals on this topic has been published. Most of the existing books are dedicated to specific microprocessor families, such as the Intel 80x86 or Motorola M68000. Moreover, most of the existing books dedicate the greater part of their text to the lower members of the above microprocessor families, such as the 8086 (80x86 family), or the MC68000 (M68000 family). Relatively little is said about the top-level members of the above families, such as the 80386, 80486 (80x86 family), or MC68030, MC68040 (M68000 family), and rarely in the same text. The new 80486 and MC68040 microprocessors have not been described at all, with the exception of being mentioned in a short paragraph here and there. The National Semiconductor microprocessors of the NS32000 family are not mentioned at all in the majority of existing texts. The new reduced instruction set computers (RISC) type microprocessors (Intel 80860, Motorola M88000, Sun SPARC, AMD 29000, and others) are described primarily in a small group of books dedicated to the RISC topic, but very rarely (and even then in insufficient detail) under the same cover with other microprocessors.

It is the purpose of this book to fill the gaps just described in the following manner.

By presenting the several microprocessor families of Intel 80x86, Motorola M68000, National Semiconductors NS32000, in considerable detail, under the same cover, along with some notable RISC-type microprocessors.

By stressing, and dedicating most of the space to, the top-level member of each family (as opposed to most existing texts).

By describing the top-level microprocessor, in detail, first and mentioning the other, lower members of the family (amply described in other books) subsequently.

The primary goal of this book is to serve as a concentrated reference of the top-level, advanced microprocessors of the most prominent (in the author's opinion) microprocessor families. An effort has been made to include as many details as possible within a limited space. This book includes details on hardware, software, architecture, organization, and realization aspects of the included microprocessors.

The book can also serve as a text on advanced microprocessors at the senior and first year graduate level. A modest number of problems for students were added to the text. Solutions to the problems are available in an instructor's manual. The preprint of this book was successfully used by the author in the graduate course (also open to advanced seniors) ECE 516 Advanced Microprocessors, offered in the spring of 1990 at George Mason University (GMU), Department of Electrical and Computer Engineering (ECE).

The primary intended audience of this book are electrical and/or computer engineers and students majoring in the above disciplines. It can also be used by professionals and students of computer science or other engineering areas, if they have sufficient basic knowledge of computer hardware. This book is not intended for beginners. It is assumed that the reader has had a basic course on digital design (including both combinational and sequential logic circuitry) and on computer organization. It is also assumed that the reader has a basic knowledge of the simpler 8-bit microprocessors, such as the Intel 8085 or Zilog Z80.

Advanced Microprocessors is divided into six parts. Part 1 serves as a basic introduction, presenting a brief historical overview of the development of microprocessors in Chap. 1, a general discussion of the structure of advanced microprocessors in Chap. 2, and a basic introduction to microprocessor architecture in Chap. 3. The next three parts are dedicated to specific microprocessor families: Part 2— Intel 80x86, Part 3—Motorola M68000, and Part 4—National Semiconductor NS32000. Part 5 discusses a selected number of advanced RISC-type microprocessors, and Part 6 includes a general discussion of microprocessor-based system development and a comparison between different systems covered in Parts 2 to 5. A list of abbreviations is added at the end for the convenience of the reader.

Chapter 3 in Part 1 contains basic material dealing with assembly language programming. Readers familiar with this material may wish to skip it. It was the experience of this author that many engineering students and practicing engineers do not know this material well enough. For this reason this material was included.

The author was fortunate to receive valuable information and comments from prominent professionals who played a leading role in the creation of some of the advanced microprocessors described in this book. They are listed in the order of appearance of the appropriate system in the book: Pat Gelsinger (Intel 80486), Ralph C. McGarity (Motorola MC68040), Les Kohn (Intel 80860), and Dr. H. Brian Bakoglu (IBM RS/6000). Helpful information and comments were also received from Tovey Barron and Chuck Swartley (Intel), Phil Brownfield (Motorola), Reuven Marko (National Semiconductor), and Max Baron (Sun Microsystems). The contributions of the above are highly appreciated.

Valuable comments, leading to considerable improvements in the text, were obtained from reviewers contacted by McGraw-Hill Publishing Company. The author is particularly indebted to James F. Fegen, Jr., of McGraw-Hill for continued support and encouragement in the preparation of the manuscript.

The manuscript was processed by the GMU Word Processing Unit, directed by Ms. Mary Blackwell. As in the author's previous books, the manuscript handling was timely and of high quality. The author would also like to express his appreciation to his wife Pnina for her understanding and patience.

Daniel Tabak

General Principles

1

Introduction

In the mid-seventies it was easy to define a microprocessor. At that time one could say that a microprocessor is a *central processing unit* (CPU) realized on a *large-scale integration* (LSI) (50,000 or more transistors) chip, operating at a clock frequency of 1 to 5 MHz, constituting an 8-bit system (with very few initial 16-bit systems around), with two to seven general-purpose CPU 8-bit registers. Because of their relatively low cost and small size, the microprocessors permitted the use of digital computers in many areas where the use of the preceding mainframe and even minicomputers would not be practical and affordable. The advent of the microprocessor permitted placing a digital computer in practically every home.

The microprocessor has come a long way since the seventies [Alex 93, Clem 92, LiGi 86, Prot 88, Rafi 84, SBNe 82, Uffe 91, Wake 89]. The microprocessors of the mid-nineties are full-scale 32- or 64-bit systems, operating at clock frequencies of 25 to 200 MHz, realized on over 3 million transistors *very large-scale integration* (VLSI) chips. We are promised by some manufacturers over 50 million transistors on a chip by the year 2000. Most of the new microprocessors issue more than one instruction per cycle. Practically all modern microprocessors have an on-chip *floating-point unit* (FPU). Most systems have 16 to 32 general-purpose CPU registers. Many have a separate 32-register file for their *integer unit* (IU), and a separate 32-register file for the FPU. Many microprocessors have a number of integer and floating-point operational units. Most microprocessors have a considerable (up to 36 kbytes) on-chip cache. In most cases the cache is dual: it is divided into an *instruction cache* (Icache) and a separate *data cache* (Dcache). The performance of today's advanced microprocessors matches and sometimes exceeds that of mainframe computers and even supercomputers. Under the circumstances, the distinction between microprocessors and other computing systems will gradually disappear. The main point of distinction, still prominent, is the microprocessor's compact VLSI realization and its relatively low price for the attained high performance. This permits placing a digital computer endowed with supercomputer capabilities in cars, airplanes, boats, and eventually in every home.

There are a number of manufacturers which developed whole families of microprocessors, featuring essentially the same architecture within each family. There are two most widely used microprocessor families, started in the late seventies: the Intel 80x86 (or simply x86), and the Motorola M68000 families. Since the early eighties we have witnessed a parallel development of a new architectural trend: the *reduced instruction set computer* (RISC) [HePa 90, Tabk 90]. The Intel x86 and Motorola M68000 families belong to what is considered to be the opposite class of RISC: the *complex instruction set computer* (CISC). The chronology of development of Intel, Motorola, and some RISC systems is summarized in Table 1.1.

Intel started with a 4-bit microprocessor 4004 in 1971. Such a microprocessor was particularly useful in calculators, easily realizing *binary*

TABLE 1.1 Microprocessor Chronology

Year	Intel	Motorola	RISC
1971	4004		
1972	8008		
1974	8080	6800	
1975			IBM 801 start
1976	8085		
1977		6809	
1978	8086		
1979	8088	68000	
1980	80186	68008	UCB RISC announced
1981	80188		Stanford MIPS
1982	80286		IBM 801 announced
1983		68010	Transputer
1984		68020	
1985	80386		
1986			MIPS R2000, IBM ROMP, HP-PA
1987		68030	Am29000, SPARC
1988	80376		M88000, R3000
1989	80486	68040	i860, R6000
1990			Am29050, RS/6000
1991			R4000, MC88110
1992			Alpha, R4400, SuperSPARC
1993	Pentium		PowerPC 601
1994		68060	

Notes: HP-PA = Hewlett-Packard Precision Architecture; MIPS = microprocessor without interlocked pipeline stages; RS = RISC system; UCB = University of California, Berkeley.

coded decimal (BCD) calculations. The 4004 was followed by an 8-bit 8008 in 1972. A more powerful 8-bit microprocessor, the 8080, was produced in 1974, followed by the top Intel 8-bit product, the 8085, in 1976 [Uffe 91]. Part of the 8080/8085 architecture, in particular the register file, is still perpetuated in Intel's x86 family. Intel started its x86 family with the 16-bit 8086 in 1978. Practically all the top manufacturers of 16-bit microprocessors of the late seventies and early eighties featured a less expensive chip with an 8-bit external data bus but with the same 16-bit internal architecture as the 16-bit counterpart. Thus came to be the 8-bit outside, 16-bit inside Intel 8088 in 1979. Such a chip, demanding about 50 percent less interface chips for its 8-bit data bus, compared to a 16-bit, offered a lower cost alternative for the 16-bit architecture. Indeed, the 8088 became the processor of the original IBM *personal computer* (PC) and all its subsequent clones, achieving a massive worldwide spread of implementation. Subsequently, more advanced members of Intel's x86 family became the CPUs of higher-level PCs and workstations.

The 8086 was followed by the highly integrated 80186 [LiGi 86], which included the 8086 CPU along with a number of *input-output* (I/O) interface and other logic units on the same chip. A small number of new instructions were added, but upward compatibility with the 8086 was preserved. As a matter of fact, there is also an upward compatibility with the 8080 (at the machine language level), preserved in the 8086. The 80186 also has an 8-bit data bus counterpart, the 80188. While still a 16-bit microprocessor, the 80286, announced in 1982, constitutes a significant step forward in microprocessor development, exemplified by the introduction of the protected mode (to be elaborated on in detail in Part 2). The protected mode was continued in all subsequent products of the Intel x86 family.

Intel launched its first full-fledged 32-bit microprocessor, the 80386 (or i386), in 1985. It was actually not quite the first 32-bit system announced by Intel. Intel had announced a 32-bit system, the 432 (unrelated to the x86 family), in the early eighties. It was never commercially marketed, and it was withdrawn in 1984. Intel also features the 80376, which is a 32-bit processor (80386 architecture) with a 16-bit external data bus, intended for implementation as an embedded, lower-cost processor. The 80386 was followed in 1989 by the 80486 (or i486), which reached 66 MHz operation in its i486 DX2 model. The i486 has an on-chip FPU, an on-chip 8-kbyte cache, while maintaining the same architecture as the i386. The last announced (in 1993) member of the Intel x86 family is the Pentium. In the past, during the development stages, it was sometime referred to as i586 or P5, but Intel decided to name it Pentium, without any numerical assignment. While maintaining the x86 family architecture, the Pentium was extended to be a two-issue superscalar (two instructions fetched, decoded, and executed in parallel), has a double 64-bit data bus in and out of chip, and a dual 16-kbyte on-chip cache (8-kbyte instruction, 8-kbyte data). Intel is currently working on the development of subsequent products of the x86 family. The next in line is currently referred to as P6, promised to double (at least) the performance of the Pentium.

The development of Motorola microprocessors proceeded in a manner similar to that of Intel. Motorola started its 8-bit 6800 family in 1974, culminating with its top 8-bit product, the 6809, in 1977. Motorola came out with its M68000 family, launching the 16-bit MC68000, in 1979 (according to the Motorola notation, M68000 denotes the 68000 family, while MC680x0 denotes a particular microprocessor in that family). There is one particular feature of the M68000 family worth emphasizing. The MC68000 microprocessor is basically a 16-bit system with a 16-bit data bus. However, its 16 general-purpose CPU registers and the program counter are 32-bit. From the beginning the M68000 family designers have been aiming toward an eventual 32-bit system, achieved by the MC68020 in 1984. In analogy to Intel's 8088, Motorola introduced an 8-bit data bus 68008 microprocessor in 1980. Motorola has also produced enhanced versions of the MC68000: the MC68010 and the MC68012 (later discontinued). The MC68010 is fully pin compatible with the MC68000. Thus, in a system designed around the MC68000, the MC68000 chip can be replaced by the MC68010, improving performance, without changing the design and configuration of the rest of the system.

Motorola started its 32-bit branch of the M68000 family by announcing the MC68020 in 1984. It was followed by the MC68030 in 1987, and the MC68040 in 1989. The 68020 had a small 256-byte Icache on chip. The 68030 had a modest dual cache of 256 bytes of instruction, 256 bytes of data on a chip. The 68040 has a 4-kbyte instruction, 4-kbyte data dual cache on chip. The next forthcoming member of the M68000 family (in 1994, announced in 1993), the MC68060, will be a two-issue superscalar, with 8-kbyte instruction, 8-kbyte data dual cache on chip (as the Pentium). Interestingly enough, both the Pentium and the MC68060 were promised to start operation at 66 MHz.

There exist, of course, several other microprocessor families, not described in this book for lack of sufficient space. The Zilog Co. features a very popular and extensively used 8-bit microprocessor Z80 [Uffe 91], announced in 1975. The Z80 was developed by some former Intel professionals, who developed the 8080 and the 8085. The Z80 architecture is indeed very similar to that of the Intel 8080. The Z80 contains all the 8080 instructions, and although the assembly language mnemonics are different, there is machine language compatibility between the two. The Z80 contains, however, many more instructions than the 8080. Zilog announced its 16-bit family, the Z8000, in parallel with Intel's 8086 at about the same time. It had also announced a 32-bit product, the Z80000, but it was not actually produced.

National Semiconductor launched its 16-bit family NS16000 in 1979, and its 32-bit family NS32000 in 1984, culminating with its top product NS32532 in 1988. The NS32000 architecture is continued in a National Semiconductor embedded processor product, called "Swordfish" [HiTa 92]. The Swordfish is a two-issue superscalar micro-controller.

As can be seen in Table 1.1, RISC microprocessors have been steadily developing in parallel with the Intel and Motorola families. In fact, Intel and Motorola launched their own RISC families in the late eighties. The start of the RISC idea is debated by some professionals. Officially, looking at the printed literature, one finds the first documented mention of the "RISC" term in a note by Patterson and Ditzel [PaDi 80] with the University of California, Berkeley in 1980. However, according to a private communication by David Ditzel, the term "RISC" was coined by another Berkeley faculty member, Carlo Sequin, coauthor of the first journal paper on RISC [PaSe 82]. A design of a RISC-type computing system, without using the term "RISC," started in effect much earlier, in the mid-seventies at IBM, under the name of IBM 801. It was publicly announced, although only in 1982, after the Berkeley RISC announcement [Patt 85, Tabk 87, Tabk 90b]. Stanford University came out with its RISC-type system right after Berkeley in 1981, the MIPS (microprocessor without interlocked pipeline stages).

Subsequently, MIPS Computer Systems Co. was founded, designing commercial versions of the Stanford MIPS, the Rx000 family ($x = 2, 3, 4, 6$), shown in Table 1.1. Some features of the basic Berkeley RISC design were used in the SPARC of Sun Microsystems. The latest top-level SPARC is the SuperSPARC, produced by cooperation of Sun with Texas Instruments (TI). Intel started its i860 RISC family and Motorola its M88000 RISC family. IBM continued with RISC development with its ROMP (Research Office products division MicroProcessor), and culminating with its RISC System 6000 (RS/6000). The RS/6000 architecture is also partially featured in the new family of IBM, Motorola, and Apple, called PowerPC. The first product of PowerPC is the 601, announced in 1993, with a more powerful 620 to follow in the future. Other RISC products are the INMOS Transputer, Hewlett-Packard Precision Architecture (HP-PA), AMD (Advanced Micro Devices) Am29000 family, and the Digital Equipment Corporation (DEC) Alpha AXP.

This text is subdivided into five parts. The first part presents a brief presentation of some basic principles common to most microprocessors, independent of any particular families or products. It includes Chaps. 1 to 6. Chapter 2 presents a general structure of a modern advanced microprocessor, specifying its essential subsystems, found in most contemporary microprocessors. Chapter 3 presents an overview of some architectural features of microprocessors, such as instruction sets, instruction formats, data formats, and addressing modes. Chapter 4 is dedicated to the discussion of the microcomputer memory hierarchy, including topics such as the register file, cache, virtual memory, paging, and segmentation. Chapter 5 discusses some basic principles of pipelining. The discussion includes such basic topics as pipeline hazards, *instruction level parallelism* (ILP): superscalar and superpipelined execution, and the *very large instruction word* (VLIW) concept. Chapter 6 presents a basic introduction to the principles and practices of RISC, its advantages and shortcomings.

Part 2 describes in detail the Intel x86 family. It includes Chaps. 7 to 10. Chapter 7 describes the basic Intel x86 architecture, featured by all members of the family. Chapter 5 describes the Pentium, while Chap. 6 describes the i386 and i486. Chapter 10 describes the 16-bit members of the family: the 8086, 80186, and 80286.

Part 3 describes the Motorola M68000 family. It contains Chaps. 11 to 13. Chapter 11 describes the M68000 architecture, common to all members of the family. Chapter 12 describes in detail the top-level members of the family: the MC68060 and MC68040. Chapter 13 discusses the earlier members of the M68000 family: the 68000, 68010, 68020, and 68030.

Part 4 is dedicated to a detailed description of a selected group of RISC microprocessors. Chapters 14 to 20 feature, respectively, the DEC Alpha AXP, the IBM RS/6000 and PowerPC 601, the Sun SPARC, and the SuperSPARC, the MIPS Rx000 family (stressing the R4000 and R4400), the Intel i860, the Motorola M88000, and the HP PA-RISC family.

The concluding Part 5 includes system development in Chap. 21, system comparison in Chap. 22, and concluding comments in Chap. 23. Some details such as instruction specifications, pin assignments, and electrical data are assigned to appendices at the end of some parts. The text concludes with a list of references, a list of abbreviations, and an index.

2

General Structure of Microprocessors

This chapter describes the structure of a general microprocessor manufactured in the mid-1990s. The structure does not represent directly any specific microprocessor. It contains, however, features to be found in the majority of microprocessors of this period.

A block diagram of a general advanced microprocessor of the mid-1990s is shown in Fig. 2.1. Most microprocessors, for the time being, are 32-bit systems. However, there are several 64-bit systems on the

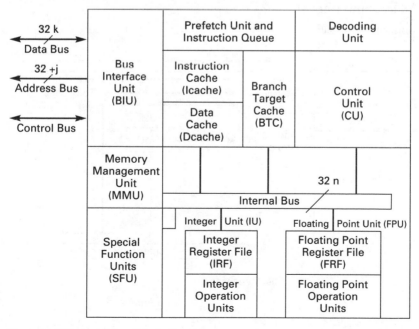

Figure 2.1 Microprocessor block diagram.

market. The number of 64-bit systems is expected to increase. Even the 32-bit systems have multiple data buses in and out of the chip. In 32-bit systems the IU and the registers in the *register file* (RF) are 32-bit. The data buses may be 32-bit, but in quite a few systems they may be 32×2 (64), 32×3 (96), or 32×4 (128). In 64-bit systems the integer unit and the registers in the register file are 64-bit. The regular data bus for the 64-bit system is naturally 64-bit. However, some 64-bit systems have multiple 64-bit buses, such as 64×2 (128) or 64×4 (256). The possibility of multiple data buses is reflected in Fig. 2.1 in the values of $32k$ outside, and $32n$ inside the chip (k, n are integers). The address bus in most systems is 32-bit for a directly addressable space of 4 gbytes. However, some new systems are being designed to accommodate a larger address space. This is reflected in Fig. 2.1 in the $32+j$ addressing lines (j is an integer). It should be noted that the internal $32n$-bit bus may include address parts, separate from the data parts. For instance, in a microprocessor with a dual cache system, there may be two internal address buses: one to the Icache and the other to the Dcache.

The *bus interface unit* (BIU) is a buffer subsystem between the internal units of the microprocessor and the external systems. It is connected to the system bus. A more detailed diagram of the BIU is shown in Fig. 2.2. The BIU is subdivided into three main parts:

1. Data interface

2. Address interface

3. Control interface

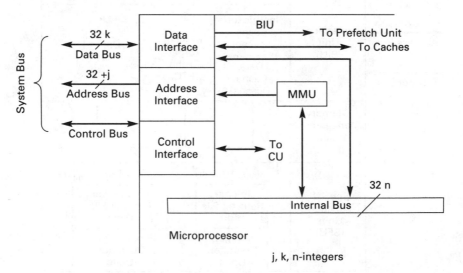

Figure 2.2 The bus interface unit.

The data interface subunit interconnects the system data bus and the internal units of the microprocessor. Generally, data transmissions between the data interface and other internal microprocessor units are conducted through the internal bus. There may be some additional direct interconnections. Usually, the data interface is directly connected to the prefetch and instruction queue unit and to the caches.

The address interface sends the internally generated address of an instruction or a data item to the address part of the system bus. The address is generated by the *memory management unit* (MMU), which is directly connected to the address interface.

The control interface sends and receives a number of control and status signals from the processor to the external systems and from external devices to the processor. Most of the control lines of the control interface are usually connected to the *control unit* (CU); however, the CU is also interconnected with other units. The emanating control lines indicate the internal status of various microprocessor operations, activate commands for read or write operations, differentiate between accesses to various types of address spaces, acknowledge interrupts and bus grant requests, and may perform other tasks. The incoming control lines indicate the status of external devices, alert the microprocessor in case of possible bus or system faults, request the servicing of an interrupt or a bus use grant, acknowledge a completed bus cycle, disable internal cache, and may serve other purposes. Detailed examples of the control lines of various microprocessors will be presented in the subsequent parts of this text.

The prefetch unit contains logic circuitry that prefetches instructions from the instruction cache, and lines them up in a *first-in first-out* (FIFO) instruction queue. A typical instruction queue may contain 8 to 32 bytes. The lined-up instructions are transmitted to the decoding unit. The decoding unit decodes the instructions and transmits appropriate intermediate control signals to the CU. Most of today's advanced microprocessors are superscalar. This means that more than one instruction at a time is forwarded for decoding. This will be discussed further later in this chapter.

Some microprocessors may include a *special function unit* (SFU) in addition to the IU and the FPU. An SFU may be one of the following (or any other application-specific) units:

1. Graphics unit
2. Signal processing unit
3. Image processing unit
4. Vector and matrix processor

A microprocessor may contain more than one identical SFU. The current technology permits the placing of over 3 million transistors on a single chip. This number may exceed 50 million by the year 2000. With such a density one may plan on having multiple operational units on the same chip. Having SFUs on the same chip yields minimal delays in the SFU's interface with other units and speeds up the whole operation.

The *cache* is a fast-access memory located between the CPU and the main memory. The availability of a sizable cache can significantly improve overall performance, since it permits the CPU to access information much faster than from the main memory [HaVZ 90, Hays 88]. Today's advanced microprocessors have up to 32 kbytes or more on-chip cache. Practically all modern microprocessors are pipelined. This means that for an n-stage instruction pipeline, the processor is dealing with different stages of up to n instructions at a time. Thus, while the CPU may access memory to fetch an instruction, it may have to access memory for a data item for another instruction at the same time. If there is just a single cache, such a simultaneous access may prove impractical [HePa 90]. For this reason, the practice of a dual cache became pervasive in advanced microprocessors. There is a separate instruction cache (Icache) and a separate data cache (Dcache), accessed by separate data paths, as shown in Fig. 2.3. In most microprocessors both caches have the same size; however, in some they are of a different size. Caches receive information from main memory through the data bus and the BIU in quanta of lines or blocks. In most modern microprocessors the line size is 16 or 32 bytes. The instruction cache is directly connected to the instruction prefetch unit, into which it transfers one or more instructions per cycle. Both caches are connected to the microprocessor internal bus. In many microprocessors, the data cache is connected to the operation units through an additional *operation data bus* (ODB), as shown in Fig. 2.3. The ODB is usually wide enough (128 or 256 bits) to carry a number of operands at a time. This certainly speeds up the operation of the microprocessor. More details on the cache will be given in Chap. 4, and on pipelining in Chap. 5.

Many modern microprocessors were designed for optional operation with a secondary cache. The cache on-chip is called in this case *primary cache*. The secondary cache is located between the primary cache and main memory in the memory hierarchy. The secondary cache access time is shorter than that of main memory and, being outside the chip, it can be considerably larger than the primary on-chip cache. The availability of a secondary cache may improve the overall performance of the system by providing the CPU with a relatively large and fast access memory [HePa 90, Przy 90]. In several microprocessors there is

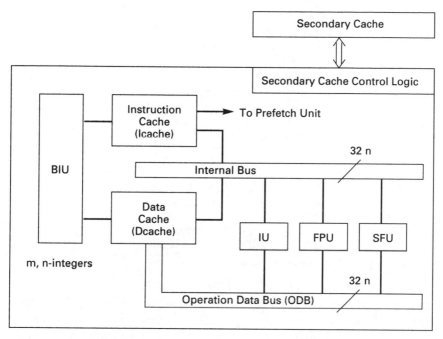

Figure 2.3 Microprocessor caches.

a special on-chip secondary cache control and interface logic unit, directly connected to an optional secondary cache (Fig. 2.3). The size of the secondary cache may be several megabytes (mbytes). Additional discussion on secondary caches can be found in Chap. 4.

A number of modern advanced microprocessors include a branch target cache (BTC) on chip. In a pipelined system, when a branch instruction is encountered, the instructions following the branch instruction must be taken out of the pipeline, and the target instructions (to the first of which the branch is executed) must be fetched into the pipeline. If the target instructions must be fetched from memory, it will introduce an extra delay in the instruction execution sequence. If, however, the first target instructions are in a special on-chip BTC, they can be fetched much faster into the pipeline. Obviously, the presence of a BTC tends to improve the overall performance of the processor. In some systems the BTC contains only the target addresses and not the target instructions.

As stated earlier in this chapter, most modern microprocessors feature instruction level parallelism (ILP). In most cases, superscalar operation is practiced. In a superscalar system, the prefetch unit issues i instructions at a time and forwards them to the decoder, as

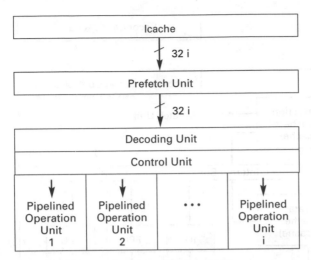

Figure 2.4 Superscalar issue of instructions.

illustrated in Fig. 2.4. The control unit subsequently generates activation commands to a number of pipelined operation units. Generally, the number of pipelines may not be equal to the number of instructions issued i. However, for better performance, it is desirable that if i instructions are issued at a time, that there should be a capability to execute i instructions at a time as well. Figure 2.4 illustrates a case where the number of operation pipelines is equal to the number of instructions issued i. In practice, such a condition is not always possible. Because of silicon area limitations it is not always possible to place a sufficient number of operation units on the same chip. The availability of operation units does not necessarily guarantee a smooth operation at all times. There is always the possibility of pipeline hazards [HePa 90], such as data dependencies among subsequent instructions, and control instructions (such as branches), which may introduce pipeline stalls and delays. These aspects of pipelined execution will be discussed in more detail in Chap. 5.

The CU can be either hardwired or microprogrammed [HaVZ 90, Hays 88]. In the more traditionally designed CISC microprocessors, the CU is usually microprogrammed. In the more recent and more pervasive trend of reduced instruction set computer (RISC) microprocessors, the CU is usually hardwired. This is done to achieve higher speed and single cycle execution of all or most instructions [Patt 85, Tabk 90b]. More details on RISC will be given in Chap. 6. Parts 2 and 3 describe in detail two major CISC families of microprocessors: Intel x86 and Motorola M68000, respectively. Part 4 describes in detail a number of advanced RISC microprocessors.

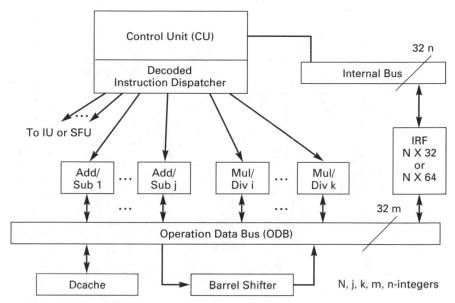

Figure 2.5 Integer unit.

The details of the microprocessor IU are illustrated in Fig. 2.5. As argued before, it is important to have a number of parallel, even some identical, operation units in a superscalar system. Thus, we may have a number of add/subtract and multiply/divide units in the IU. There is an IU register file (IRF) consisting of N 32- or 64-bit registers (depending on whether it is a 32- or 64-bit system). CISC systems usually have 8 to 16 registers. RISC systems usually have at least 32 registers. A few RISC systems have more than 100 registers (see Part 4). The information flows in two ways to the operation units. The instruction dispatcher receives decoded commands from the control unit and directs these commands to the appropriate operation units. Integer instruction commands are thus forwarded to the integer operation units, floating-point instruction commands to the FPU, and special function commands to the SFU (if available; so far only a minority of chips has an SFU). Data are forwarded to the operation units from the Dcache through the ODB (Fig. 2.5). Most systems have a barrel shifter permitting efficient execution of multiple-bit shift instructions in a single cycle.

The FPU datapath is illustrated in Fig. 2.6. The information flow is similar to that in the IU, as discussed in the previous paragraph. There is a separate floating-point register file (FRF), consisting of N registers. In CISC systems the number of FRF registers is usually 8, while in RISC systems it is usually 32. Practically all modern designs feature the Institute of Electrical and Electronics Engineers (IEEE)

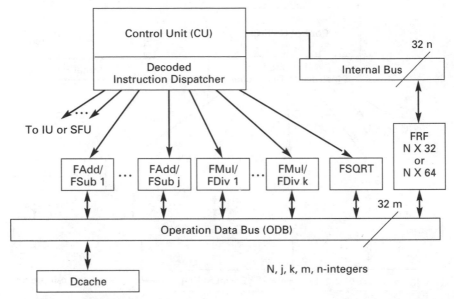

Figure 2.6 Floating-point unit.

754-1985 floating-point standard [HaVZ 90, Hays 88, IEEE 81, IEEE 85]. This includes the 32-bit single-precision and the 64-bit double-precision representation of floating-point numbers. For this reason, some FRFs feature 64-bit registers, even if the system is 32-bit. Some systems feature an additional 80-bit extended floating-point format. For this reason, some FRF registers are 80-bit in such systems. The DEC Alpha features both IEEE standard floating-point formats as well as some VAX floating-point formats for compatibility (see Chap. 14).

The primary functions of the MMU are:

1. Translation of the virtual (logical) address into a physical (real) address, and the transmission of the physical address to the cache or through the BIU and the address bus to the external devices.

2. Provide for the paging mechanism involved in the virtual memory organization. This is done by the paging unit.

3. Provide for the segmentation mechanism (if implemented) by the segmentation unit.

4. Provide for memory protection. This is usually done within the paging or segmentation unit, or both.

5. Inclusion and management of a fast-access *translation lookaside buffer* (TLB), or *address translation cache* (ATC), for virtual to physical page number translation.

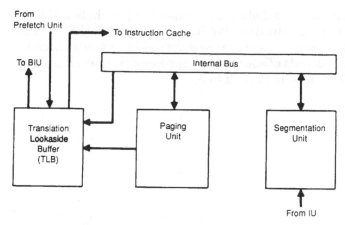

Figure 2.7 The MMU.

A block diagram of an MMU is shown in Fig. 2.7. Practically all advanced microprocessors have a paging unit and a TLB (or ATC). Only the Intel x86 family has a segmentation unit (Part 2). In case of a TLB miss, the MMU contains the logic that supervises the access to the appropriate page directories and tables in main memory, and the loading of the TLB with the missing page number. Details of specific MMUs will be presented in the subsequent parts of the text. Details on MMU basic principles will be given in Chap. 4.

Modern microprocessors may have hundreds of interconnection pins. This permits having multiple data buses, separate data and address buses, separate interface to the system bus and secondary cache, and a great variety of control and status signals. One can also have multiple power and ground interconnections on all sides of the chip. This permits having shorter internal power and ground interconnections, simplifying the chip design and thus reducing its cost. Generally, shorter interconnections have shorter delays of signal propagation.

A very general microprocessor structure, not necessarily identified with any specific microprocessor, was presented in this chapter. Some basic principles of memory hierarchy, pipelined execution, and other aspects of computer architecture and organization were used in this presentation. For the convenience of the readers not fully versed in certain aspects of computer architecture and organization, some basic principles are discussed in the subsequent chapters of Part 1. Chapter 3 discusses some basic principles of microprocessor architecture, including a general discussion of instruction sets, instruction and data formats, and addressing modes. Chapter 4 discusses memory hierarchy and organization, including caches, virtual memory, paging, and segmentation. Chapter 5 is dedicated to pipelining, including ILP and pipeline hazards. RISC principles and its advantages vs. CISC are discussed in detail in Chap. 6. Readers well versed in computer architecture, interested in details of specific microprocessors, may skip Chaps. 3 to 6 and follow directly to Parts 2 to 5.

3

Microprocessor Architecture

3.1 Introduction

When one discusses the concept of "computer architecture," one has to provide a definition, since there is no standard definition acceptable to all professionals. Looking through the literature one can find a number of definitions and conceptions of computer architecture. Myers provides the following definition [Myrs 82], based on the original IBM formulation:

> Computer architecture is an abstraction of a physical computing system as seen by a machine language programmer or a compiler writer. In other words, computer architecture refers to those aspects of a computing system that are visible to the most sophisticated programmer, operating at the most privileged protection level (if any such levels are implemented). This includes all registers accessible by any instruction (including the privileged instructions), the complete instruction set, instruction and data formats, addressing modes and other details that are necessary in order to write any program on the system under consideration.

A number of people have commented that the above definition is outmoded; however, so far, nobody has provided an alternative clear-cut definition. Until that happens, it is suggested that the above definition, as formulated by Myers, be used. The acceptance of this definition dictated the choice of topics to be covered in this chapter, dealing with the architecture of microprocessors: instruction set, data formats, instruction formats, and addressing modes.

As opposed to computer architecture, *computer organization* refers to the physical arrangement and interconnection of the components in a computing system, and to the characteristics of those components. The details of the computer organization may be completely transparent to the programmer. For example, a prefetch instruction queue and its size in any microprocessor is an organization item; it is completely transparent to the user.

There are computer subsystems that may or may not belong to the architecture in different computing systems. In many systems the cache is managed by hardware and there isn't even one instruction that controls the cache or affects it directly in any way. It is completely transparent to the user. In such a case the cache is definitely a part of the organization and not of the architecture. On the other hand, some modern systems, such as the M88000, do have instructions that affect the cache. In this case, the cache becomes a part of computer architecture, visible to the programmer. In fact, any part of a computing system can be made part of the architecture by an appropriate design of the instruction set. In such a case, the instruction set should include instructions which permit the programmer to control the cache and to affect its content. The instruction set is a very important aspect of any computing system, since it clearly defines and specifies what exactly the system is capable of doing. The microprocessor instruction set is discussed in the following section. Readers familiar with the basic aspects of computer architecture may skip this chapter.

3.2 Instruction Set

The instruction set is indeed one of the key features of computer architecture, defining and describing the capabilities of any computing system, including microprocessors. It constitutes a specific set of operations that a given system can perform. We can usually distinguish the following principal types of instructions:

1. Data movement instructions
2. Integer arithmetic and logic instructions
3. Shift and rotate instructions
4. Control transfer instructions
5. Bit manipulation instructions
6. System control instructions
7. Floating-point instructions
8. Special function unit instructions

The above types of instructions will now be discussed in more detail in the subsequent paragraphs.

Data movement instructions perform the transfer of information from one location in a computing system to another. The transfer can be

Register to register

Memory to register or register to memory

Memory to memory

By "register" we understand any register in the CPU register file. By "memory" we understand any location in the main memory addressing space.

Data movement instructions are usually denoted in the following assembly form:

```
MOV source, destination
```

or

```
MOV src, dst
```

where src (source) is the address of the source operand and dst (destination) is the address of the destination operand. The above addresses may be either a reference to a register in the CPU register file or an address in memory.

In some systems there is a distinction between instructions operating on different sizes of data. The most often used data sizes are

B Byte	8 bits
H Halfword	16 bits
W Word	32 bits
D Doubleword	64 bits

For each of the above the MOV instruction can be expressed:

MOVB for moving a byte

MOVH for moving a halfword

MOVW for moving a word

MOVD for moving a doubleword

It should be noted that in some systems (particularly those that evolved from 16-bit products) the term *word* applies to a 16-bit data item, and a 32-bit item is called a *doubleword* or a *longword*. These distinctions will become apparent during the discussion of particular microprocessor families in Parts 2 to 4.

The data movement instructions also include input and output operations. In many microprocessors a part of the memory addressing space is dedicated to *input-output* (I/O) interface units and devices. This constitutes memory-mapped I/O. In this case I/O operations are performed by the regular memory access instructions, such as MOV. Other systems (such as Intel, see Part 2) also have a separate I/O addressing space and special I/O instructions, such as IN and OUT (see Part 2), in addition to the memory-mapped I/O capability.

Some systems accomplish memory access from the CPU by using LOAD and STORE instructions. The LOAD instruction

```
LOAD madr, ri
```

loads a data item from a memory location, at the address "madr" into a CPU register ri. The STORE instruction

```
STORE ri, madr
```

stores the content of the CPU register ri into a memory location at the address "madr."

Instructions involving the manipulation of stack, such as PUSH and POP, also fall into the category of data movement instructions. In some systems, such as Motorola, stacks are manipulated with MOV instructions, using a particular combination of addressing modes (see Part 3). Other variants of data movement instructions will be discussed in conjunction with particular examples of microprocessors in Parts 2 to 4.

Integer arithmetic instructions are structured as follows:

1. Two-operand instructions:

```
ADD src, dst; (dst) + (src) → dst
SUB src, dst; (dst) - (src) → dst
MUL src, dst; (dst) x (src) → dst
DIV src, dst; (dst) / (src) → dst
```

In case of the divide instruction (DIV) a double destination is usually specified to store the quotient and the remainder.

2. Three-operand instructions:

```
ADD src1, src2, dst; (src2) + (src1) → dst
SUB src1, src2, dst; (src2) - (src1) → dst
MUL src1, src2, dst; (src2) x (src1) → dst
DIV src1, src2, dst; (src2) / (src1) → dst
```

where ";" denotes the beginning of a comment, (x) denotes the content of x, src, src1, src2 are the locations of the source operands, and dst denotes the destination. The addresses of the operands can be expressed in a variety of addressing modes, discussed in Sec. 3.5.

The use of three-operand instructions results in a more compact code. This can be seen from the following simple example. Take a HLL expression:

```
C: = A + B; C ← (A) + (B)
```

To accomplish the same with a two-operand instruction, we need:

```
ADD A,B; (A) + (B) → B
STORE B,C; (B) → C
```

With a three-operand instruction, the same can be done in a single line:

```
ADD A,B,C
```

In some processors there is a separate version of arithmetic instructions for different data types, such as bytes, halfwords, and words. As in the MOV instructions, the notation can be of the type ADDB, ADDH, and ADDW, respectively.

In many systems there are separate signed and unsigned multiply and divide instructions. The notation in these cases can be MULS and DIVS for signed, and MULU and DIVU for unsigned operations.

In some cases, different options of operand sizes are provided for multiply and divide instructions. In the multiply case we may have

16×16-bit operands, 32-bit product

32×32-bit operands, 64-bit product

64×64-bit operands, 128-bit product

Modern advanced microprocessors arc 32- or 64-bit systems with 32- and 64-bit CPU registers, respectively. If there is a 64-bit product in a 32-bit system, or a 128-bit product in a 64-bit system, a double destination location is to be provided.

Example   In the Motorola MC68030 the 32×32-bit, 64-bit product signed multiplication is (see Part 3)

```
MULS.L src, Dh:Dl; (src) x (Dl) → (Dh,Dl)
```

where .L indicates a 32-bit (longword) operation, and Dh and Dl are 32-bit data registers. The most significant 32 bits of the 64-bit product go into Dh, and the least significant 32 bits go into Dl (which contained the multiplicand at the beginning of the operation).

Division can also be featured for different operand sizes, such as

32/16-bit operands, 16-bit quotient, 16-bit remainder

32/32-bit operands, 32-bit quotient, 32-bit remainder

64/32-bit operands, 32-bit quotient, 32-bit remainder

64/64-bit operands, 64-bit quotient, 64-bit remainder

128/64-bit operands, 64-bit quotient, 64-bit remainder

Example   In the MC68030 (see Part 3), a 64-bit dividend, stored in two 32-bit data registers (Dr,Dq), is divided signed by the divisor at the src address:

```
DIVS.L src, Dr:Dq;  (Dr,Dq)/src → Dr (remainder), Dq (quotient)
```

In all cases when the size of an operand or the result is greater than the size of the standard CPU register, the architecture should provide appropriate locations, as illustrated in the previous example, or in any other way. In some systems, a number of predefined CPU registers may be permanently assigned to contain multiplication and division operands and results, as it is done in the Intel 80x86 family (see Part 2).

Logical two-operand instructions are usually designed in a similar manner to the ADD or SUB instructions:

```
AND src, dst; src AND dst → dst
OR src, dst; src OR dst → dst
XOR src, dst; src XOR dst → dst
```

where XOR is the exclusive OR operation. Similarly to the ADD or SUB, the logical instructions may also have three operands in some systems such as

```
AND src1, src2, dst; src1 AND src2 → dst
```

In many systems there is also a complement or NOT single-operand instruction:

```
NOT dst; dst# → dst
```

where dst# denotes the logical complement of dst, such that 0# = 1, 1# = 0.

The compare instruction is usually classified as a logical instruction, although a subtraction is actually performed. Its basic expression is

```
CMP src1, src2; (src2) - (src1), flags affected.
```

The result of the subtraction is not stored anywhere, and the original values compared, in src1 and src2, are not affected. The flags are affected as follows:

If (src2) = (src1), the zero flag Z is set.

If (src2) < (src1), the sign flag S is set, since the difference is negative.

The flags can be checked after the compare instruction by using conditional branch instructions, to be discussed later in this section.

In earlier generations of computers [Hays 88, Chap. 1] shift and rotate operations were performed one bit at a time, left or right. Many modern microprocessors have an additional logic subsystem in the CPU, such as a barrel shifter [Prot 88], permitting fast shifts by a specified number of bits (usually up to 31) at a time. There are two basic types of shifts [Wake 89]:

1. *Logical shift:*   The operand is shifted left or right by n bits, and zeros are transferred into the n vacated positions. Usually the carry flag C is a part of the operation as illustrated in Fig. 3.1.

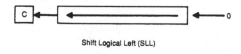

Shift Logical Left (SLL)

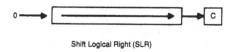

Shift Logical Right (SLR)

Figure 3.1 Logical shift operations.

2. *Arithmetic shift:* The shifted operand is treated as a signed 2's complement number. An arithmetic shift left by each bit is equivalent to a multiplication by 2, and an arithmetic shift right by each bit is equivalent to a division by 2. The arithmetic shift operation is illustrated in Fig. 3.2. The *shift arithmetic right* (SAR) extends the sign bit [the *most significant bit* (MSB)] to the right for the number of bits shifted. The left shift is basically the same as the logical left shift.

There are generally two operands specified for a shift instruction. One specifies the number of bits to be shifted, and the other specifies the operand to be shifted.

Examples

SLL Dc, Ds; shift logical left the content of register Ds by the number of bits equal to the content of register Dc.

SLR count, Ds; shift logical right the content of register Ds by the immediate number of bits specified by "count."

In many systems one can specify a variety of addresses instead of Ds, including locations in memory. One can also specify the size of the operand to be shifted (byte, halfword, word) by appending the appropriate letter to the instruction, such as SARB, SARH, SARW.

Shift Arithmetic Left (SAL)

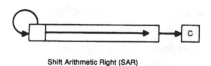

Shift Arithmetic Right (SAR)

Figure 3.2 Arithmetic shift operations.

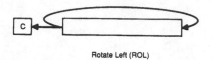

Rotate Left (ROL)

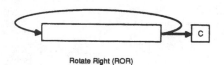

Rotate Right (ROR)

Figure 3.3 Rotate instructions.

Rotate instructions, illustrated in Fig. 3.3, are denoted in a similar manner:

ROL Dc, Ds; rotate Ds left by (Dc) bits.
ROR Dc, Ds; rotate Ds right by (Dc) bits.

Control transfer instructions consist primarily of

1. Jumps or branches
2. Calls to subroutines
3. Returns from subroutines

All the above can be conditional or unconditional. Unconditional branches and subroutine calls are of the form

```
BRA addr
CALL subname
```

where addr and subname are addresses in memory. Both addresses may be symbolic; subname is usually also the name of the subroutine called. The return from subroutine instruction is usually of the form

```
RET
```

Some systems also have special *return from exception* (RTE) instructions.

Conditional branches are usually based on the status of the flags, such as

C, Carry   Set if there is a carry out from an ALU operation.

Z, Zero   Set if the result of the last operation is zero.

S, Sign   Set if the result of the last operation is negative.

V, Overflow   Set if an overflow was caused by the last operation.

Conditional branches can then take the form:

```
BZ addr; branch to addr if Z = 1
BNZ addr; branch to addr if Z = 0
```

And similarly for other flags: BC, BNC, BS, BNS, BV, BNV. There may also be compound branch conditions such as

```
BBE addr; branch on below or equal, branch to addr if C OR Z = 1.
```

The BBE is usually given after the CMP (compare) instruction (see above). If (src2) < (src1) and a subtraction (src2) − (src1) is performed, a borrow will be needed and C will be set. The S flag will also be set. The flag Z will be set if (src2) = (src1). Complete sets of conditions for different systems will be presented in Parts 2 to 4.

Bit manipulation instructions operate on specified fields of bits, usually from 1 to 31 bits (32 bits usually constitutes a whole word). A bit field within any word can be specified by two parameters, as illustrated in Fig. 3.4:

Width, in bits

Offset, in bits

The width specifies the field size in bits, and the offset specifies the placement of the field within a word. The offset is measured starting with the *least significant bit* (LSB) 0. Of course, in some systems the offset can be defined in a different way. In the example illustrated in Fig. 3.4, the bit-field width is 5 bits and the offset is 10 bits. In assembly, this may be denoted:

```
5<10>
```

Alternately, one may preload the above two parameters (5 and 10 bits) into specific fields of a CPU register, say, Rb, and then use Rb as an operand in a bit-field manipulation instruction. The width and offset fields in the Rb register can be just 5 bits each, since $2^5 = 32$, and the width and the offset are limited to be no higher than 31. There are systems, however, where the offset is measured from a specified address, in which case its value may be much higher.

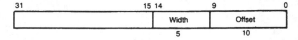

Figure 3.4 Bit-field specification.

The instructions performed on bit fields are usually of the following type:

Test, affecting flags

Clear

Set

Extract and transfer to another destination

Extract and sign extend

and others. Specific bit-field instructions will be presented in conjunction with actual systems (see Parts 2 to 4), wherever implemented.

System control instructions permit the user to influence directly the operation of the processor and other subsystems such as the MMU and the cache. Practically all systems have a HLT (halt) instruction that stops the operation of the processor. A number of systems have instructions that flush or invalidate entries in the cache or the TLB. In some systems there are instructions permitting the loading of TLB entries. Some systems have a special LOCK instruction that prevents other systems from using the system bus. It causes a LOCK control signal to be activated. The bus is locked to other systems until the LOCK signal is deactivated. The deactivation of the LOCK signal can be done either by the hardware or by using a special UNLOCK instruction, as implemented on the Intel 80860 (see Part 4). Data movement instructions to or from control registers are considered to be system control instructions in some cases.

Floating-point instructions are implemented on the on-chip FPU. Earlier generations of microprocessors used an off-CPU floating-point coprocessor. This instruction group usually consists of floating-point data movement (FMOV), arithmetic (FADD, FSUB, FMUL, FDIV), comparison (FCOMP), square root (FSQRT), absolute value (FABS), and a variety of transcendental functions (FCOS, FSIN) and others, depending on the system (see Parts 2 to 4). We distinguish between *diadic,* two-operand (FADD, FMUL), and *monadic,* single-operand (FABS, FSIN) instructions.

Special function instructions perform particular operations of the *special function unit* (SFU), depending on its nature. For instance, the Intel 80860's SFU (see Part 4) is a graphics unit. It has 10 instructions associated with it, such as

```
faddp     add with pixel merge
pfaddp    pipelined add with pixel merge
```

A special type of instructions called atomic instructions are used in the management of semaphores, which control the access to critical sections in multiprocessors [AlGo 89, Hwan 93, Ston 93, Tabk 90a]. Operationally, these instructions constitute a combination of data movement and arithmetic or logic instructions. Many modern microprocessors have been designed in advance for multiprocessor implementation, and their instruction set includes atomic instructions. In order to ensure the correct implementation of semaphores, the operations handling them must be *uninterruptible*. Such instructions are called *atomic*.

A general form of an atomic instruction implemented on a number of actual systems is the *Test And f* (TAf) operation [AlGo 89], where f(V,e) is an arithmetic or logical function of two values V and e. A particular case of the TAf operation, widely implemented, is f = OR(V,TRUE), e = TRUE. This operation is often called *Test And Set* (TAS). The TAS is an uninterruptible (atomic) instruction, operating with a *Read-Modify-Write* (RMW) bus cycle. It reads the value of the semaphore located in memory, tests whether it is 0 or 1, and sets the condition code flags accordingly (Read part of cycle). It then proceeds to set the semaphore value to one (Modify part of the cycle) and store it back in its original location in memory (Write part of the cycle). No other processor can access the system bus and the semaphore during the uninterruptible RMW cycle. Had such an access been possible, the test could yield wrong results, thus disrupting the operation.

Another form of an atomic instruction is the *Compare and Swap* (CAS), used to update shared data in a multiprocessor. A shared data item is locked at the beginning of the execution of CAS, compared with the content of a compare register, setting condition code flags accordingly, and is then updated by moving into it the content of an update register. After that, the access to the shared data item is unlocked, and the execution of CAS is complete. The implementation of instructions such as TAS and CAS in a computing system provides the system with synchronization capabilities for multiprocessing operation [AlGo 89, Hwan 93, Ston 93, Tabk 90a]. Instructions of this type have indeed been implemented on a number of advanced microprocessors (see Parts 2 to 4).

3.3 Data Formats

A typical set of integer data formats as practiced in modern microprocessors is shown in Fig. 3.5. They are

1. Signed and unsigned

```
Byte            8 bits
Halfword       16 bits
Word           32 bits
Doubleword     64 bits
```

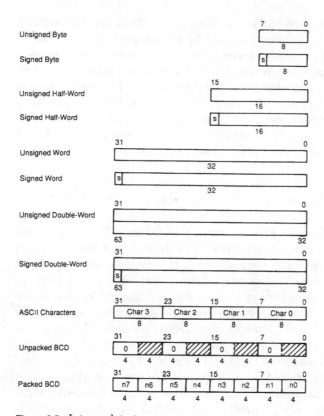

Figure 3.5 Integer data types.

2. ASCII characters, 8 bits each

3. Packed and unpacked *binary coded decimal* (BCD) numbers, 4 bits each.

It should be noted, however, that in some systems, particularly those which developed from 16-bit products (see Parts 2 to 4), a word is 16 bits and a 32-bit data item is called a doubleword or a longword. In that case the 64-bit data item is called a quadword. The recent trend, however, is to use the term "word" for 32-bit data items, as illustrated in Fig. 3.5. Appropriate terminology will be used in conjunction with each particular microprocessor family in the subsequent parts.

A 32-bit data word contains 4 bytes and takes up 4 addresses in memory. There are two ways of ordering the byte addresses within a word:

1. *Little-endian,* where the LSB has the lower address. In other words, the bytes are addressed 3, 2, 1, 0, from the MSB down.
2. *Big-endian,* where the LSB has the upper address. In other words, the byte are addressed 0, 1, 2, 3, from the MSB down.

American Standard Code for Information Interchange (ASCII) characters take up 8 bits each as a standard. Four such characters can be packed in a 32-bit word, as shown in Fig. 3.5. A table of ASCII characters is given in Appendix 1.A.

BCD numbers can be stored in two ways (see Fig. 3.5):

1. Unpacked, with the 4-bit BCD number taking up the 4 lower bits of each byte (usually each byte is separately addressed).
2. Packed, two BCD numbers per byte. Eight such numbers can be packed into a 32-bit word.

Whenever necessary, an extra sign bit may be placed in front of a string of BCD numbers. In some cases, the upper 4 bits of the word (n7 in Fig. 3.5) may just contain a sign bit, with 3 other bits unused (usually cleared).

The floating-point data formats are shown in Fig. 3.6. The single- (32-bit) and double-precision (64-bit) formats are according to the IEEE 754-1985 standard [HaVZ 90, Hays 88, IEEE 81, IEEE 85]. The 80-bit extended precision format may vary with different manufacturers. The one shown in Fig. 3.6 is implemented by Intel (Part 2).

The most significant sign bit (s) denotes the sign of the whole floating-point number: 0, positive, 1, negative. Let us assume the following notation:

e   Biased exponent; the number actually appearing in the exponent field

E   Actual exponent

ne   Number of bits in the exponent field

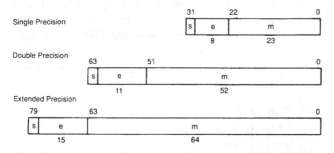

Figure 3.6 Floating-point data types. Note that e = biased exponent and m = mantissa or significand.

Precision	ne, bits
Single	8
Double	11
Extended	15

The biased exponent is defined as follows:

$$e = E + [2^{(ne - 1)} - 1] \qquad (3.1)$$

or the actual exponent expression, in terms of the biased exponent, is

$$E = e - [2^{(ne - 1)} - 1] \qquad (3.2)$$

The *bias* is $b = 2^{(ne - 1)} - 1$. For the three precision types, shown in Fig. 3.6, the bias values are

Precision	ne	Bias
Single	8	127
Double	11	1023
Extended	15	16383

The general IEEE standard floating-point number representation is

$$f = (-1)^s 1.m \times 2^{e - b} \qquad (3.3)$$

where s is the sign bit value (0 for positive, 1 for negative) and m is the mantissa or significand binary fraction bits:

$$m = .m_1 m_2 m_3 ... m_p \qquad (3.4)$$

where p is the total number of bits in the mantissa field. Thus, for

single precision $p = 23$
double precision $p = 52$

The "1." in Eq. (3.3) is implicit and does not take up a bit in the mantissa field. On the other hand, the "1" does take up bit 63 in the extended format (Fig. 3.6) and therefore $p = 63$ in this case (and not 64 as it would be if there were complete adherence to the IEEE standard in the extended format).

Translating the number ranges into decimal, we have approximately:

Single precision 10^{-38} to 10^{38}
Double precision 10^{-308} to 10^{308}
Extended precision 10^{-4932} to 10^{4932}

for both positive and negative numbers.

Example [Inte 87]   Represent the number 178.125 (decimal) in single-precision floating-point format.

$$178.125 = 1.78125 \times 10^2 \text{ (base 10)}$$

Converted to binary: $178.125 = 1.0110010001 \times (10)^{111}$ (base 2). The bias is 127 (decimal) or 0111 1111 (binary). The biased exponent is

$$b = 0111\ 1111$$
$$\underline{E = \qquad +\ 111}$$
$$e = 1000\ 0110$$

Since the "1." is implicit and not shown in the format, the final binary representation of 178.125 in the single-precision floating-point format is

Sign	Biased Exponent	Mantissa
0	1000 0110	01100100010000000000000
	8 bits	23 bits

The leading zero in the mantissa field may seem to represent an unnormalized number [Hays 88, HaVZ 90]; however, one should always remember the implicit "1.", which is not stored but is assumed to be there, as shown in Eq. (3.3). The normalized numbers in this notation are higher or equal to 1 and less than 2.

From this point on, unless otherwise stipulated, when mentioning "IEEE Standard," the IEEE 754-1985 standard will be assumed [IEEE 85].

3.4   Instruction Formats

Instruction formats vary considerably among different microprocessors. In some systems, instructions are formed in units of bytes, in some in units of 16-bit halfwords (however, in some systems they are called "words"), and in others in units of 32-bit words. The basic information included in an instruction format consists primarily of

1. Instruction opcode

2. Addresses of operands

Depending on the system considered, other details may be included. This can be clearly seen only when discussing specific systems, as will be done in Parts 2 to 4.

The opcode usually consists of 8 bits (a byte), yielding 256 different function encodings. In some systems, up to 2 bytes may be allocated, although the second byte would not be fully utilized for opcode (there is no system with 2^{16} opcodes). In fact, usually only a few bits of the second opcode byte would be utilized for opcode extension (see Part 2).

Some instructions, such as "clear accumulator" or "return" (from subroutine), do not need to specify any operands; a single opcode byte is sufficient for their specification. Many instructions need to specify operands, though. Modern microprocessors may have up to two or three operands specified by an instruction. In the two-operand systems, one of the operands serves as a source (src) and the other both as a second source and a destination (dst) in which the result of the operation is stored. In three-operand systems two operands constitute two sources, src1 and src2, and the third, the destination. As a particular case, a source may be identical to a destination. For instance, consider an addition instruction in a two-operand system:

```
ADD r1, r1; (r1) + (r1) → r1 = 2 x (r1)
```

This is equivalent to multiplying the content of register r1 by 2, with the result remaining in r1. Similarly, for a three-operand system:

```
ADD r1, r1, r1; (r1) + (r1) → r1 = 2 x (r1)
```

Another example:

```
ADD r0, r1, r1; (r0) + (r1) → r1
```

This is equivalent to the following operation in a two-operand system:

```
ADD r0,r1
```

As pointed out earlier in this chapter, the use of three-operand instructions yields a more compact code (same operations can be performed with fewer instructions).

The operands may be either CPU registers, as was the case in the previous examples, values stored in memory, or immediate values. Since the number of CPU registers is usually small, such as 8, 16, or 32, one needs only a field of 3, 4, or 5 bits to specify them. There are only two examples of RISC-type microprocessor families [*scalable processor architecture* (SPARC)], Chap. 16 and Am29000 [Tabk 90b]), where the number of directly addressable CPU registers is higher than 32. The maximum is 192 registers on the Am29000, for which an 8-bit field is sufficient.

The situation is different for memory addressing. Modern microprocessors usually have a full 32-bit address (and some have more), spanning 2^{32} = 4-Gbyte addresses. They are usually byte-addressable. Some systems, such as the Intel 80386 (see Part 2), offer a 2^{46}-byte virtual address space. With the appearance of new 64-bit systems (DEC Alpha, MIPS R4000; see Part 4), it will not take long before 64-bit addresses are featured. However, restricting our discussion to 32-

bit addresses, it is clear that we need 32 bits for a direct representation of a memory address within an instruction. Such an addressing mode is certainly used in many systems; however, it is clear that its use tends to increase considerably the size of instructions and programs. This in turn causes the use of more memory to store programs and increases the system's cost. For this reason, a number of indirect memory addressing modes were developed. These are discussed in the next section.

A number of typical instruction formats for 32-bit systems are shown in Fig. 3.7. A three-operand system is assumed. The formats shown constitute a general example and do not correspond exactly to any formats of actual systems. Formats for specific systems will be presented in Parts 2 to 4.

The register format, shown in Fig. 3.7(a), is intended primarily for *register-to-register* operations, that is, operations for which both the operands and the result destination are in the CPU register file. With

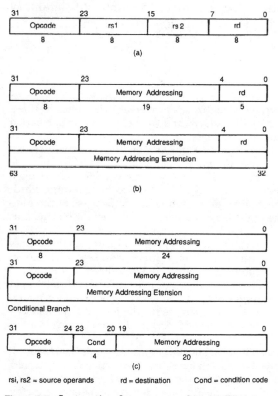

Figure 3.7 Instruction format examples. (*a*) Register format; (*b*) register-memory formats; and (*c*) branch and call formats.

8 bits for each register field rs1, rs2, and rd, there is the capability of addressing up to 256 registers (maximum implemented so far, 192 on Am29000, described in [Tabk 90b]). Most microprocessors have no more than 32 CPU general-purpose registers, requiring only 5 bits for their direct addressing. The other 3 bits of the 8-bit register field could be used for other purposes, such as addressing mode encoding (3 bits to encode 8 addressing modes). If there are only 16 registers, then the register field can be equally divided: 4 bits for register addressing and 4 bits to encode up to 16 addressing modes.

The register format can, of course, be used for indirect memory addressing, to be discussed in the next section. The register-memory formats, shown in Fig. 3.7(b), are intended for *register-to-memory, memory-to-register,* and *memory-to-memory* operations. In register-to-memory operations the source is in a CPU register and the destination is in memory: for example, a store instruction that transfers data from a CPU register into a location in memory. In a memory-to-register instruction, the source is in memory and the destination in a CPU register, as in a load instruction, for instance. In memory-to-memory operations both the source and the destination are in memory.

Example

```
ADD madr1, madr2; (madr1) + (madr2) → madr2
```

Both madr1 and amdr2 are symbolic addresses in memory. The data in these locations are transferred to the CPU, added, and the result stored in location madr2. The data in madr1 remain unchanged.

The memory addressing field (19 bits) in Fig. 3.7(b) may contain another register field, encodings for an addressing mode, and/or a numerical value to be used in calculating the operand address in memory. If direct addressing using a 32-bit address is to be implemented, one may add a 32-bit extension word, which will contain the complete and explicit address in memory. Because of the indirect addressing capability, the register format can also be used for instructions involving memory operands.

The branch (or jump) and call formats, shown in Fig. 3.7(c), are single-address instructions. The 32-bit format can be used for indirect addresses or for direct addresses not exceeding 24 bits (up to 16 mbytes). If 32-bit direct addresses are to be used, one can add the 32-bit extension word. For conditional branches or subroutine calls an extra condition field is to be added. Most modern microprocessors feature up to 16 branch conditions, requiring a condition field of 4 bits. Naturally, an extra 32-bit word can be added to conditional branch instructions as well, if required.

There are a number of tradeoffs involved in addressing operands in memory. If we allow memory addressing of operands in operation instructions (such as add, multiply), the CPU must access memory for the operand values during the execution cycle. On the other hand, if the operation is strictly register-to-register, the operands are fetched from the CPU register file, whose access is much faster than that of the memory. Even if the operands are in an on-chip cache, one needs an extra cycle to provide the complete address for cache access (also see discussion in Chap. 6). In systems working strictly with a register-to-register operation (RISC systems, Chap. 6), memory access is performed only by load and store operations. Of course, if register-to-memory or memory-to-register operations are permitted, some load or store operations may be saved, allowing for a better code compaction. An even better code compaction may be achieved if we allow memory-to-memory operations. However, in this case, we will have large variations in instruction size and execution duration. Moreover, a memory access bottleneck may develop. For this reason, memory-to-memory operations are practically not used in real-life systems. Most of the modern systems (which are predominantly RISC-type; see Part 4) use register-to-register operation and strictly load/store memory access.

3.5 Addressing Modes

The addressing modes found in most modern microprocessors are described in the following paragraphs.

Register

The operand is stored in one of the CPU registers. This mode is also called *register direct*. The assembly notation is Ri or ri. An example of a two-operand instruction using the register mode is

```
ADD r1, r2; (r1) + (r2) → r2.
```

Immediate

The operand is a part of the instruction. In some systems, such as the M68000 family (Part 3), an immediate value is denoted by writing the "#" sign in front of it. For example,

```
ADD #50, D1; 50 + (D1) → D1
```

The decimal number 50 is added to the content of the CPU data register D1. In most systems the numbers are considered to be decimal by default. For other number representations an extra letter is added:

H for hexadecimal

B for binary

O for octal

For example, the following operations are equivalent:

ADD 255, r1

ADD FFH, r1

ADD 1111 1111B, r1

ADD 377O, r1

Although widely used, the above notation is not universal. For instance, the symbol "$" in front of the number is used in the M68000 family (see Part 3) for hexadecimal values. A symbolic notation, defined elsewhere in the program, may be used instead of an explicit number in many assemblers.

Direct (or absolute)

The address of the operand in memory is a part of the instruction. Usually a symbolic name is used, although explicit addresses can be given as well. Some examples:

LOAD madr, r2;   (madr) → r2

CALL subr1

JMP AAF0H

Memory locations are specified by the symbolic names "madr" and "subr1" in the first two instructions, while an explicit hexadecimal address AAF0 is given in the third.

Register indirect (or indirect)

The address of the operand in memory is stored in one of the CPU registers. A possible assembly notation is

```
(Ri) or (ri)
```

For example,

```
MOV (r7), r1
```

which means: move a data item whose address in memory is in the CPU register r7, into register r1. In this instruction the source operand is specified by the register indirect, and the destination by the register addressing mode.

Register indirect with postincrement

Basically this mode works like the register indirect, except that the content of the register is *incremented after* the use of the address in it by an integer equal to the number of bytes of the operand. The notation is

```
(Ri)+ or (ri)+
```

Example

```
MOVW (r1)+, r3; memory[(r1)] → r3, then, (r1) + 4 → r1
```

Move a 32-bit (4-byte) word from memory at the address, equal to the content of r1, to register r3. After the address in r1 is used to access memory, the content of r1 is incremented by 4, since the operand is 4 bytes long. The register indirect with postincrement mode permits automatic advancing in loops, accessing byte, halfword, or word arrays of operands. Assume in the above example that r1 is pointing to an element in an array of word-size (4 bytes) operands. After the execution of the above instruction, r1 will point to the next element in the array.

Register indirect with predecrement

The content of the register is decremented by an integer equal to the number of bytes in the operand and is then used as an indirect address. The notation is

```
(Ri) or (ri)
```

Example

```
MOVH -(r6), r11; (r6) - 2 → r6, then, memory[(r6)] → r11
```

The content of register r6 is decremented by 2 (since the operand is a halfword; 2 bytes), and then it is used as an address in memory, from which a halfword operand is transferred to the register r11.

This mode can be used to scan an array in the direction of decreasing indices. That is, after dealing with element j of the array, element j − 1 will be dealt with.

Register indirect with displacement

The *effective address* (EA) of the operand is the sum of the content of a CPU register and a value, called *displacement,* specified in the instruction. The notation is

```
d(Ri) or d(ri)
```
, where d is the displacement.

The displacement may be an immediate constant or a symbolic value, defined elsewhere in the program.

Examples

```
ADD 125(r2), r4; memory[125 + (r2)] + (r4) → r4
```

The content of a memory location, whose address is the content of register r2 plus 125, is added to the content of register r4. The sum will be in r4.

```
MOV DISP(r1), r3; memory[DISP + (r1)] → r3
```

In this case the displacement is represented by a symbolic value DISP, which should be defined elsewhere in the program. The assembler will place the appropriate specific value instead of DISP. The EA of the source operand is obtained as in the preceding example.

In some systems this mode is denoted as the *base* mode and the register involved is called the *base register*.

Indexed and scaled indexed

The indexed mode works essentially as the indirect. Similarly to the indirect with displacement, there is also an indexed with displacement mode, defined in an identical way. The CPU register containing the address in the indexed mode is called *index register,* or simply *index*. The main difference between the indirect and the indexed is that the content of the index register can be scaled by a factor of 1, 2, 4, 8, or 16, depending on the system. If the index is scaled, the mode is called *scaled indexed*. The scaling number is called the *scale factor*. Of course, when the scale factor is 1, there is no difference between the indirect and the indexed modes. In the scaled indexed mode the content of the index register is multiplied by the scale factor to form the EA.

Example

```
MOV (r2*4), r5; memory[(r2)*4] → r5
```

The content of a memory location, whose address is four times the content of r2, is moved to register r5. In this example, r2 is the index register, and the scale factor is 4.

The motivation for the indexed mode is the extra flexibility given to the user when it is used in conjunction with the indirect mode, as exemplified by the next two addressing modes. The availability of the scale factor, along with the index, permits scanning of data structures of any size, at any desired step.

Indirect scaled indexed

The EA is the sum of the content of the indirect (or base) register and the scaled content of the index register. The notation is

```
(Ri,Rj*SC) or (ri,rj*SC)
```

where Ri or ri is the indirect (or base) register, Rj or rj is the index register, and SC is the scale factor.

Example

```
ADD (r5, r7*2), r9; memory[(r5) + (r7)*2] + (r9) → r9
```

In this example, r5 is the indirect register, r7 is the index register, and the scale factor is 2. The EA of the source operand in memory is the sum of the content of r7 multiplied by 2, and the content of r5.

Indirect scaled indexed with displacement

This mode is essentially as the indirect scaled indexed, except that a displacement is added to form the EA. The notation is similarly

```
d(Ri, Rj*SC) or d(ri, rj*SC)
```

where d is the displacement.

Example

```
MOV AAH(r1, r5*4), r7; memory[AAH + (r1) + (r5)*4] → r7
```

The EA of the source operand in memory is

```
EA = AAH + (r1) + (r5)*4
```

The displacement is AA (hexadecimal) = 170 (decimal).

PC-relative

The PC-relative mode is similar to the register indirect with displacement. The difference is that the *program counter* (PC) serves as the register indirect (or the base register). After an instruction is fetched, the PC is incremented to point to the next instruction (in systems with 32-bit, or 4 byte uniform instruction size, the PC will be incremented by 4). This is what we call the *updated PC,* and it is the content of the updated PC that is used in the PC-relative mode. The PC-relative mode is used automatically with program control instructions in many systems.

Example Given a system where the PC-relative mode is used automatically with any *branch* (BR) instruction. Thus, only the displacement must be given explicitly. It is assumed that all instructions are 4 bytes long. The instruction with its address is:

4000 BR 400H; 4000 is the hexadecimal address of the instruction.
4004 next instruction

After the branch instruction is fetched, the updated PC value is 4004. The hexadecimal displacement value 400H is added to the updated PC value to form the branch target address:

```
4004 + 400 = 4404
```

The addressing modes defined above are found in most modern microprocessors. Some variations of the above may exist, and in some systems a different terminology may be used. The details of the addressing modes, as implemented in actual systems, are presented in Parts 2 to 4.

3.6 Concluding Comment

Main architectural features of microprocessors, such as instructions sets, data formats, instruction formats, and addressing modes, were presented in the preceding sections of this chapter. The presentation was general; the architectural features discussed were not associated with any specific product, except in a few examples. Later on, in subsequent parts, when architectural features of specific products are presented, the reader will recognize the features presented in this chapter, or some features which are very similar. This chapter was primarily intended for the benefit of readers not sufficiently familiar with architectural features of microprocessors.

Architectural features, discussed in this chapter, are closely connected with the concepts of data storage in memory and its addressing. The next chapter will discuss in detail the subject of memory organization in microprocessors, stressing the concept of memory hierarchy and its different levels, such as the register file, the cache, main memory, and secondary memory. Naturally, readers familiar with the above aspects of mcmory organization may skip the next chapter.

4

Memory Hierarchy

4.1 Introduction

The interface between the memory and the CPU plays an important role in establishing the overall performance in any computing system [HaVZ 90, Hays 88, HePa 90]. Memory access operations are always slower than internal CPU operations, particularly in microprocessors, where the CPU is usually on one chip, while the main memory is distributed among a number of other chips. Naturally, access to resources on the same chip is much faster than access to resources outside of the chip. In modern computing systems, including microprocessors, we can establish a number of distinct levels of the memory hierarchy. Five notable levels of the memory hierarchy, numbered 0 through 4, that can be found on a number of recent microprocessors, are illustrated in Fig. 4.1. The levels are:

0. CPU register file. Any register (or any flip-flop for that matter) can be viewed as a particular case of memory. Thus the CPU register file, the closest to the CPU and easiest to access by the CPU, may be regarded as level 0 of the overall memory hierarchy. The register file is actually a part of the CPU.

1. Primary cache. The primary cache is the fastest access memory level outside of the CPU. In most modern microprocessors the primary cache is on the same chip with the CPU. In many systems the primary cache is dual: a separate cache for instructions and a separate cache for data. In case of a dual cache, both the instruction and the data cache are considered to be on the same level in the memory hierarchy.

2. Secondary cache. The secondary cache is considerably larger than the primary cache. On current systems where it is implemented it is outside of the CPU chip and it is a unified cache, containing both instructions and data. With the increase in chip densities, this situa-

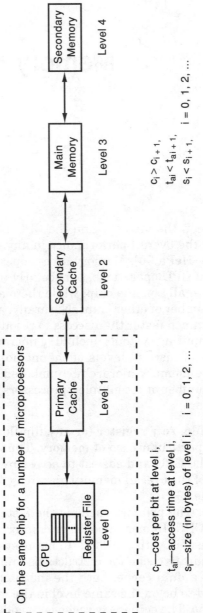

Figure 4.1 Memory hierarchy.

tion may change in the future: in some systems, the secondary cache may be included on the same chip. A dual secondary cache is feasible, however, so far, it has not been implemented on any existing system.

3. Main memory. The main memory is the one actually addressed by the CPU. It contains the code and data of currently running programs. Some of this information may also be in the cache. The main memory is in general of insufficient size to contain all information needed by the users. This is why another level of memory is needed.

4. Secondary memory. The secondary memory is much larger than the main memory. It is used as repository storage of information in any computing system. Magnetic disks belong to the category of secondary memory; they are very useful in information porting from system to system.

The closer the memory hierarchy level is to the CPU, the more expensive (per bit of storage) it is, the faster is its access by the CPU, and the smaller is its size (in bytes). We can express these rules of memory hierarchy quantitatively. Introducing the notation:

c_i—cost per bit at level i

t_{ai}—access time at level i

s_i—size (in bytes) of level i $i = 0, 1, 2,...$

we have the following inequalities:

$$c_i > c_{i+1},$$

$$t_{ai} < t_{ai+1},$$

$$s_i < s_{i+1}, \qquad i = 0, 1, 2,...$$

The higher cost for the lower levels, such as the cache, is the primary reason why the cache size is limited. Another factor limiting the size of the primary cache is the finite number of resources that can be put on a chip. The total size of the on-chip primary cache, in microprocessors of the mid-nineties, is measured in tens of kilobytes. This size will increase in time. However, increasing the cache size beyond a certain limit, increases the access time [Przy 90].

The CPU usually accesses the register file register by register, requesting the information stored in the whole register. In some systems, a register may be accessed for a part of its information only. For instance, in the Motorola M68000 family (Part 3) the registers are 32 bits wide. They may be accessed for all 32 bits, for 16 bits (least significant in the register), and the data registers, even for a single byte (LSB in the register). The CPU usually accesses memory for the same data structures. However, information between memory and cache is transmitted in quanta of lines, or blocks. In most modern microprocessors the line size is 16 or 32 bytes, but there are a few exceptions. Information between the main and secondary memory is transmitted in quanta of pages. In a large number of computing systems there is a standard page size of 4 kbytes. Some systems offer larger page sizes as an option, in addition to the 4-kbyte page. More on line and page size considerations will be presented in Secs. 4.3 and 4.4, respectively.

Details of the properties of the different levels of the memory hierarchy will be presented in the next sections of this chapter.

4.2 Register File

A register file is a small, fast access memory. It is really a part of the CPU. It can be accessed within a fraction of a CPU cycle. It has been denoted here as level 0 in the memory hierarchy. CISC systems of today have a register file of 8 to 16 registers. Many of these registers may be committed to special tasks, such as serving as a *program counter* (PC) or a *stack pointer* (SP), and others. RISC systems (see Chaps. 1 and 6) usually have 32 registers, but some have more (SPARC has 136; see Part 4). Most modern systems have two sets of register files: one for the IU, and one for the FPU. The registers in the register file are usually denoted: r0, r1, r2,... or R0, R1, R2,..., and so on. An example representation of a register file of n registers is shown in Fig. 4.2. Specific examples of register files of actual microprocessors are shown throughout Parts 2 to 4.

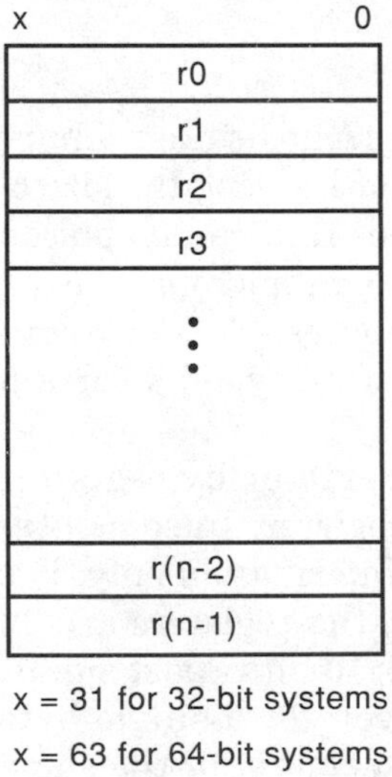

```
x                       0
┌─────────────────────────┐
│           r0            │
├─────────────────────────┤
│           r1            │
├─────────────────────────┤
│           r2            │
├─────────────────────────┤
│           r3            │
├─────────────────────────┤
│            ·            │
│            ·            │
│            ·            │
│                         │
├─────────────────────────┤
│          r(n-2)         │
├─────────────────────────┤
│          r(n-1)         │
└─────────────────────────┘
```

x = 31 for 32-bit systems

x = 63 for 64-bit systems

Figure 4.2 CPU register file (n registers).

Each register is 32 bits wide for 32-bit systems, and 64 bits wide for 64-bit systems. In most systems the bits are numbered starting with the LSB: the LSB is bit 0 and the MSB is bit 31 in 32-bit systems, and bit 63 in 64-bit systems, as shown in Fig. 4.2. Of course, in other systems the bits may be numbered the other way around, denoting the MSB as bit 0.

Most of the registers in the register file may be freely used by the programmer. In most microprocessors, a minority of registers is assigned special tasks by the designers. Some of the mostly used special tasks are as a SP, or a return address (from a subroutine) register. Many of the RISC-type systems assign register r0 (in a few exceptions it is r31) to be permanently hardwired to a zero value. Even if any nonzero value is moved to a so designed r0, the register r0 will remain cleared. This feature is used to generate other features not offered by the design in a direct manner.

Example In a three-address system, in which r0 = 0 on a permanent basis, a register-to-register move instruction is not offered. It may be artificially generated by an add instruction, using r0 as one of its operands:

```
add r0, ri, rj; (r0) + (ri) → rj, since (r0) = 0, equivalent to (ri)
→ rj.
```

The above instruction is equivalent to a move instruction from register ri to register rj.

As noted above, many modern microprocessors have a separate floating-point register file, containing 32 registers in most cases. These registers are naturally denoted in a different manner, such as f0, f1, f2,... and so on, or other similar notations. Practically all FPUs implement the IEEE standard (see Chap. 3), featuring the 32-bit single-precision and the 64-bit double-precision floating-point formats. For this reason all floating-point registers are 64 bits wide, even in 32-bit systems. Some FPUs feature 80-bit-wide floating-point registers to accommodate the 80-bit extended format (see Chap. 3).

Design considerations in selecting the size of the CPU register file will be given in Chap. 6, in conjunction with the general RISC architecture design considerations. The next section takes up the discussion of the next level of the memory hierarchy: the cache.

4.3   Cache

Cache operation is based on the principle of locality [Denn 72, Smit 82]. We can distinguish between two types of locality:

1. Temporal locality (in time). If an information item is accessed by the CPU, there is a high probability that it will be accessed again in the near future (possibly in a few nanoseconds).

2. Spatial locality (in program space). If an information item is accessed, there is a high probability that other items nearby in the program will be accessed in the near future.

It should be noted that the principle of locality will work in well-structured programs where most of the instructions are executed sequentially, where far target jumps are avoided, and near target jumps are kept to the minimum. Although Fig. 4.1 illustrates two levels of cache, the primary and the secondary caches, the discussion in this section will start with a simpler model, featuring just a single-level cache, as shown in Fig. 4.3.

The distinction between the primary and secondary cache levels will be introduced later in this section.

As noted in Fig. 4.3, data between the cache and the CPU are transferred in basic data type quantities such as words, halfwords, and bytes. Information between the cache and main memory is usually transferred in quanta of *lines* (or *blocks*). The lines stored in cache do not have a separate address. They are referred to by their addresses in main memory. The question arises, how then can one know where exactly in the cache is a given line stored (it is assumed that its main memory address is known)? The answer to this question can be known if we know the method of address mapping between the main memory and the cache.

Address mapping

In most existing systems the cache is subdivided into sets. Each set may contain a number of lines. The mapping between the main memory and a cache containing K sets is illustrated in Fig. 4.4.

The mapping works in the following way: line 0 from main memory is stored in set 0 in the cache (if it is in the cache), line 1 in set 1, line 2 in set 2, and so on until the number of sets in the cache is exhausted, with line $K - 1$ from main memory stored in set $K - 1$ in the cache. After that, the count starts again: line K from main memory in set 0, line $K + 1$ in set 1,..., line $2K - 1$ in set $K - 1$, line $2K$ in set 0, and so on. In general, we can say that line X from main memory will be stored in set X(modulo K) in the cache. The value X(modulo K) is obtained by

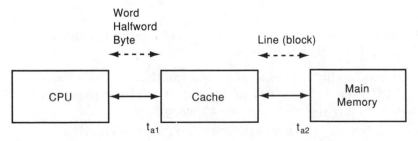

Figure 4.3 CPU, cache, main memory interconnection.

Cache   Main Memory

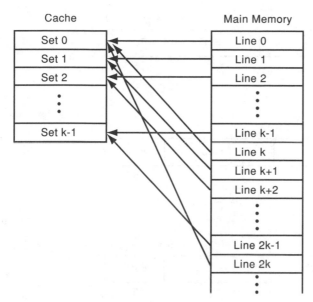

Figure 4.4 Cache-memory mapping.

dividing X/K and taking the integer remainder (both X and K are integers by definition).

Example Given line $X = 270$ in main memory. The number of sets in the cache is $K = 128$. Line 270 will be mapped as follows: $270/128 = 2 + 14/128$, that is, in set 14.

This method of mapping, practiced in most existing systems, is called set-associative mapping [HaVZ 90, Hays 88, HePa 90]. As mentioned before, each set in the cache may contain several lines. Let us say that there are L lines in each set. The main memory line, mapped to a given set, may be stored in any of the L lines in the set. The set-associative mapping with L lines/set can also be called L-way set-associative. So for two lines/set it will be two-way set-associative, for four lines/set: four-way set-associative, and so on.

We can tell which line from main memory resides in which particular line space in its assigned set, by looking at the complete address of the accessed item in the line. This is best illustrated by an example.

Example Given a main memory of 4 Gbytes = 2^{32} bytes. The cache size is 16 kbytes = 2^{14} bytes. The line size is 32 = 2^5 bytes. The cache is two-way set-associative, that is, two lines/set. The complete address of any data item will naturally be 32-bit. The least significant 5 bits of the address identify the byte within any line.

The total number of lines in the cache is $2^{14}/2^5 = 2^9 = 512$ lines.

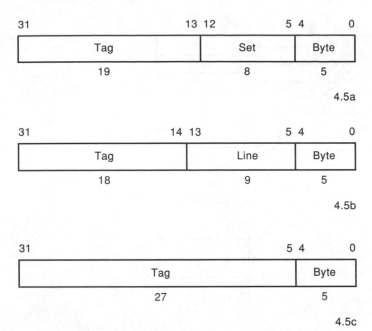

Figure 4.5 Address for a (a) two-way set-associative cache; (b) direct-mapped cache; (c) fully associative cache.

The total number of sets in the cache is $512/2 = 256 = 2^8$ sets.

Hence, the field denoting the set in the cache will have 8 bits. It is the field next to the byte field, as illustrated in Fig. 4.5(a).

The number of main memory lines that may be stored in any set in the cache is $2^{32}/2^5 = 2^{27}$ lines in main memory, $2^{27}/2^8 = 2^{19}$ lines in main memory/set.

Thus, the memory line residing in the cache may be uniquely identified by the 19-bit upper part of the address. The field containing this part is called the tag field, as illustrated in Fig. 4.5. Each line in cache has a tag value associated with it, stored in cache memory, indicating which line in main memory is stored in the specific line space in the cache.

We can note two particular boundary cases of the set-associative mapping:

1. Each set contains only one line ($L = 1$). This is called *direct mapping*. In this case each line in main memory can be stored in only one specific line slot in the cache.

2. There is only one single set in the cache; it includes all the cache lines. This is called *associative* or *fully associative* mapping. In this case, any line in main memory may be stored in any line slot in the cache.

Example Assuming the basic data of the previous example (two-way set associative), assume now a direct mapping. There are $512 = 2^9$ lines in the cache. The address will now be expressed as shown in Fig. 4.5(b). If we assume full associativity, the address will be expressed as in Fig. 4.5(c). Since any line in main memory can be stored in any line slot in cache, then naturally all the 2^{27} lines of main memory can be stored in any cache line, and the tag field is now 27 bits long. The only other field needed in this case is the byte field, indicating the byte in line being accessed.

The higher the level of associativity, the more complex the logic. This is one of the reasons why the level of associativity in practical systems is 2 or 4 in most cases. In some systems (notably the Alpha realization 21064; see Chap. 14) the designers accepted a lower potential hit ratio in exchange for simpler logic, and implemented the direct mapping method.

To summarize, the higher the level of associativity, the higher the hit ratio (not spectacularly, though) on the average (it is very much program dependent). On the other hand, the higher the level of associativity, the higher the hardware complexity and cost. The resulting compromise is somewhere in between. Therefore, mappings used in most systems are four-way, two-way set-associative, and direct.

In general, only a part of a given program can fit into the cache, which is usually much smaller than the main memory. Thus, when a CPU attempts to access any item of information, there are two possible outcomes:

1. The item is in the cache. This event is called a *hit*. The probability of a hit is called the *hit ratio,* denoted by H. Naturally we want to increase H as much as possible.

2. The item is not in the cache. This event is called a *miss*. The probability of a miss is called the *miss ratio.* Since a hit and a miss form a complete probability space (there may be only a hit or a miss, no other event exists), the miss ratio is equal to $(1 - H)$. Naturally, we strive to reduce the miss ratio as much as possible. When a miss occurs, the line containing the missing item is loaded into the cache, replacing another line.

We can now discuss mapping methods in light of the hit and miss ratios. If direct mapping is used, each line from main memory is restricted to be placed in one single line slot in the cache. The line residing in the cache in that slot will be replaced on any miss on a line mapped to that slot. This tends to limit the hit ratio because of the lack of flexibility of line mapping. On the other hand, if we increase the level of associativity (increase the number of lines per set), each line in main memory can be placed in any of the line slots within the set. This increases the flexibility of line mapping and tends to increase the hit ratio. Of course, one of the lines in the set must be replaced. The method according to which it is decided which line will be replaced is called the *replacement algorithm*. There are several replacement algorithms:

1. Random. The line to be replaced is chosen at random. This is the simplest algorithm to implement; however, since the line to be accessed next may be replaced, it tends to increase the miss ratio.

2. First-in first-out (FIFO). The first line placed in the set will be the first to be replaced. This will yield a lower miss ratio than the random method. A circuit keeping track of successive line storage in the set is needed. This method does not keep track of the actual access of the lines by the CPU.

3. Least recently used (LRU). There is a logic circuit which keeps track of actual access of the lines in each set by the CPU. According to the principle of locality, if a line has not been accessed recently, chances are that it will not be accessed any more, and therefore it is a better candidate for replacement. The LRU method is implemented on many systems. It certainly assures a better hit ratio (on the average) than other methods of replacement, although it needs more complicated logic for realization.

Average access time

As mentioned earlier in this section, the hit ratio is highly program dependent. Thus, for different programs, at different instances, we may get different access times by the CPU. We can, however, calculate a mean or average access time [Hays 88], using the individual access times of the cache (t_{a1}), the main memory (t_{a2}), and the hit ratio (H). The probability of having a hit and accessing the cache is H, the probability of having a miss and accessing the main memory is $(1 - H)$. Thus, the average access time is given by the expression of calculating a discrete mean value:

$$t_a = Ht_{a1} + (1 - H)t_{a2} \qquad (4.1)$$

We can also define a line (or block) transfer time t_b, as the time needed to transfer a line from main memory to the cache on a miss:

$$t_b = t_{a2} - t_{a1}$$

or

$$t_{a2} = t_b + t_{a1} \qquad (4.2)$$

A miss consists of the sequence of an attempted reference to a cache line, detection of the fact that the data is missing, the load of that data into the cache, and then the reinitiation of the access.

Substituting expression (4.2) into (4.1), we obtain:

$$t_a = t_{a1} + (1 - H)t_b \tag{4.3}$$

Usually t_{a1} is much smaller than t_b or t_{a2}, while the last two are close to each other in their values. The memory access time, in case of a miss, that is, the time it takes to replace a line in the cache plus the time to deliver the missing item to the CPU, is also called the miss penalty [HePa 90].

Let us now expand the discussion to include a second-level cache (see Fig. 4.1).

Second-level cache

The motivation for using a second-level cache is improved performance. The primary cache (usually on the CPU chip) is limited in size. With the current technology it is several tens of kbytes. An off-chip secondary cache has several Mbytes in today's technology. Thus, the total cache fast-access storage is considerably improved if we use a second-level cache. A block diagram of a system with two levels of cache and a main memory is shown in Fig. 4.6.

The respective access times of the primary cache, secondary cache, and the main memory are t_{a1}, t_{a2}, and t_{a3}, respectively. The hit ratios of the primary and secondary cache are H_1 and H_2, respectively. H_2 is the probability that a miss in the primary cache will hit in the secondary cache. It is assumed that if there is a miss on both cache levels, the missing item is in the main memory. Thus, the event space is such that an accessed item is either in the primary cache, or in the secondary cache, or main memory. The hit probability in the primary cache is H_1. The probability of a miss in the primary cache and a hit in the secondary cache is $(1 - H_1)H_2$. The probability of a miss in both cache levels is $(1 - H_1)(1 - H_2)$. Therefore, the average access time for the system with a two-level cache is:

$$t_a = H_1 t_{a1} + (1 - H_1)H_2 t_{a2} + (1 - H_1)(1 - H_2)t_{a3} \tag{4.4}$$

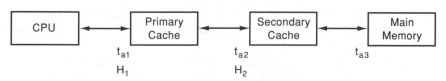

Figure 4.6 Two-level cache.

The miss ratio in the second cache level is also referred to as the *local miss rate,* and the product of the two miss ratios of the two cache levels is referred to as the *global miss ratio* [HePa 90].

Example   Given $H_1 = 0.96, H_2 = 0.50$;
   The local miss rate $= 1 - 0.50 = 0.50 = 50$ percent
   The global miss rate $= 0.50(1 - 0.96) = 0.50 \times 0.04 = 0.02 = 2$ percent

Some actual values of primary and secondary cache parameters are given in the following table [HePa 90]:

Parameter	Primary cache	Secondary cache
Size (kbytes)	1–256	256–4096
Line size (kbytes)	4–128	32–256
Hit time (cycles)	1–4	4–10
Miss penalty (cycles)	8–32	30–80
Miss ratio (%)	1–20	15–30 (local)

Cache operation

Consider the single-level cache model shown in Fig. 4.3. We will now look at the four possible events of access: hit, miss, and read and write operations in each case [HaVZ 90, HePa 90, Smit 82].

I. Hit
   A. Read operation
       The accessed item is transferred from the cache into the CPU. The memory is not involved.
   B. Write operation
       There are two options open to designers:
       1. Write through method (also called store through). The memory location is updated together with the cache. This method is simple to implement, but it involves multiple bus transfers and memory write operations.
       2. *Write back* method (also called *copy back*). Memory is updated only when the updated line (block) is replaced. This method has a reduced memory bus traffic, but it may have lengthy periods when there are different values at the same addresses in memory and cache (in other words, we have memory incoherency).
           In some systems both the write through and write back methods are offered as options to the user for selection.

II. Miss
  A. Read operation
     The line containing the missing item that was accessed is transferred from memory into the cache, replacing another line. In this case there is a speedup option called load-through (or load bypass [Smit 82]), where the missing item is forwarded to the CPU immediately upon arrival in the cache, without waiting for the loading of the whole line. This will speed up the operation, particularly if the missing item is close to the beginning of the line. In well-structured programs this will usually be the case. There is also a technique called wraparound load [Smit 82], by which the target bytes are fetched first. Then, a wraparound is performed, reading in the rest of the line to the end and then around to the beginning.
  B. Write operation
     There are two design options available:
     1. *Write-allocate* (also called *fetch on write*). The line containing the missing item is loaded into the cache. This makes sense considering the principle of locality. Chances are that the line in question will be accessed again by the CPU in the near future. If the write back method is also used, the memory bus traffic will be considerably reduced.
     2. *No write-allocate*. The item written into is modified in the memory. The missing line is not transferred into the cache.
        In some systems both write-allocate and no write-allocate options are offered to the user for selection.

Design considerations

Some basic design considerations in establishing cache parameters, such as cache size, line size, and associativity, will be discussed in the following paragraphs [Przy 90, Smit 82].

Cache size

The larger the cache, the more information it will contain, and therefore the higher will be the hit ratio. Instinctively, we may strive to increase the cache size indefinitely. There are, however, some practical restrictions on increasing the cache size. First of all, the larger the cache, the higher its cost. As argued above, cache is composed of a more expensive semiconductor material than the main memory. There is, however, another and probably more important restriction. The larger the cache, the more complicated will be its access logic. The complexity will increase with the increase of the cache size to a point at which any further cache size increase may cause a longer access time, thus defeating the main cache advantage of lower access time. The cache size is of course limited by the size of the board or the chip area. For these reasons the cache size is kept up to several hundreds of kbytes for primary caches, and several Mbytes for secondary caches.

Line size

The relative advantages of a small and a large line (block) size are summarized in the following table [Smit 82]:

Small Line	Large Line
Shorter line transmission time	More information transmitted on each line fetch
Less likely to contain unneeded information	Smaller number of lines in cache; less extra logic (tags, etc.)
Smaller bus data width needed	Decreased miss ratio

As we can see, there are arguments for and against a large line size. Thus, the optimal solution is somewhere in between. Experimental studies have shown [Smit 87] that best performance is attained for cache line sizes within the interval of 16 to 64 bytes. In most modern microprocessors the line size is either 16 or 32 bytes.

Associativity

The higher the associativity [HiSm 89, Smit 78], that is, the more lines in a set, there is more flexibility in placing a line within a set, while replacing a least recently used line (and hence most probably not needed anymore). This tends to increase the hit ratio, or decrease the miss ratio. On the other hand, increasing the associativity tends to increase circuit complexity and cost (both material and design cost). Therefore, again, a tradeoff is in order. Modern systems implement either four-way, or two-way associativity, or direct mapping. It has been a decision of some designers to give up somewhat on the hit ratio in favor of logic circuit simplicity, by featuring direct mapping (DEC Alpha AXP, Chap. 14; MIPS R4000 and R4400, Chap. 17). Furthermore, it was argued in [Hill 88] that as caches grow larger, the difference between miss ratios of direct-mapped and set-associative caches decreases, and set-associativity loses its relative advantage while still retaining the disadvantage of higher complexity. Because of the direct-mapped cache's reduced complexity and simpler access path, its cost is lower and its cache hit and average access times are shorter [Hill 88].

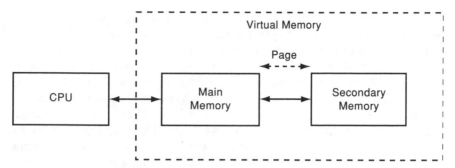

Figure 4.7 Virtual memory.

4.4 Virtual Memory and Paging

We shall concentrate now on the interface between the CPU, main memory, and secondary memory, as illustrated in Fig. 4.7. Information between main and secondary memory is transferred in quanta of pages. Page sizes range from 0.5 to tenths of kbytes. Some systems have options for even larger pages of several Mbytes. However, most actual systems use a 4-kbyte page.

The interface between the main and the secondary memory is managed by the *operating system* (OS) in a manner that is completely transparent to the user [HaVZ 90, Hays 88, HePa 90]. The overall memory system appears to the user as a very large memory whose size is that of the secondary memory. This is what we call the *virtual memory*. It may be formally defined as follows [HaVZ 90]:

> *Virtual memory* is a hierarchical storage system of at least two levels, managed by an operating system (OS) to appear to a user as a single, large, directly addressable main memory.

The virtual memory is addressed by *virtual* (or *logic*) addresses. The hardware-based main memory, identified as the physical (or real) memory, is addressed by the *physical* (or *real*) *address*. Some virtual memory parameters are given in the following table [HePa 90]:

Page Size	0.5 kbytes–4 Mbytes
Main memory size (Mbytes)	4–4096
Hit time (cycles)	1–20
Miss penalty (cycles*10^5)	1–6
Miss rate (%)	0.000001–0.0001

The CPU usually accesses memory by producing a virtual address. However, the physical memory must be addressed by physical addresses. Therefore, the virtual address must be translated into a physical address. In most systems the virtual address is 32 bits long. There are systems featuring a larger virtual address. Such systems will be described in chapters dedicated to specific systems, particularly in Part 4. The current discussion will concentrate on 32-bit addresses, both virtual and physical. Not all of the 32 bits of the address must be translated. For instance, if the page size is 4 kbytes = 2^{12} bytes, the least significant 12 bits of the address (virtual and physical) refer to the internal address within the page, called the *offset* (in some systems it is called *displacement*). This is illustrated in Fig. 4.8.

The starting address of the page is called the *page base* (or the *page base address*). The offset is the distance in bytes between the page base and the data item within the page which is being accessed (see Fig. 4.8). Naturally, since the offset is the same for both the virtual and the physical address, it does not need to be translated. It is only the field containing the upper bits of the virtual address (20 bits for 4-kbyte pages) that must be translated. This field is called the *virtual page number* (vpn) in the virtual address, and the *physical page number* (ppn) in the physical address (see Fig. 4.8). The system pages are stored in *page frames* in the main memory. The page frames start from address zero, and after that they are allocated consecutively in mem-

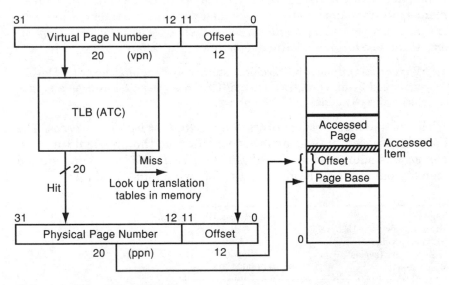

TLB—translation look aside buffer
ATC—address translation cache

Figure 4.8 Virtual to physical address translation.

ory without skipping locations, and without overlapping. Thus, if the page size is 4 kbytes, page 0 will start at location 0 and end at location 4095 (4 kbytes–1), page 1 will start at location 4096 (4 kbytes), page 2 will start at location 8192 (8 kbytes), and so on. The page base address will be divisible by $4096 = 2^{12}$, in other words, its least significant 12 bits will always be zero. Therefore, the upper 20 bits of the page base address are sufficient for unambiguous specification of any page base (see Fig. 4.8).

In general, the complete information for translating any virtual vpn into its physical counterpart, the ppn, is stored in *page tables* in the main memory. The structure of page tables and the procedure of their access will be discussed later in this section. Suffice it to say that this access is very time consuming. For this reason, all modern systems, microprocessors included, feature a shortcut for virtual to physical address translation. The CPU memory management subsystem (see also Chap. 2) includes a relatively small special-purpose cache, structured like a dictionary for translation of vpn into ppn. Such a cache is called *translation lookaside buffer* (TLB), or *address translation cache* (ATC). The term TLB is used in the majority of systems. The term ATC is used primarily by Motorola in its M68000 family (see Part 3). Obviously, the term ATC describes more clearly the actual task of this subsystem. Since the TLB is relatively small (up to about a couple of hundred entries, and usually 32 to 64 entries), not all page addresses may be included in it. Thus a hit or a miss is possible, as in any regular cache (see Sec. 4.3). Some actual TLB parameters are summarized in the following table [HePa 90]:

Entry size (bytes)	4–8
TLB size (bytes)	32–8192
Hit time (cycles)	Less than 1
Miss penalty (cycles)	10–30
Miss rate (%)	0.001–0.02

It follows from the above data that the probability of a TLB hit ranges from 0.98 to 0.999. Thus, a page base translation will usually take a single CPU cycle. In those few cases (probability of 0.001 to 0.02) when we have a miss, page tables will have to be looked up in main memory. The mechanism of page tables lookup will be discussed next.

Page table mechanism

Each page in memory has a *page table entry* (PTE) corresponding to it. The PTE contains:

1. The upper bits of the page base address (or the ppn, as specified earlier)

2. Status and protection bits related to the page

The PTEs are located in *page tables*. A system may have several levels of page tables. At this point it is better to explain the details of the paging mechanism by example. The paging mechanism of the Intel i386 and i486 will be used (a similar mechanism is used in other systems). The Intel paging mechanism is illustrated in Fig. 4.9 (see also a more detailed discussion in Part 2).

Each page has a PTE corresponding to it, located in a page table, as shown in Fig. 4.9(*c*). The PTE format is shown in Fig. 4.9(*a*). The upper 20 bits of the PTE contain the page base address (its upper 20 bits), called by Intel the page frame address. Bits 0 through 6 are used to store status and protection information for the page (to be discussed in

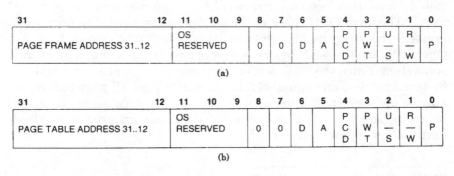

(a)

(b)

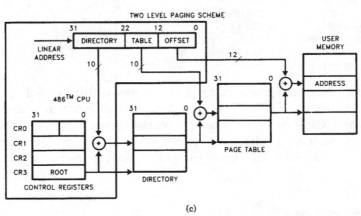

(c)

Figure 4.9 Paging mechanism. (*a*) Page table entry (points to page); (*b*) page directory entry (points to page table). (*Courtesy of Intel Corp.*)

Part 2). The PTE is 32 bits (4 bytes) wide. The Intel page size is 4 kbytes. Thus, the address offset is 12 bits, as shown in Fig. 4.9(c). The page table contains 1024 (1 kilobyte) entries. Therefore, the size of a page table is 4 kbytes, as any other page in the system. Since there are $1024 = 2^{10}$ entries in the page table, we need a 10-bit field to point to each entry. This is the 10-bit field (bits 12 to 21) in the address, as shown in Fig. 4.9(c). The base of each page table is pointed to by a *page directory entry* (PDE), located in the page directory. Each PDE is 32 bits (4 bytes), and the directory contains 1024 PDEs. Thus, the directory is also page size—4 kbytes. Each PDE is pointed to by the entry in the upper 10-bit field of the address, the directory field, as shown in Fig. 4.9(c). The PDE format is shown in Fig. 4.9(b); it is practically identical to the format of the PTE. The base of the directory is pointed to by *control register* 3 (CR3), located in the CPU.

The Intel x86 family paging system is very convenient to implement, because of the standard, uniform size of all data structures involved: page frames, page tables, directory—all 4 kbytes. This permits a uniform treatment of these structures and simplifies the hardware which supports the paging mechanism. It also simplifies the part of the OS software in charge of managing the paging mechanism. Other computer manufacturers have adopted similar paging mechanisms. Even Motorola, which has a three-level paging mechanism in its M68000 family (see Part 3), has adopted a two-level paging mechanism, similar to that of Intel, in its RISC-type MC88110 system (see Part 4).

Example   Combined page mechanism and cache addressing, given the i486 system, whose page table mechanism is illustrated in Fig. 4.9, and was discussed in the preceding paragraphs. The i486 has an on-chip 8-kbyte unified cache (code and data in the same cache). The cache is four-way set-associative with 16 bytes/line.

Number of lines in the cache: $2^{13}/2^4 = 2^9 = 512$

Number of sets in the cache: $512/4 = 128 = 2^7$

Given a virtual address: 00F0 10A4 H

The binary equivalent of this address is placed in a 32-bit address format, shown in Fig. 4.10. The three cache fields, shown on top, are: BYTE, 4 bits (0 to 3), SET, 7 bits (4 to 10), and TAG, 21 bits (11 to 31). The three fields of the page table mechanism are: OFFSET, 12 bits (0 to 11), TABLE, 10 bits (12 to 21), and DIRECTORY, 10 bits (22 to 31).

The byte in line value is 4, the set is 0A H (10 decimal), and the tag is 001E02 H.

The offset is 0A4 H. The table entry is 301 H. The internal address in the page table is 301 H $\times$ 4 = C04 H. The directory field entry is 003 H. The internal address in the directory table is 003 H $\times$ 4 = 00C H.

It should be noted that bit 11 of the address plays a double role: it is the lowest bit of the TAG in the cache access address, and it is the upper bit in the page offset field. Therefore it is directed to two places in the system:

1. To the TLB as the lowest bit.

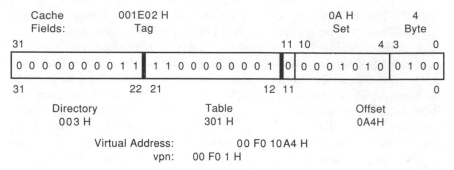

Figure 4.10 Combined addressing example.

2. As bit 11 (upper page offset bit) of the physical address forwarded to the hardware memory. As argued before, the whole offset field does not undergo translation.

From the description of the page table mechanism one can easily see that in case of a TLB miss, the CPU has to access memory two or three times (depending if the number of page table levels is two or three, respectively) to obtain an address translation. This explains the large miss penalty (10 to 30 cycles) incurred in such a case. Fortunately, it does not happen too often (less than 2 percent of all accesses). A TLB miss is not the only possible cause for a significant delay in processing. For very large programs, not all of the program or data may be placed in the main memory at any time. Thus, it is possible that when a CPU attempts to access any information item, the page containing it may not be in the main memory. This phenomenon is called a *page fault*. The OS handles page faults by bringing the missing page into the main memory, replacing another page. Fortunately, the probability of a page fault is just 0.001 percent.

One can, of course, reduce the TLB miss ratio by increasing the size of the TLB. This can be helpful up to a point. Increasing the TLB too much will eventually cause an increase in its access time, defeating the benefit of a reduced miss ratio. A fast access to the TLB is extremely important in modern systems. The reason is that most caches are addressed by the physical address to avoid aliasing (when more than one virtual address corresponds to the same physical address). The TLB should be able to produce a translation while the cache is accessed.

Another issue is the memory size encompassed (mapped) by the TLB. This memory amount is called the *mapping size* of the TLB. The mapping size is equal to the product of the number of TLB entries multiplied by the page size.

Example The TLB size of the Intel i486 is 32 entries. The page size is 4 kbytes.

The mapping size of the TLB is $2^5 \times 2^{12} = 2^{17}$ bytes = 128 kbytes.

On the other hand, Motorola MC68040 has two ATCs, one for code and one for data, 64 entries each. The total number of ATC entries is 128. The MC68040 has a choice of either 4-kbyte or 8-kbyte pages.

For 4-kbyte pages the mapping size is $2^7 \times 2^{12} = 2^{19}$ bytes = 512 kbytes.

For 8-kbyte pages the mapping size is double: 1 Mbyte.

The mapping size may of course be increased by increasing the page size, and not just the TLB number of entries. But is it advantageous to increase the page size indefinitely? Some design considerations are discussed in the following.

Page size design considerations

If we increase the page size for the same size of memory, the total number of page frames will be smaller. Hence we will have smaller page tables and a smaller amount of such tables, thus saving on overhead space in memory. Since a larger page contains more information, any page transfer will be more efficient. These arguments support a large page size.

On the other hand, a smaller page can be transferred faster on the same bus, saving time and reducing the page fault penalty. If any information structure is not equal to an integer number of pages, there will be less memory space wasted in unfilled pages for smaller pages. These arguments support a smaller page size.

Clearly, the answer is somewhere in between, although the question of optimal page choice is still an active research subject and has not been fully resolved. In any case, it is quite clear today that DEC's choice of a 0.5-kbyte page for the VAX (dictated by compatibility considerations with the PDP-11) was far from optimal (although it should be said that this was corrected in the recent Alpha which offers several page sizes, starting with 8 kbytes). As stated earlier, most manufacturers use the 4-kbyte page as an unofficial standard, while some offer more than one page size. This has become a trend in recent designs. Some recent research studies [ChBJ 92, Tall 92], although not quite conclusive, point to an advantage in having more than one page size in a system. Clearly, for very large program packages it may be of advantage to offer a larger page option. This was indeed realized in the recent Intel design of the Pentium (see Part 2), which offers a very large page option of 4 Mbytes (for large OS software), in addition to its regular 4-kbyte page.

Paging is not the only way of subdividing the memory. There exists also another way of memory partitioning called *segmentation*. It is practiced primarily by Intel in its x86 family of microprocessors. Some basic principles of segmentation will be discussed in the next section. Details of the Intel segmentation mechanism will be presented in Part 2.

4.5   Segmentation

The concept of a *segment* can be defined [Hays 88] as a set of logically related contiguous words generated by a compiler or a programmer. Thus, *segmentation* is a memory management mechanism that allocates main memory by segments and supervises any segment-related activities.

The main differences between pages and segments may be summarized as follows:

Page	Segment
Fixed size, or a finite number of fixed sizes	Variable size, from a single byte to the whole physical memory
Page frames allocated contiguously in memory, no overlapping	May be placed anywhere in memory and may overlap
Not characterized by type of information	Each segment defined by the type of information in it

Segments are pointed to in memory by a segment table mechanism similar to that of paging. There is also CPU caching of segment information to speed up segment access. Details of segmentation features and mechanisms, as implemented in the Intel x86 family, will be given in Part 2.

The advantages of segmentation are:

1. Segmentation supports well-structured software, since segments can contain meaningful information units.

2. Supports more compact code, since references within a segment can be shorter.

3. Segmentation lends itself to efficient implementation of sophisticated memory management, virtual memory, and memory protection techniques. This is because segmented information is encapsulated in segments classified according to the type of information stored in them. Each segment may be handled efficiently as a single unit, regardless of its size.

4. Segmentation supports efficient typing (classification) of data, thus contributing to the narrowing of the semantic gap [Myrs 82].

Segmentation disadvantages are:

1. Increased hardware complexity

2. Execution time overhead in handling segmentation resources

3. Extra memory need for segmentation tables

4. Added memory fragmentation for small segments

In spite of some advantages of segmentation, it is not practiced by most computing systems designers because of the extra complexity and overhead involved. It is practiced primarily by Intel's x86 family. However, in the latest models, starting with i386 (see Part 2), segmentation is an option; the system may be run without it.

4.6   Concluding Comment

The importance of the memory system and memory management cannot be overemphasized. No matter how fast and efficient is the processor (CPU), if the memory system is inefficient and slow, memory access bottlenecks will arise and the overall system performance will be low. Therefore, in any system design adequate attention should be paid to the design of memory hierarchy subsystems, such as memory interface, cache, TLB, CPU memory hierarchy support registers, paging mechanism, and other related features. Various aspects of memory hierarchy design are a part of active research and development conducted in numerous universities, government agencies, and industrial companies.

5

Pipelining

5.1 The Instruction Pipeline

Pipelining is one of the most commonly used features of parallel processing [Hays 88, HePa 90, Hwan 93]. An operation is subdivided into a number of elementary suboperations, say k. We then form a k-stage system and execute the above suboperations in each stage, one after the other. Thus, a k-stage pipeline is formed. If we continue to send data into the pipeline, then at any time the k-stage pipeline will handle k sets of data simultaneously, performing an elementary suboperation on each of the k sets of data at each of the k stages. The pipeline is analogous to a manufacturing production line, consisting of a number of stages. At each stage some operation is performed on a product, until the final product is obtained at the last stage.

We differentiate between two basic types of pipelines:

1. *Instruction pipeline*—where different stages of instruction fetch and execution are handled in a pipeline.

2. *Arithmetic pipeline*—where different stages of an arithmetic operation are handled along the stages of a pipeline.

The discussion in this chapter will concentrate on the details of the operation of the instruction pipeline. This topic is crucial in the design considerations of modern microprocessors. The arithmetic pipeline is also very important. However, it involves the details of design of the *arithmetic logic unit* (ALU), a topic completely outside of the scope of this text. A good coverage of arithmetic pipelines can be found in [Hays 88].

The discussion of the instruction pipeline will start with a simple example of a four-stage pipeline ($k = 4$). The four stages are:

1. F—fetch. The instruction is fetched into the CPU.

2. D—decode. The instruction is decoded. The register file is accessed for operands.

3. E—execute. The instruction is executed.

4. W—write back. The result of the execution is written back, that is, stored.

A time diagram of the above pipeline is illustrated in Fig. 5.1. A similar pipeline is actually practiced in the integer unit of Intel 860 (see Part 4). The diagram shows four stages of execution of four subsequent instructions: i, i + 1, i + 2, and i + 3. It similarly continues in both directions prior to instruction i, and after instruction i + 3. One can observe that at any time the pipeline is busy on some aspect of four consecutive instructions simultaneously.

The diagram in Fig. 5.1 is drawn under idealized conditions, assuming that the pipeline execution proceeds in a smooth manner. This is not always the case. It is assumed in the diagram that every one of the four stages can be completed in a single cycle. While it may be assumed that the decoding will not take more than one cycle, it may not be assumed about the other operations. If an instruction consists of more that one word, and the data bus is only one word wide, the fetch may take more than one cycle. Only in systems where all instructions are no longer than the width of the data bus, may we assume a single-cycle fetch at all times. The most difficult thing is to achieve single-cycle execution. This is certainly not always possible, particularly for instructions such as multiply, divide, and floating-point operations. As will be argued later in Chap. 6, only in RISC systems can we achieve single-cycle execution for most instructions (Intel 860 is RISC-type). Completing the storage of the result in a single cycle W is not a problem if the result is stored in one of the CPU registers. However, if the result is to be stored in an off-chip memory, it may take more than one

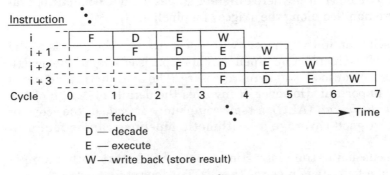

Figure 5.1   An example of a four-stage pipeline.

cycle. As will be seen in Chap. 6, all operations in RISC systems are register-to-register. This means that the result is always stored in a CPU register, and therefore, there is no problem to complete stage W in a single cycle.

There are other potential problems that may disrupt the smooth execution of a pipeline. If we have to access memory to fetch an instruction (stage F) for one instruction, and access memory to store the result of the other (stage W) at the same time, we have a conflict. If an instruction depends on the result of the preceding one, that is, it uses as an operand the destination of the previous instruction, we have another problem. If one instruction happens to be a branch or a jump instruction, we have to interrupt the flow of instructions in the pipeline, and switch to another flow of instructions. All these phenomena which tend to disrupt the smooth execution of a pipeline, as shown in Fig. 5.1, are generally called *pipeline hazards* [HePa 90]. Pipeline hazards, and ways to minimize their effect, will be discussed in detail in the next section.

5.2   Pipeline Hazards

We can define three types of pipeline hazards [HePa 90]:

1. Structural hazards arise from resource conflicts when the hardware cannot support all possible combinations of instructions in simultaneous overlapped execution in different pipeline stages.

2. Data hazards arise when an instruction depends on the results of a previous instruction in a way that is exposed by the overlapping of instructions in different stages of the pipeline.

3. Control hazards arise from the appearance of branch, jump, and other control flow change instructions in the pipeline.

We shall discuss each pipeline hazard type separately.

Structural hazards

Let us check which system resources are used in each stage of the pipeline in Fig. 5.1:

Pipeline Stage	Resources Needed
F	PC, MAR, address and data bus
D	Decoder, internal bus
E	ALU, MAR, address and data bus
W	Internal bus

Note: ALU = arithmetic logic unit; MAR = memory address register; PC = program counter.

Looking at the fourth cycle in Fig. 5.1, we can see that the MAR and the address and data buses are needed in the E stage of instruction i + 1 (to possibly fetch an operand from memory during the execution stage), and at the same time they are needed in stage F of instruction i + 3 (to fetch the instruction). The internal bus is needed during the D stage of instruction i + 2 (to transfer data from the register file), and at the same time, during the W stage of instruction i (to store the final result in the register file). We definitely have a conflict of resources in this case, that is, a structural hazard.

The effects of structural hazards can be alleviated by replication of resources. For instance, if we have a dual cache, one for code, and one for data, with separate access buses and MARs, the CPU can access the code cache at stage F of one instruction, and simultaneously the data cache in stage E of another instruction. Using multiple internal buses with a multiport register file will alleviate the problem of the simultaneous use of internal buses and register file access by different instructions, at different pipeline stages. Such features are indeed practiced by modern microprocessors. Another problem that may arise is due to the necessity of incrementing the PC after the instruction was fetched. If we attempt to use the ALU, there will be a conflict with another instruction at its E stage. This problem can be solved by featuring a special incrementing logic circuit dedicated to the PC, as practiced in most systems.

Data hazards

Dealing with data hazards can be best explained by example. Consider a sequence of two arithmetic instructions:

```
add r3, r2, r1; (r3) + (r2) → r1, register r1 is the destination
sub r4, r1, r5; (r4) - (r1) → r5, register r1 is a source
```

Let us look more closely at the pipelined execution of the two instructions:

Cycle	1	2	3	4	5
Add	F	D	E	W	
			New (r1) calculated here	(r1) stored here	
Sub		F	D	E	W
			(r1) read here		

Looking at the above sequence we can easily see that while the new value in r1 will be stored during cycle 4, an attempt to read it will be made in cycle 3 during stage D of the sub instruction. If nothing is done about it, the sub instruction will use the old, stale value in r1, and a wrong result will be obtained. This is the essence of a data hazard.

The simplest way to deal with it is to stall the pipeline for two cycles:

Cycle	1	2	3	4	5	6	7
Add	F	D	E	W			
Sub		F	ST	ST	D	E	W

Note: ST = stall in the pipeline.

Doing it this way, and this is a very simple solution, we lose two cycles, but make sure that the new value in r1 is used for the sub instruction.

There is of course a more sophisticated solution, requiring a hardware investment. The new value of r1 is actually calculated at cycle 3 in the E stage of add. It is then needed in cycle 4, which is the E (execution) cycle of the sub. What can be done in this case is to capture the new value to be stored in r1 in a special register at the output of the ALU, and forward it back to the ALU input as an operand for the subsequent sub instruction. The storage of the new value in the r1 register can then proceed in the W stage of add (cycle 4) as usual. Meantime, the sub instruction receives the new value at the beginning of its E stage and a correct result is now obtained. This method is called *for warding.* A block diagram illustrating the forwarding method is shown in Fig. 5.2 [HePa 90].

Not all types of data hazards can be resolved by forwarding. It was possible to do so in the previous case because the actual calculation in the first instruction was completed during the E cycle. The reason for

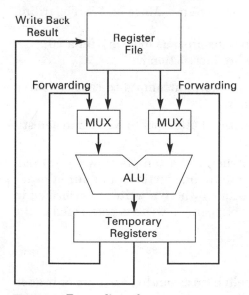

Figure 5.2 Forwarding scheme.

that was that it was a register-to-register operation, executable within a single cycle. This will not be the case in a memory access instruction, such as a load. Even if the data item is loaded from an on-chip data cache, one needs a cycle to calculate the effective address and complete the virtual-to-physical address translation, and then, a cycle to access the data cache, bring in the data item, and store it in the destination register. The result of a load instruction is not available at the end of the E stage, and no forwarding can be accomplished.

Example   Given the instruction sequence

```
load memr,r1; (memr) → r1, r1 is the load destination
add r1,r2,r3; (r1) + (r2) → r3, r1 is the add source
```

   The add instruction will have to be stalled until the end of the W stage of the load, when the new value will be stored in r1. The total delay in the pipeline will be two cycles.

When the detection of a pipeline hazard and the subsequent pipeline stall are performed by the hardware, the mechanism is called a *pipeline interlock*. Of course, this may also be accomplished by software, as it is done in the MIPS family of microprocessors [HePa 90, Tabk 90b; see Part 4 of this text]. One method of hardware-based keeping track of data hazards is by *scoreboarding*. There is a special scoreboard register whose bits represent CPU registers. If register ri is a destination in a certain instruction, bit i of the scoreboard register is set. It is cleared when the instruction is completed. As long as bit i is set, no other instruction may use register ri. Scoreboarding was recently used mainly in RISC-type systems. More on scoreboarding will be said in Chap. 6 on RISC.

In general, data hazards may be classified as follows [HePa 90]:
Assume instruction i occurs before instruction j.

1. RAW—read after write. Instruction j attempts to read a source before instruction i writes into it.

2. WAR—write after read. Instruction j attempts to write into a destination before instruction i reads it.

3. WAW—write after write. Instruction j attempts to write into an operand before it is written into by instruction i (an out of order write). This can happen in pipelines where a write is performed in more than one stage. It cannot happen if a write is performed in the W stage only.

Control hazards

When an instruction within a pipeline turns out to be a jump or branch instruction, the instructions subsequent to it in the pipeline must be flushed, and a new target instruction fetched. This is the essence of a control hazard in a pipeline. In order to minimize the effect of a control hazard (i.e., minimize the number of instructions flushed from the pipeline) the following steps must be taken:

1. Detect the branch as early as possible in the pipeline. If it is an unconditional branch, this can easily be done in stage D. For a conditional branch it is more difficult, since a condition must be tested after the instruction is decoded, taking up more time.

2. Attain the target address and load it into the PC as early as possible. If the target address must be computed following some addressing mode, an extra cycle may be required.

One of the ways of reducing the delay caused by a control hazard is to have a *branch target cache* in the CPU. A branch target cache contains a set of some first instructions of possible branch targets. If a branch target happens to be present in the branch target cache, it takes much less time to fetch it into the pipeline, than if one had to access memory for it.

The most simple way of handling a control hazard is to stall the pipeline until the arrival of the target instruction, while flushing the prefetched instructions following the branch instruction. If the branch was detected in the D stage, the PC loaded fast, and the target instruction was readily available in the branch target cache, only a single cycle may be lost. Otherwise, the pipeline may be stalled for more cycles.

Another way of dealing with control hazards is by *branch prediction*. One can design the hardware assuming (predicting) that the branches will be *not-taken*. In this case, instructions following the branch will continue to be executed. When and if it turns out that the branch is actually taken, the pipeline is stopped, results of the instructions following the branch flushed, and the correct target instruction is fetched. Similarly, one can design the system with a branch *taken* prediction performing in a similar manner. The outcome of a branch prediction policy is strongly program dependent. On the average, it may work either way, and nothing may be gained by branch prediction. There are systems, such as Intel i960 microcontrollers [HiTa 92], which offer the user an option of taking up the policy of branch taken or not-taken prediction. This is helpful if the user has specific and reliable information about the branching behavior of the program.

A method of reducing the penalty of control hazards, practiced in many RISC-type systems, is the method of the *delayed branch*. In this case, the instruction following the branch is always executed, while the branching is delayed for a whole cycle. Thus, no cycles are usually lost, and at least, the number of lost cycles is minimized. More on delayed branch is given in Chap. 6 on RISC. Although the delayed branch method was widely utilized in the first generations of RISC-type systems (the Motorola M88000 family features the delayed branch as a user option), its use becomes too complicated in the new superscalar systems where two or more instructions are fetched and processed simultaneously. Superscalar systems and *instruction level parallelism* (ILP) are discussed in the next section.

5.3 Instruction Level Parallelism

Instruction level parallelism can be defined as a technique of simultaneous issue and processing of multiple instructions within a single processor (CPU). In an n-issue ILP system, n instructions are issued per CPU cycle, and n results per cycle may be attained.

A more formal definition of ILP is given by Rau and Fisher [RauF 93]:

> ILP is a family of processor and compiler design techniques that speed up execution by causing individual machine operations, such as memory loads and stores, integer and floating-point operations, to execute in parallel.

We distinguish the following main types of ILP:

1. *Superscalar,* where a number of instructions are issued simultaneously each cycle. A two-issue superscalar execution, in a four-stage pipeline, is illustrated in Fig. 5.3(*a*).

2. *Superpipelined,* where a number of instructions are issued within a cycle, but not simultaneously. In an n-issue superpipelined system, a new instruction is issued every $1/n$ of a cycle. A two-issue super-pipelined system is illustrated in Fig. 5.3(*b*). It is in analogy with running the pipeline at a double frequency. Similarly, an n-issue superpipelined system is in analogy with running the system at a frequency n times as fast [Joup 89, JoWa 89].

    Some authors [HePa 94] characterize superpipelined execution as "*superpipelined* processors, an informal term suggesting a deeper pipeline than the five-stage model." It is felt that defining the superpipelined concept by the depth of the pipeline is a rather vague way of doing it. The definition in [Joup 89] is more precise and clear.

3. *VLIW—very long instruction word,* where an instruction contains multiple operation codes with their operand specifications. The design of a notable VLIW system is described in [Colw 88]. The details of VLIW operation are outside the scope of this text.

Most modern microprocessors implement superscalar operation, two-issue in most cases (see Parts 2 to 4). There are some exceptions

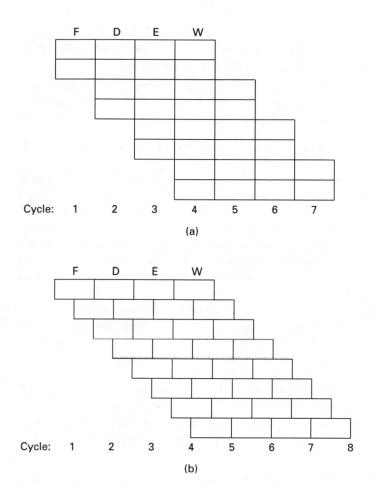

Figure 5.3 Superscalar and superpiplined execution. (*a*) Two-issue superscalar; (*b*) two-issue superpipelined.

where three-issue (SuperSPARC), or four-issue superscalar operation is implemented (IBM RS/6000; see Part 4). Some computer manufacturers are currently working on even higher-issue superscalar systems (such as eight-issue). The only superpipelined system produced to date, is the MIPS R4000 and R4400 (see Part 4).

Rau and Fisher provide a more profound classification of ILP systems [RauF 93]:

1. *Sequential architectures,* for which the program is not expected to convey any explicit information regarding parallelism. Superscalar systems belong to this class.

2. *Dependence architectures,* for which the program explicitly indicates the dependencies that exist between operations. Dataflow processors [ArCu 86, Papd 91] belong to this class.

3. *Independence architectures,* for which the program provides information as to which operations are independent of one another. VLIW systems belong to this class.

A comparison of the above classes of ILP is given in Table 5.1, originally published in [RauF 93].

In an n-issue superscalar system, n instructions are fetched and decoded simultaneously. Subsequently, they must also be executed simultaneously in order to keep the advantage of a superscalar operation and attain n results per cycle. In order to execute n instructions simultaneously, we need to have n operation resources, some of which may be replicated. For instance, if we fetch two integer arithmetic instructions at the same time, we need two integer units to process them simultaneously.

Suppose we have an adequate number of resources. This alone will not solve all possible problems arising in superscalar operation. There is also a problem of data dependence. As argued earlier in this chapter, data dependence poses a serious hazard even in regular pipelines. It is much more serious in superscalar systems. If two or more dependent instructions are fetched simultaneously, forwarding will not be applicable, and stalling one or more of the pipelines might be unavoidable.

In a two-issue superscalar it is not difficult to group most programs into pairs of independent instructions. This can be done either by the

TABLE 5.1   Comparison of ILP Architectures

Feature	Sequential	Dependence	Independence
Additional information required in the program	None	Complete specification of dependence between operations	A partial list of independences
Type of ILP	Superscalar	Dataflow	VLIW
Analysis of dependences by	Hardware	Compiler	Compiler
Analysis of independent operations by	Hardware	Hardware	Compiler
Operation scheduling by	Hardware	Hardware	Compiler
Role of compiler	Rearranges the code	Replaces some analysis hardware	Replaces all analysis and scheduling hardware

SOURCE: [RauF 93]

user or the compiler. In three- or four-issue systems such grouping is more difficult, but still possible. However, more stalling should be expected. If we keep increasing the issue, the class of programs where instructions can be grouped into larger independent groups, will get smaller and smaller. In a large-issue system (eight-issue and up), dependence among instructions issued together should be expected, multiple pipeline stalling may occur, and the resources may be under-utilized most of the time. For this reason, most of the existing super-scalar systems are two-issue, and very few three- or four-issue. There are some manufacturers however, who are working on higher-issue systems. Their success remains to be seen in the future.

A feature alleviating some of the data hazard problems in super-scalar operation is register renaming. *Register renaming* prevents stalling of the completion of load operations by allowing the load oper-ation to complete even though some previous operation using the same destination register as an operand has not yet moved into the decode stage and accessed that register. Even though it appears that the orig-inal contents of the register are destroyed by allowing the load opera-tion to complete before the previous instruction has accessed the regis-ter, the load operation is actually loading temporarily another register. To be more specific, let us say that the load destination regis-ter is rd. Then there would exist temporarily two rd registers: the orig-inal and the new. All previous instructions to the load, specifying rd, that are waiting in a queue to execute, access the register containing the original rd content, and all instructions subsequent to the load instruction specifying rd, access the register containing the new con-tent of rd. Register renaming is accomplished by implementing more physical registers than what appear to be available to the program-mer. For instance, the user model may officially contain 32 registers, while in reality there are 40 registers; 8 extra registers for register renaming implementation. The above description of the register renaming implementation corresponds to such a feature in the FPU of the IBM RS/6000 [BaWh 90] (see Chap. 15). There actually exist a number of ways of realizing the same concept which serves to solve the problem of possible out-of-order execution in superscalar systems [BuPa 93, HwPa 87, MPVa 93, SmPl 85].

One of the proposed methods of dealing with out-of-order execution is the *reorder buffer* [SmPl 85]. Instructions are put into a FIFO queue in the order in which they were issued. When an instruction completes execution, possibly out-of-order, its result is placed in the appropriate slot in the FIFO queue, rather than in the register file. The queue, in turn, updates the register file in-order, at a time when a possible error, resulting from an out-of-order execution, can be avoided. Another method, proposed by the same authors [SmPl 85], is the use of the *his-*

tory buffer. In this method, instructions update the register file upon completion of execution; however, the previous value of the register is maintained in a *last-in first-out* (LIFO) queue. The LIFO queue is arranged with a slot for each instruction in the order in which they were issued, with the head of the queue containing the oldest instruction. The LIFO queue stores the history of the use of the register file. Any corrections of possible errors resulting from out-of-order execution can be accomplished using the LIFO queue.

Another method recently proposed to deal with out-of-order execution, is the method of *register mapping* [MPVa 93]. Only one set of physical registers is maintained. The registers defined by the system architecture are mapped into a subset of the available physical registers. This mapping changes as the instructions are issued. The association of the architected registers with physical registers is maintained in a mapping table. Correct results are maintained by repeated references to the mapping table [MPVa 93].

5.4   Concluding Comment

Pipelining and ILP can improve significantly system performance. These features are therefore implemented in a vast majority of new systems created in the past years. The implementation of these features is not limited to RISC (see Chap. 6). For many years pipelining has been implemented in numerous CISC systems, such as the DEC VAX and the Motorola M68000 families. As will be discussed in more detail in Chap. 6 on RISC, efficient handling of even a regular pipeline depends strongly on the ability to fetch an instruction in a single cycle, and then execute it in a single cycle. Such a requirement is even more important in superscalar systems, when a number of pipelines are run in parallel. In CISC systems we have multiword instructions which must be fetched in more than one cycle. Many CISC instructions execute in more than one cycle because of their complexity and memory access for operands during execution. On the other hand, with the uniform single-word instruction size in RISC systems, all of the instructions can be fetched in a single cycle. Because of RISC simplicity and restricted memory access, most of RISC instructions execute in a single cycle. Therefore, RISC systems will handle pipelining and ILP more efficiently than CISC systems. The details of RISC principles and its properties will be discussed next in Chap. 6.

6

Reduced Instruction
Set Computer
Principles

6.1   RISC Versus CISC

The microprocessor families Intel x86 and Motorola M68000, mentioned in Chap. 1, and described in detail in Parts 2 and 3, are known for their abundant instruction sets, multiple addressing modes, and multiple instruction formats and sizes. Their control is microprogrammed, and different instructions execute within a different number of cycles. The control units of such microprocessors are naturally complex, since they have to distinguish between a large number of opcodes, addressing modes, and formats. This type of system belongs to the category called complex instruction set computer (CISC). Although many CISC microprocessors are pipelined, there exists an inherent difficulty in managing a pipeline (see Chap. 5) in a system with a variety of instruction sizes and different instruction execution lengths.

As opposed to the traditional CISC design, in the early eighties there emerged a new trend of computer design called RISC—reduced instruction set computer [HePa 90, PaDi 80, PaSe 82, Patt 85, Tabk 87, Tabk 90b]. What is "reduced" in a RISC? Practically everything: the number of instructions, addressing modes, and formats. In an ideal RISC all instructions have the same size (usually 32 bits) and execute within a single CPU cycle. In practice, only the majority of the instructions (over 80 percent in most RISC systems) execute in a single cycle.

The relative properties of CISC versus RISC systems will now be elaborated in more detail.

A CISC system with a large menu of features implies a larger and more complicated decoding subsystem, preceding the complex control logic. Logic signals will usually have to propagate through a considerable number of gates, increasing the duration of delays and slowing down the system. In a microprogrammed environment (and most CISCs are microprogrammed), increased complexity will directly result in longer microroutines and therefore their longer execution to produce all necessary microoperations and their corresponding control signals to execute an instruction.

One of the ways to increase the speed of execution on any computer is to implement pipelining (see Chap. 5). For a pipeline with n stages, we can get the system to deal with n subsequent instructions simultaneously. Consider a simple two-stage instruction pipeline:

Stage One: Fetch, F

Stage Two: Execute, E

Assume a simple model where each of the above stages takes just a single CPU cycle to complete. We get the following instruction time (in CPU cycles) layout for three subsequent instructions:

Cycle	1	2	3	4
Instr.				
i	F	E		
i + 1		F	E	
i + 2			F	E

All three instructions are fully taken care of in four cycles. It should also be mentioned that instruction i − 1 is executed during cycle 1, while instruction i + 3 is fetched during cycle 4. At any cycle, two instructions are being worked on in this simple two-stage pipeline.

The above streamlined pipeline model does not occur in CISC systems. The instructions are of different length; while some can be fetched in a single cycle, others need more. Different instructions are executed in a different number of cycles. A more realistic example on a CISC can be the following:

Cycle	1	2	3	4	5	6	7
Instr.							
i	F	E	E	E			
i + 1		F			E		
i + 2			F	F		E	E

It takes now seven cycles to execute three instructions (in the previous model it would take six cycles to handle all three instructions without a pipeline). Because of the disparity in instruction lengths and execution times, some instructions have to be suspended and wait for a few cycles within the pipeline. Instruction $i + 3$ can be fetched starting with cycle 5, and its execution can begin no earlier than cycle 8.

The above example illustrates that there is a difficulty in implementing an instruction pipeline efficiently in a CISC-type system. In actual systems, instruction pipelines have more than two stages (usually three to six for integer operations and more for floating-point). If there are considerable differences between lengths and execution cycles of different instructions, which can appear in the CPU in any order, the pipeline design and utilization will be much more complicated. This complication will be even more severe for superscalar or superpipelined systems (see Chap. 5).

The complexity of a CISC system would imply a long design time with a significant probability of design errors. In a complex system the errors will take a long time to locate and correct. By the time a CISC system is designed, built, and tested, it may become obsolete from the standpoint of the state of the art of the current computer technology, in which significant advances occur on a quarterly basis (sometimes even more frequently).

A large instruction set presents too large a choice for the compiler of any *high-level language* (HLL). This in turn makes it more difficult to design the optimizing stage of a CISC compiler. This stage would have to be longer and more complicated in a CISC system. Furthermore, the results of this "optimization" may not always yield the most efficient and the fastest machine language code.

Some CISC instruction sets contain a number of instructions particularly specialized to fit certain HLL instructions. However, a machine-language instruction that fits one HLL may be redundant for another and would constitute an excessive effort for the designer. Such a machine may have a relatively low cost-benefit factor.

Considering the pipeline operation example discussed above, one can see that an efficient system operation can be attained if all instructions take the same number of cycles for the fetch and execution stages. If the above take a single clock cycle, the operation will naturally be the speediest for a given technology. The designer should therefore strive to achieve uniform, single-cycle fetch and execute operations for each instruction implemented on the computing system being developed.

A single-cycle fetch can be achieved by keeping all instructions at a standard size. The standard instruction size should be equal to the basic word length of the computing system, which is usually equal to the number of data lines in the system bus, connecting the memory (where the program and data are stored) to the CPU. At any fetch cycle, a complete single instruction will be transferred to the CPU. For instance, if the basic word size is 32 bits and the data part of the system bus (the data bus) has 32 lines, then the standard instruction length should be 32 bits, as it is today in most systems. Some systems have a double bus, in and out of chip (64 bits), thus being able to fetch two instructions at a time.

Achieving uniform (same time duration) execution of all instructions (desirably in a single cycle) is much more difficult than achieving a uniform fetch. Some instruction executions may involve simple logical operations on a CPU register (such as clearing the register) and can be executed in a single CPU clock cycle without any problem. Other instructions may involve memory access (load from or store to memory, fetch data) or multicycle operations (multiply, divide, floating-point) and may be impossible to execute in a single cycle. In order to attain better performance the designer should strive to achieve a situation where most of the featured instructions are executable in a single cycle.

Ideally, we would like to see a streamlined and uniform handling of all instructions, where the fetch and the execute stages take up the same time for any instruction (in the two-stage pipeline model)— preferably a single cycle. This is one of the first and most important principles inherent in the RISC design approach. All instructions go from the memory to the CPU, where they are executed, in a constant stream. Each instruction is executed at the same pace, and no instruction is kept waiting. The CPU is kept busy all the time. Having thus introduced the basis of the RISC idea, the RISC properties will be discussed in detail in the next section.

6.2   RISC Properties

As argued in the preceding section, some of the necessary conditions to achieve a streamlined operation in a RISC-type system, are

1. Standard, fixed size of the instruction, equal to the computer word length and to the width of the data bus (with the stipulation that in some new systems the word length and/or the data bus may be an integer multiple of the instruction size, as it is in the new 64-bit systems and in some 32-bit systems with a 64-bit data bus).

2. Standard execution time of all instructions, preferably within a single CPU cycle (with the stipulation that a minority of instructions, such as divide, will have to be executed in more than a single cycle).

One might raise the following argument: why not pack two instructions into a single word, transferred into the CPU on the data bus? Considering a 32-bit system, why not have some simple instructions of 16-bit length? In a 32-bit system, two 16-bit instructions can be packed into a single 32-bit word and fetched together. This would seem to enhance the speed of operation. On the other hand, having instructions of both 16 and 32 bits is contrary to the principle of uniformity in size of all instructions and does not permit a continuous streamlined handling at all times. Having more than one size of instructions also tends to complicate the decoding and other logic. Then why not have a standard 16-bit instruction? A 16-bit standard instruction is not practical in modern computers. Of course, a considerable number of simple single-operand instructions could be 16 bits in length. The trend in modern computer design is to have three-operand instructions. This permits efficient encoding of operations with different source operands and a different destination in a single instruction (with a two-operand format, two instructions would be required). A 32-bit format is needed for three-operand instructions. This will also permit a larger range for immediate values and address displacements (see Chap. 3). Therefore, since it is impractical having all instructions of halfword (16-bit) length, they all should be a full word (32-bit) long, as they are in most modern systems.

Requiring all of the instructions to be of the same length is not in itself sufficient to ensure streamlined handling for all cases. It is also essential to have relatively simple decoding and control subsystems. A complex control unit will introduce extra delays in producing control signals, which in turn will tend to interfere with the expected streamlined and uniform handling of all instructions. An obvious way of significantly reducing the complexity of the control unit is to provide a reduced number of choices (a reduced "menu") of instructions, data and instruction formats, and addressing modes. A reduction in the number of operation possibilities will first of all simplify the design and speed up the operation of the decoding subsystem, since it will have many fewer items to distinguish. Since there are fewer instructions and addressing modes, the control unit needs less logic circuitry to implement them. For a reduced menu, the control unit will be simpler and less costly to design, manufacture, and test. The reduction of the operations menu has been one of the primary points made by the original proposers of the RISC idea [PaDi 80].

How much is reduced? There is no definite answer to this question. One can only inspect the menu of the existing RISC-type systems (see Part 4), comparing them to the CISCs (such as VAX with 304 instructions, 16 addressing modes, and over 10 different instruction lengths). Based on a number of existing systems one can tentatively adopt the following constraints for a RISC menu:

Number of instructions: less than or equal to 128

Number of addressing modes: less than or equal to 4

Number of instruction formats: less than or equal to 4

Realizing that it might not be practical to hope that all instructions will execute in a single cycle, one can request that at least 80 percent should.

Which instructions should be selected to be on the reduced instruction list? The obvious answer is, the ones used most often. A number of earlier studies [Fair 82, Kate 85, PaDi 80] established that a relatively small percentage of instructions (10 to 20 percent) takes up about 80 to 90 percent of execution time in an extended selection of benchmark programs. Among the most often executed instructions were data moves and arithmetic and logical operations. Another criterion for selection is the general support of HLL. This is an important consideration, supporting the reduction of the semantic gap [Myrs 82] between the basic machine design and the HLLs, particularly since over 90 percent of all programming is done in HLL. The term *general support* is stressed, as opposed to the support of a particular HLL. In other words, one should strive to provide features that tend to support HLLs in general (such as support for procedure handling, parameter passing, and process management), as opposed to a particular HLL (such as Pascal or FORTRAN).

It was mentioned earlier that one of the reasons preventing an instruction from being able to execute in a single cycle is the possible need to access memory to fetch operands and/or store results. The conclusion is therefore obvious: we should minimize as much as possible the number of instructions that have to access memory during the execution stage. This consideration brought forward the following RISC principles, adopted on all of the existing systems of this category:

1. Memory access, during the execution stage, is done by load and store instructions only.

2. All operations, except load and store, are register-to-register, within the CPU.

Systems featuring the above two rules are said to be adhering to a load/store memory access architecture.

Most of the CISC systems are microprogrammed, because of the flexibility that microprogramming offers the designer [Hays 88]. Different instructions usually have microroutines of different length. This means that each instruction will take a different number of cycles to execute. This contradicts the principle of a uniform streamlined handling of all instructions. Uniform handling of instructions can be achieved by using hardwired control, which is also faster. In hardwired control, a set of input signals is passed through a logic network to produce a set of control signals [Hays 88]. In such a system, uniformity of instruction handling is easily achieved. Therefore, RISC-type systems should have *hardwired control*.

In order to facilitate the implementation of most instructions as register-to-register operations, a sufficient number of CPU general-purpose registers has to be provided. A sufficiently large register set will permit the temporary storage of the intermediate results needed as operands in subsequent CPU operations. This, in turn, will reduce the number of memory accesses by reducing the number of load and store operations in the program, speeding up its run time. A minimal number of 32 general-purpose CPU registers has been adopted by most industrial RISC systems designers. Design considerations for establishing the size of the CPU register file are discussed in Sec. 6.4.

A summary of the basic points of RISC definition is given in Table 6.1. It is not a rigorous, acceptable by all, definition. In fact, some systems, advertised as RISC-type, violate some of the points in Table 6.1. The above points should be viewed as guidelines, explaining the nature of RISC. Loosely speaking, a system satisfying the majority of these points could be accepted as a RISC. Naturally, the more of these points are satisfied, the closer is the system to being recognized as a "full-fledged RISC."

In addition to the basic properties, forming the essence of RISC systems, summarized in Table 6.1, a number of features are practiced in many actual RISC systems. These features are not necessarily unique to RISC-type systems, and do not constitute a part of its definition. In fact, they might well be adopted by any CISC-type system, and indeed some of them have been.

One of the most important practices in modern computer design to be discussed here is the machine support of HLL. Although most programming is done in HLLs, the basic computer design of earlier generations did not provide any hardware-based support for HLL features, such as array management, handling of procedure parameter passing, typing and classifying information, process, and memory management. Usually, a wide *semantic* gap existed between the HLL and the

TABLE 6.1   RISC Definition

A RISC system satisfies the following properties:
1. Single-cycle execution of all (or at least most, over 80 percent) instructions
2. Single-word standard length of all instructions
3. Small number of instructions, not to exceed about 128
4. Small number of instruction formats, not to exceed about 4
5. Small number of addressing modes, not to exceed about 4
6. Memory access by load and store instructions only
7. All operations, except load and store, are register-to-register, within the CPU
8. Hardwired control unit
9. A relatively large (at least 32) general-purpose CPU register file

machine design [Myrs 82]. In the early generations of computers this gap had to be narrowed and bridged by software. A wide semantic gap would in general cause more complicated and hence more costly and less reliable system software. Lately, many systems (such as the VAX [LeEc 84, Leon 87] have started to incorporate features supporting HLLs in their basic design, thus narrowing the semantic gap.

The support of HLL features is mandatory in the design of any computing system, be it RISC or CISC. It is a rather complicated matter and has to be approached very carefully. Loading the design with a great number of HLL features and instructions (some of which may happen to be used rather rarely) may result in a very complex and low-throughput system. A better approach is to investigate statistically the frequency of usage of various HLL features and to run a substantial number of benchmark programs written in HLL. Such experimental work has indeed been performed by the Berkeley team during the design of their RISC I and RISC II [Kate 85, PaSe 82, Patt 85]. This investigation suggested that the procedure call-return is the most time-consuming operation in typical HLL programs. The percentage of time spent on handling local variables and constants turned out to be the highest, compared to other variables. Based on that, the Berkeley team decided to support HLLs in their RISC design by supporting efficiently the handling of local variables, constants, and procedure calls, while leaving less frequent HLL operations to instruction sequences and subroutines. In other words, the Berkeley team decided to support HLLs by enhancing the performance of the most time-consuming HLL features and operations. Many other subsequent RISC designs followed this policy, obtaining reasonable HLL support and narrowing the semantic gap, while maintaining the simplicity and low complexity of the designed system.

One of the mechanisms supporting the handling of procedures, and their parameter passing in particular, is the feature of the *register window*. It was adopted by the Berkeley RISC designers and later featured on the Pyramid [Tabk 87, Tabk90b] and the Sun SPARC (see Part 4).

The register file is subdivided into groups of registers, called windows. A certain group of i registers, say r0 to r(i − 1), are designated as *global registers*. The global registers are accessible to all procedures running on the system at all times. On the other hand, each procedure is assigned a separate window within the register file. The window base (first register within the window) is pointed to by a field called *current window pointer* (CWP), usually located in the CPU's *status register* (SR), as illustrated in Fig. 6.1. If the currently running procedure is assigned the register window J, taking up registers K, K + 1,...,K + W − 1 (where W is the number of registers per window), the

REGISTER FILE

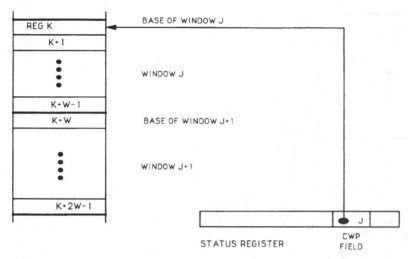

Figure 6.1   Register windows. (*Courtesy of RSP Ltd.*)

CWP contains the value J, thereby pointing to the base of window J. If the next procedure to execute takes up window J + 1, the value in the CWP field will be incremented accordingly to J + 1.

Register windowing can be particularly adapted to efficient parameter passing between calling and called procedures by partial overlapping of the windows, as illustrated in Fig. 6.2. The last N registers of window J are the first N registers of window J + 1. If the procedure taking up window J calls a procedure, which in this design will necessarily be assigned the next window J + 1, it can pass N parameters to the called procedure by placing their values into registers (K + W − N) to (K + W − 1). The same registers will be automatically available to the called procedure without any further movement of data. Naturally, the procedure call will cause the CWP field to be incremented by one. In a computer with a small register file, parameters are passed by placing them on stack or any other data structure in memory. Extra traffic on the CPU to memory bus is necessarily involved, taking up additional time.

Although register windowing has been implemented primarily on RISC-type systems, the concept is not directly connected with RISC principles, listed earlier. Theoretically, register windowing could be implemented on any system. However, an important point has to be noted. Modern implementations involve the use of VLSI chips. A CISC control unit takes up a large percentage of the chip area, leaving very little space for other subsystems and basically not permitting a large

REGISTER FILE

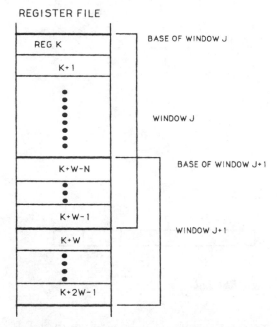

Figure 6.2 Partially overlapping windows. (*Courtesy of RSP ltd.*)

register file, needed for an efficient implementation of windowing. A RISC control unit takes up a much smaller percentage of the chip area, yielding the necessary space for a large register file. The overall size of a CPU register file is a debatable matter, to be discussed in more detail in Sec. 6.4.

Most modern computers use a number of parallel processing approaches to speed up operations [Hwan 93]. In particular, the pipelining technique is widely used (see Chap. 5). Pipelining was already featured in the third-generation computers such as the CDC 6600, and it became widely implemented later on (VAX 8600, MC68040, and many others). Pipelining was widely used on various CISC systems even before the RISC approach became popular, and the concept is not really a part of RISC definition. However, as argued earlier in this section, a streamlined RISC can handle pipelines more efficiently. Pipelining is indeed implemented in practically all modern high-performance RISC systems. Moreover, the most recent systems, implement more than one pipeline by using the superscalar approach (see Chap. 5).

Another design feature, associated with pipelining, which became very popular on some RISC systems, is that of the delayed branch. The problem occurs in systems where instructions are prefetched (they are always prefetched in an instruction pipeline), right after a branch. If the branch is conditional, and the condition is not satisfied (an unsuccessful branch), then the next instruction, which was prefetched, is executed, and since no branch is to be performed, no time is lost. If on the other hand the branch condition is satisfied (a successful branch), or the branch is unconditional, the next prefetched instruction is to be flushed and another instruction pointed to by the branch address is to be fetched in its place. The time dedicated to the prefetching of the flushed instruction is lost. Such loss of time is remedied by using the delayed branch approach.

Consider the following example:

```
CLR r2 ; clear register r2
CMP r1, 10 ; check the difference (r1) - 10
BZ adr1 ; if (r1) - 10 = 0, branch to adr1
Next instruction; otherwise, execute next instruction
```

where (r1) is the content of register r1. The nature of the "next instruction" is immaterial (can be any) and ";" represents the beginning of a comment. The next instruction was prefetched and will be flushed if (r1) = 10. The first instruction can be placed between the branch and the next instruction, as follows:

```
CMP r1, 10
BZ adr1
CLR r2
Next instruction
```

The reshuffling of the instructions does not change the result. The instruction "CLR r2" is now prefetched following the branch. Applying the delayed branch principle and assuming a successful branch, the execution of the branch [placing the value of the branch address "adr1" into the program counter (PC)] is delayed, until the following prefetched instruction (CLR r2) is executed. No time is lost, and there is no change in the intended program operation since r2 had to be cleared anyway before the branch, and it did not influence the branch condition (check whether the value stored in another register r1, is equal to 10).

The delayed branch technique may be implemented on any system, be it RISC or CISC. It so happens that it was implemented on some RISC systems. The reason why it was not implemented on CISCs is that CISCs have more serious problems associated with handling the pipeline (see Sec. 6.1), and the extra complexity associated with the introduction of a delayed branch, is not going to be of significant help and is not worthwhile. Implementing the delayed branch in superscalar systems, with multiple parallel pipelines, is too complicated, and for this reason not practiced in such systems.

There is another problem associated with the handling of instruction pipelines. It is the problem of data dependency, discussed in Chap. 5. Consider the following sequence of instructions:

```
LOAD memr, r1; load CPU register r1 from memory location memr
ADD r1, r2, r3; (r1) + (r2) → r3
```

The register r1, loaded from memory by the first instruction, is needed as an operand in the execution of the next instruction. It is important that the add instruction should use the new value in r1, attained after the completion of the load. Assuming both instructions can be fetched in a single fetch cycle (F), the load from memory would usually require an extra execute cycle (E). We have the following pipeline scheme:

Load	F	E	E
Add		F	E

The add is ready to execute before the new value in r1 is available. Unless appropriate steps are taken, the old value in r1 may be used, yielding a possibly incorrect result.

A method currently used to deal within such a case is called *scoreboarding.* A special CPU control register, called the *scoreboard register,* is set aside for this purpose. Assume that there are 32 CPU registers, as is the case in most RISCs. The scoreboard register will then be 32 bits long. Each of its bits represents one of the 32 CPU registers. For instance, bit 0 represents r0, bit 1 represents r1, and so on. In general, if register ri (i = 0,1,...,31) is involved as a destination in the execution of any instruction, bit i in the scoreboard register will be set. As long as bit i is set, any subsequent instruction in the pipeline will be prevented from using ri in any way until bit i is cleared. This will happen as soon as the execution of the instruction, which caused bit i to be set, is completed.

In the previous example, bit 1 of the scoreboard register will be set until the load is complete. The execution of the add instruction will be held (H cycle) until bit 1 is reset:

Load	F	E	E	
Add		F	H	E

A cycle may be lost, but the final result is correct. Scoreboarding is used in a number of RISC-type systems, but it is not a property characterizing RISC.

Another feature, implemented in a number of RISC-type systems, is separate data and code caches, or the dual cache (see Chap. 4). Some manufacturers refer to this feature as "Harvard architecture." It should be borne in mind, however, that in the original Harvard design the separation of data and code referred to the *main memory*. In most RISC-type systems of today, and even in some CISCs, only the primary cache is usually separated into data and code parts; the secondary cache (if implemented; see Chap. 4) and the main memory store both code and data. Dual cache is practiced today both on RISC and CISC systems; it is not a part of the RISC definition.

Practically all most recent RISC systems and some CISCs (see Parts 2–4) practice ILP, superscalar design in most cases, and super-pipelined design in a few (see Chap. 5). This practice, although implemented in most recent RISCs, is not a part of the RISC definition.

To summarize, the features implemented in in RISC systems, but not necessarily constituting the basic principles of RISC, are:

1. HLL support
2. Implementation of register windows
3. Pipelining
4. Delayed branch
5. Scoreboarding
6. Dual cache
7. ILP

After discussing the basic principles and properties of RISC, an evaluation of RISC, presenting its advantages and disadvantages, will be taken up in the next section.

6.3   RISC Evaluation

Advantages of RISC

The advantages of RISC will be discussed from a number of points of view:

VLSI realization

Computing speed

Design cost and reliability

HLL support

RISC shortcomings will be presented subsequently.

RISC and VLSI realization.   The VLSI viewpoint argumentation was one of the principal points presented by the original RISC proponents at Berkeley in 1980 [PaDi 80]. As argued earlier, a RISC has relatively few instructions, few addressing modes, and few instruction formats. As a result, a relatively small and simple (compared to CISC) decoding and executing hardware subsystem of the CPU is required. This yields the following results when we contemplate the realization of a computing system by VLSI chips.

1. The chip area, dedicated to the realization of the control unit (the so-called control area), is considerably reduced. For example, the control area on RISC I took up 6 percent of the chip area [PaSe 82]; on RISC II, 10 percent; and on the CISC Motorola MC68020, 68 percent. In general, the control area for CISCs might take up over 50 percent of the chip area. Therefore, on a RISC VLSI chip, there is more area available for other features. There is a higher chance of fitting a whole CPU and some additional features on a chip (cache, FPU, part of the main memory, memory management unit, I/O ports).

2. As a result of the considerable reduction of the control area, the RISC designer can fit a large number of CPU registers (138 on RISC II) on the chip. This in turn enhances the throughput for a large class of programs.

3. By reducing the control area on a VLSI chip and filling the area by numerous identical registers, we actually increase the regularization factor of the chip. The *regularization factor* is defined [Latt 81] as the total number of devices on the chip, excluding ROMs, divided by the number of drawn devices (such as registers, ALUs, counters, and other subsystems). It is the effective number of devices on the chip that we get for each device that we draw. Basically, the higher the regularization factor, the lower the VLSI design cost. While the regularization factor for MC68000 was 12, it was 25 for RISC I [Patt 82].

4. The GaAs VLSI chip realization technology is currently limited to a relatively low density compared to *complementary metal-oxide semiconductor* (CMOS). Therefore, since a RISC reduces the control area, it represents an attractive approach for GaAs, single-chip, CPU realization [Milu 86].

The computing speed aspect.   As explained earlier, the essence of a RISC is its uniform, streamlined handling of all (at least most) of the instructions. The RISC design approach is particularly suitable for a more efficient handling of pipelines (compared to CISC). As a result of the uniformity of instruction size and duration of execution, wait or hold periods in the pipeline are reduced to a minimum. These factors contribute significantly to the increase in computing speed.

A simpler and smaller control unit in a RISC has fewer gates. This results in shorter propagation paths (fewer gates to propagate through) for the control unit signals, yielding a faster operation.

A significantly reduced number of instructions, formats, and modes results in a simpler and smaller decoding system. As in the case of the simpler control unit, the decoding operation is faster on a RISC.

A hardwire-controlled system with a reduced control unit will in general be faster than a microprogram-controlled one—particularly if the later has instructions corresponding to microroutines of different lengths, some of which may be considerably long.

A relatively large (32 or more) CPU register file tends to reduce CPU-memory traffic to fetch and store data operands. Data items that are needed often can be kept in CPU registers. This tends to save computing time, particularly for programs handling large amounts of data.

A large register set can also be used to store parameters to be passed from a calling to a called procedure, to store the information of a process that was preempted by another, and to store the information of an interrupted program. Without an adequate CPU register file, all of the above information would have to be stored in memory. This would cause extra CPU memory traffic for the storage, and later, for the eventual restoration of the above information. All in all, a considerable amount of computer time can be saved by a large register set in a number of different events.

The delayed branch technique also contributes to the enhancement of speed by preventing the flushing (and thus a waste) of prefetched instructions in case of a successful branch.

From the quantitative point of view, we can say that the RISC design contributes to the reduction of the program run or to the increase of speed by reducing the number of clock cycles per instruction. This follows from the basic characterization of RISC, which minimizes the number of cycles (ideally, to one) needed to execute each instruction.

Design cost and reliability considerations. A relatively small and simple control unit in a CPU usually yields the following design cost and design reliability benefits:

1. It takes a shorter time to complete the design of a RISC control unit, thus contributing to the reduction in the overall design cost.

2. A shorter design time would reduce the probability that the end product will be obsolete by the time the design is completed.

3. A simpler and smaller control unit will have a reduced number of design errors and, therefore, a higher reliability.

4. Because of the simplicity and low number of instruction formats (usually not above 4) and the fact that all instructions have the

same standard length, instruction will not cross word boundaries and an instruction cannot wind up on two separate pages in a virtual memory. This eliminates a potential difficulty in the design of a virtual memory management subsystem.

HLL support.   Several of the modern CISC systems, such as the VAX [Leon 87], have many features in their machine design that support directly functions which are common in HLLs (procedure management, array operations, array index testing, information typing and protection, memory management, and others). Several CISC systems have machine-language instructions that are either identical or very similar to some HLL instructions. As it turns out, the RISC design also offers some features that directly support common HLL operations and simplify the design of certain HLL compilers.

1. Since the total number of instructions in a RISC system is small, a compiler (for any HLL), while attempting to realize a certain operation in assembly language, will usually have only a single choice, as opposed to a possibility of several choices in a CISC. This will make that part of the compiler shorter and simpler in a RISC.

2. The availability of a relatively large number of CPU registers in a RISC permits a more efficient code optimization stage in a compiler by maximizing the number of faster register-to-register operations and minimizing the number of slower memory accesses.

3. The "register windows" arrangement in a RISC CPU permits fast parameter passing between procedures and constitutes a direct support of HLL handling of subroutines and procedures.

4. All in all, a RISC instruction set presents a reduced burden on the compiler writer. This in turn tends to reduce the time of preparation of RISC compilers and their cost [PaDi 80, PaSe 82].

5. A simplified instruction set in a RISC provides an opportunity to eliminate a level of translation at run time in favor of translating at compile time (since the RISC compiler is simpler).

RISC shortcomings

RISC shortcomings are directly related to some of its points of advantage. The principal RISC disadvantage is its reduced number of instructions. Since a RISC has a small number of instructions, a number of functions, performed on CISCs by a single instruction, will need two, three, or more instructions on a RISC. This in turn will cause the RISC code to be longer. More memory will have to be allocated for RISC programs, and the instruction traffic between the memory and the CPU will be increased [FlMM 87]. Recent studies [PaSe 82] have shown that, on the average, a RISC program is about 30 percent longer than a CISC program, performing the same function. This is because only a minority of the instructions are used most of the time [Fair 82], and this minority is usually featured on RISC systems. This consideration has been taken seriously by most commercial RISC systems manufacturers. In fact, a number of commercial RISCs feature more than 100 instructions (see Part 4), compared to fewer than 40 on the Berkeley RISC.

The CPU register file played an important role in the discussion of RISC advantages in this section. Modern microprocessors also feature a sizable on-chip cache. The relative merits of these features will be discussed in the next section.

6.4 On-Chip Register File Versus Cache Evaluation

A controversial feature of a number of RISC systems is the large (sometimes over 100 registers) CPU register file. Some of its potential advantages are quite obvious, and have already been mentioned in the preceding section. By keeping data values to be used as operands in the program in the CPU register file, the overall data traffic between the memory and the CPU is reduced. If the register set is small, many intermediate results, even if needed later in the program, have to be stored in memory, only to be fetched again at a later time. With a large register set, intermediate results and any other data can be kept in CPU registers for as long as they are needed. A large register set also permits efficient parameter passing between procedures. With a small register set, all of the parameters have to be stored in memory and fetched from it whenever needed, increasing the CPU-memory traffic.

In a multitasking environment, the processor is often switched between different tasks. Each task has a certain set of data associated with it called the *task context* or the *task state*. The context of the interrupted task has to be saved. It is usually saved in memory and later retrieved when the interrupted task is reinstated. If the CPU has a large register file and the task context is limited to a finite subset of the CPU registers, the context can be saved in another subset of the CPU register file, saving on CPU memory traffic. The same can be said about saving the basic information of an interrupted program. A large enough CPU register file can be configured by the designer to serve, optionally, as a stack or a data queue, if needed [Tabk 90b, Chap. 5]. Without a large CPU register file, the above data structures have to be configured in the memory, and any communication with them constitutes extra CPU memory data traffic. Having a large number of identical registers on the CPU chip increases its regularization factor and thus reduces the VLSI design and manufacturing cost.

On the other hand, having a large, on-chip CPU register file is plagued with disadvantages. The register address decoding system will be more complicated for a larger register file, increasing the access time to any of its registers. The use of window pointers in some systems also tends to increase the register address decoding time. Elaborate window management policies may also complicate the CPU logic, raising the cost and slowing down the operation. A large register file will take up more space on a chip. Some designers may decide that it is more important to put other resources on the chip (cache, SFUs), and have only a modest CPU register file (32 registers).

In all systems where all of the CPU registers are saved in memory during a context switch, a large register file take more time to store and later to retrieve. In addition, some compiler techniques make more efficient use of relatively small register files (16 to 32 registers). The advantages and disadvantages of a large (over 32 registers) CPU register file, discussed above, are summarized in Table 6.2. Most RISC-type commercial processors have 32 registers for the IU and separate 32 registers for the FPU.

As can be seen, the question of the optimal size of the CPU register file is a controversial one, requiring additional research. Despite some obvious advantages, certain studies have cast doubt on the benefit of a large register set [FlMM 87, Wall 88]. The simulation experiments in these studies were conducted with somewhat contrived rather than actual models of widely used computers. For this reason, the recommendations resulting from the above are indicative rather than conclusive. More extensive experimental results are needed.

TABLE 6.2   Large CPU Register File

Advantages

1. Speedup of operations by reducing CPU memory traffic
2. Procedure parameter passing support within the CPU
3. Multitasking context switching and interrupt handling support within the CPU
4. On-chip stack and/or queue of data
5. Increase in the chip regularization factor

Disadvantages

1. Longer access time
2. If window pointers are used, longer to decode register address
3. Register file takes up more chip space
4. Elaborate window policies complicate CPU logic
5. Advanced compiler technology makes efficient use of relatively small register files
6. If all CPU registers are saved on a context switch, a large register file takes more time to store or retrieve

An alternative to a large register file is the use of a cache (see Chap. 4). The cache is now implemented in practically all modern computing systems. In order to compare the relative merits of a cache and a register file, let us look at some of their properties, listed in Table 6.3.

Since cache addresses are actually memory addresses (usually 32 bits), they would take longer to decode than register file addresses (7 bits for 128 registers, for instance). Moreover, three direct register addresses can easily be packed into a 32-bit instruction format. With memory addressing, single-word instruction length (32 bits), a three-operand addressing would necessarily be indirect (see Chap. 3). This would imply a longer access time compared to the direct mode. If the cache is on the same chip with the CPU, the access time will be comparable to that of a register file, but still at least a cycle longer because a complete 32-bit address must be calculated for the cache access, according to the specified addressing mode.

In many systems the user has no direct control over the manipulation of the cache. The cache is usually managed either by the hardware or by OS. On the other hand, the CPU register file is usually general purpose (or most of it), fully accessible by the user.

A simulation study, using VHSIC hardware description language (VHDL), on the relative merits of on-chip cache vs. register file, was recently reported [MaTA 91]. The simulation was conducted for a model of the RISC-type Intel 860XR (see Part 4) which has an 8 kbyte data cache, and 32 CPU registers (32 for IU and 32 for FPU). The simulation was conducted using the Linpack benchmark for the floating-point register file. The results are illustrated in Fig. 6.3. The horizontal axis represents the on-chip data cache, and the vertical axis

TABLE 6.3   On-Chip Cache Versus Register File

	Cache	CPU register file
1.	Addressed as locations in memory-long addresses.	Separate register addressing—short addresses
2.	Has to be tens of kbytes to be effective	About 128 registers (512 bytes) will have significant effect on performance
3.	Information loaded in units of lines (blocks)	Information can be loaded individually to each register
4.	Slower access (EA calculation, virtual to physical address translation)	Faster access
5.	Information loaded based on prefetch and replacement policies	Any information can be loaded at any time by the user
6.	Usually inaccessible by the user	Fully accessible by the user
7.	Possibility of a miss	No miss

Note: EA = effective address.

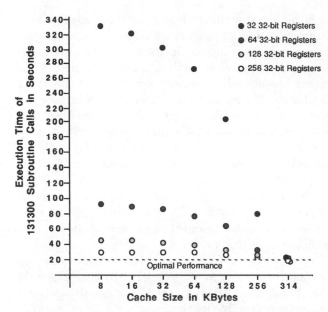

Figure 6.3   Register/cache ratio versus performance.

represents computing time. There are four curves for different sizes of the register file: 32 (the existing one), 64, 128, and 256 registers. One can see from Fig. 6.3 that doubling the register file from 32 to 64 registers improves considerably the performance. After that, only a modest improvement is achieved. This is in line with the previous discussion about the size of the register file: it is advantageous to increase it only up to a point. It can also be seen from Fig. 6.3 that one needs hundreds of thousands of bytes of cache to achieve the performance attainable by several hundreds of bytes of register file. The conclusion is that although cache can improve performance considerably, we should not give up the register file. The register file should be kept at a size of 32 to 64 registers. Most modern RISC systems have 32 IU and 32 FPU registers, for a total of 64 CPU registers.

6.5   Overview of RISC Development and Current Systems

As can be seen from the preceding discussion in this chapter, the RISC concept is not quite clearcut; it has both advantages and shortcomings. It has encountered opposition right from its inception [ClSt 80], in the same issue where it has been first publicly announced [PaDi 80]. The RISC controversy continued over a number of years [Colw 85]. Notwithstanding the controversy, an important fact is notable: there is a considerable number of commercial computer products (see Part 4) announced as RISC-type by their manufacturers. To be sure, some of them do not adhere to all the RISC properties specified in Table 6.1. One particular RISC "violation" is in the number of instructions. In some systems such as IBM RS/6000 it is close to 200. Some of the announced RISC-type systems are more "RISCy" than others; however, all of them strive to achieve a uniform and streamlined handling of all (or at least of most) instructions.

The reason for the success of the RISC idea, despite its criticism, is the proven performance of RISC-type systems, attained over the years. Some examples of experimental results, demonstrating RISC system benchmark performance, compared to some CISCs, will be presented in Part 5.

Among the RISC manufacturers there are companies which started with a RISC product, such as MIPS Computer Systems (now a part of Silicon Graphics) with its Rx000 series, and Sun Microsystems with its SPARC (see Part 4). There are other manufacturers, known for their CISC microprocessor families, who also started their own RISC system families, such as Intel, with its x86 family (see Part 2), which started the RISC 860 family; and Motorola, with its M68000 family (see Part 3), which started the RISC M88000 family.

Of particular note is IBM, which was actually the first to start with the development of an experimental RISC system, the 801 (see also Chap. 1), and now features the RISC System 6000 (see Part 4). This effort is continued jointly by the cooperation of IBM, Motorola, and Apple in creating a new RISC-type family of microprocessors, called PowerPC, with the 6xx series (see Part 4).

DEC, some of whose professionals opposed the RISC idea in the beginning [ClSt 80], now features its own RISC product, the Alpha AXP (see Part 4), considered to be one of the fastest microprocessors of the early nineties.

Practically all new RISC-type products, as well as some CISCs, are superscalar. The MIPS R4000 and R4400 are two-issue superpipelined (see Chap. 5). Of the superscalar systems, the majority are two-issue.

The application of RISC processors is widening. Generally speaking, most RISC processors are universal and their field of application is not limited. However, some of the most notable recent RISC applications are in workstations, multiprocessors, and real-time systems, primarily because of their superior performance, at a relatively low cost. The application area of RISCs is expected to widen in the future.

1

The ASCII Character Set

Character	Hex	Character	Hex
SPACE	A0	@	CD
!	A1	A	C1
"	A2	B	C2
#	A3	C	C3
$	A4	D	C4
%	A5	E	C5
&	A6	F	C6
'	A7	G	C7
(	A8	H	C8
)	A9	I	C9
*	AA	J	CA
+	AB	K	CB
'	AC	L	CC
−	AD	M	CD
.	AE	N	CE
/	AF	O	CF
0	B0	P	D0
1	B1	Q	D1
2	B2	R	D2
3	B3	S	D3
4	B4	T	D4
5	B5	U	D5
6	B6	V	D6
7	B7	W	D7
8	B8	X	D8
9	B9	Y	D9
:	BA	Z	DA
;	BB	[	DB
<	BC	/	DC
=	BD	]	DD
>	BE	†	DE
?	BF	←	DF

Problems to Part 1

1. Given a system with a 64-Mbyte main memory and a 64-kbyte cache. The cache is four-way set-associative with 32 bytes/line. Draw a block diagram of the cache and the address format for the cache access.

2. Given the following instruction sequence:

```
load memr,r10; (memr) → r10
add r10,r11,r12; (r10) + (r11) → r12
sub r4,r3,r1; (r4) - (r3) → r1
cmp r11,r2; (r11) - (r2), flags affected
bz adr1; branch on zero [if (r11) = (r2)]
dcr r12; otherwise, decrement r12
adr1 next instruction
```

Assume a four-stage pipeline IF ID EX WB, as presented in Chap. 5. Assume that any register-to-register operation can be executed in a single cycle, and a load in two cycles. Draw a detailed time-space diagram for the pipelined execution of this program. Identify any possible pipeline hazards and find efficient ways to minimize the effect of these hazards. Complete the scoreboard register setting sequence for this program.

3. Given a system with a 64-Mbyte main memory and a cache of 64 kbytes. The cache is two-way set-associative with 64 bytes/line.

   a. Find the cache access address format.

   b. Design an LRU implementation for the cache.

4. The system in Problem 3. has an 8-kbyte page size. A two-level page tabulation is required. Work out and sketch a detailed page table hierarchy and provide an address format accessing it.

5. In a 50-MHz i860 microprocessor the on-chip access time is one cycle. The memory access is five cycles. What should be the hit ratio H in order to achieve an average access time of 23 nsec?

6. Design a 32-bit instruction format for a 3-operand register-to-register operation. Assume there are 200 instructions, 12 addressing modes, and 64 CPU registers.

7. Sketch a CPU register file consisting of 128 registers, 8 of which are global, and the rest of which are subdivided into windows of 16 registers each. There is an overlap of four registers between each adjacent pair of windows. How many windows are there? What should be the size of the control register field pointing to a window?

8. The MIPS R4000 has an on-chip direct-mapped data cache of 8 kbytes. The line size is 32 bytes. The main memory implemented is 512 Mbytes. Provide a detailed format for the physical address for cache access.

9. A secondary cache was connected to the system in Problem 8. The hit ratio of the primary cache is 0.92, and its access time is 10nsec. The hit ratio of the secondary cache is is 0.90, and its access time is 50 nsec. The access time of the main memory is 500 nsec.

  a. What is the system average access time?

  b. What should be the change in the hit ratio of the secondary cache in order to achieve an average access time of 15 nsec?

10. In a system with a four-stage instruction pipeline: IF ID EX WB, the following subprogram is executed:

```
load r5,memr1;  r5 ← (memr1)
load r6,memr2;  r6 ← (memr2)
add r1,r5,r6;  r1 ← (r5) + (r6)
sub r2,r1,r3;  r2 ← (r1) - (r3)
inc r1;  r1 ← (r1) + 1
comp r2,r1;  (r2) - (r1), flags affected
bp adr1; branch on positive to adr1
dcr r2;  r2 ← (r2) - 1
adr1 next instruction
```

  a. Find and explain all possible pipeline hazards.

  b. What is the best way to alleviate the effect of these hazards?

  c. Explain the forwarding method. Is it applicable in any particular part of the above program?

11. Given a 64-bit CPU with 200 instructions, 64 CPU registers R0,R1,...,R63, and 16 addressing modes. The instructions are 64 bits long, always permitting up to three operands. Each operand can be specified by a separate addressing mode. Design in detail, specifying all fields and bits, two instruction formats: one permitting two direct memory addresses and one permitting a single direct memory address. What will be the directly addressable memory (in bytes) in each case? The memory is byte addressable.

12. Given the following sequence of instructions in a four-stage pipelined system with stages IF ID EX WB:

```
load memr,r5;  (memr) → r5
add r5,r1,r2;  (r5) + (r1) → r2
incr r6;  (r6) + 1 → r6
incr r7;  (r7) + 1 → r7
cmp r3,r4;  (r4) - (r3), flags affected
bz adr1; branch on zero, if (r3) = (r4) to adr1
next instruction; executed if condition not met
adr1 target instruction; executed if condition met
```

Identify all pipeline hazards and explain how they can be resolved in a most efficient way.

13.   Given a computing system with 64-kbyte primary cache, 1-Mbyte secondary cache, and 256-Mbyte main memory. Both caches are four-way set-associative. The secondary cache has 64 bytes/line, and the primary cache has 16 bytes/line. Given a physical hexadecimal address: 1F0 A024:

   a. What is the byte in line represented in each cache?

   b. What set is used in each cache?

   c. What is the TAG value for each cache?

   d. Draw the address formats for both caches.

Present all answers in hexadecimal.

14.   The access times in Problem 13. for the primary cache, the secondary cache, and for main memory are 10 nsec, 50 nsec, and 100 nsec, respectively. The hit ratio of the primary cache is 0.95. What should be the hit ratio of the secondary cache in order to achieve a mean access time of 12.25 nsec? Is it possible to achieve a mean access time of 12 nsec? Why?

15.   Given a computing system with a full 4-Gbyte main memory. The page size is 2 kbytes. Design a three-level page table scheme, so that all page tables at all three levels should be of equal size. Assuming a 4-byte page table entry, what will be the table size and how many entries will it contain? Draw a complete address pointing diagram. Use the address 112A 0400 (hexadecimal) in illustrating your diagram. What is the offset?

16.   Given a computing system with a 4-Gbyte main memory and a 16-kbyte page size. Design a two-level paging system so that the page table size is equal to the size of the page directory. Draw a complete diagram for the paging system. Draw the address format. Assume 32-bit page table and directory entries.

17.   The system of Problem 16 has a 64-kbyte, four-way set-associative cache with 64 bytes/line. Draw a complete cache management diagram and an access address format. Relate your answer to that of Problem 16. How do they fit?

18.   Design instruction formats for a 16-bit system with a single group of 16 CPU registers and eight addressing modes. Up to two-operand instructions are allowed. One direct address per instruction is allowed. Address size is 24 bits. There should be at least 32 instructions.

The Intel x86 Family

7

The Intel
x86 Family
Architecture

7.1 Introduction

As mentioned in Chap. 1, Intel was the pioneering manufacturer of microprocessors. After producing its successful 8-bit products (8080, 8085), it started in 1978 a new family of 16-bit microprocessors, which later evolved into a 32-bit family: the x86 (also known as the 86 or the 80x86) family. It started with the 16-bit 8086 and continued up to the latest Pentium (1993). As announced by Intel, work on additional future generations of the x86 family is underway.

One of the strongest motivations for perpetuating the x86 family of microprocessors is the fact that it is so widely used in PCs and other notable applications. There exists a vast amount of software associated with this family. Therefore, compatibility with previous generations of the family becomes a crucial design consideration. Even though Intel products were used in numerous applications, an important step in the x86 family development occurred when IBM decided to use its 8088 microprocessor as the CPU in IBM's original PC in the early eighties. The 8088 has essentially the same architecture of the 8086. The main difference is that while the 8086 external data bus is 16 bits wide (as it should be for a 16-bit system), the data bus of the 8088 is only 8 bits wide. The disadvantage of such a microprocessor is an obvious reduction in speed. The advantage is that because of the 8-bit bus there is a significant reduction in the number of auxiliary and interface chips needed in an 8088-based microcomputer (compared to an 8086-based). This permitted IBM to offer the PC at a reasonable price affordable by many users. Subsequently, a great number of IBM PC compatible and even less costly PCs were created, spreading the use of the x86 family all over the world. Today there exists a great number of systems based on the upper members of the x86 family, such as the 286, 386, 486, and Pentium.

Although each subsequent member of the x86 family differs considerably from the preceding members in its organization and realization, the basic architecture (see Chap. 3) remains essentially the same. The architecture of the Intel x86 microprocessor family will be described in detail in the subsequent sections of this chapter. The architecture of the 32-bit members of the x86 family is essentially the same (with very few differences) starting with the i386, and continuing with i486, Pentium, and possible future products. The organizational details of different members of the x86 family will be described in the subsequent chapters (8 to 10) of this part.

7.2   The Register Set

The Intel x86 architecture register set is subdivided into the following groups [Pnt3 93]:

1. Base architecture registers (or: application register set)
   a. General-purpose registers
   b. Instruction pointer
   c. Flags register
   d. Segment registers

2. System registers
   a. Memory management registers
   b. Control registers

3. Floating-point registers
   a. Data registers
   b. Tag word
   c. Status word
   d. Control word
   e. Instruction and data pointers

4. Debug registers

The base architecture and floating-point registers are accessible by applications programs. The system and debug registers are accessible only by system programs (such as OS), running on the highest privilege level (see Sec. 7.10).

Base architecture registers

The base architecture registers (or the application register set) are shown in Fig. 7.1. There are eight 32-bit general-purpose registers

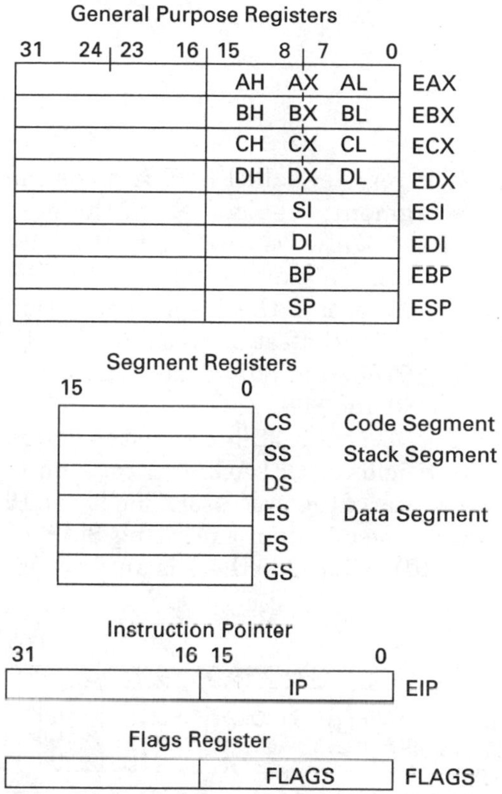

General Purpose Registers

31	24	23	16	15	8	7	0	
				AH	AX	AL		EAX
				BH	BX	BL		EBX
				CH	CX	CL		ECX
				DH	DX	DL		EDX
				SI				ESI
				DI				EDI
				BP				EBP
				SP				ESP

Segment Registers

15	0		
		CS	Code Segment
		SS	Stack Segment
		DS	
		ES	Data Segment
		FS	
		GS	

Instruction Pointer

31	16	15	0	
		IP		EIP

Flags Register

	FLAGS	FLAGS

Figure 7.1 Base architecture registers. (*Courtesy of Intel Corp.*)

EAX, EBX, ECX, EDX, ESI, EDI, EBP, and ESP. These registers hold data or address quantities, supporting data operands of 1, 8, 16, and 32 bits; bit fields of 1 to 32 bits; and address operands of 16 and 32 bits. This register set is identical for all products of the Intel x86 family, starting with the i386 and on (with the exception of a few bits in the flags register).

The least-significant 16 bits of the general-purpose registers, labeled identically to the registers of the 8086, 80186, or 80286 (see Chap. 10), AX, BX, CX, DX, SI, DI, BP, SP, can be accessed separately. The upper 16 bits of the register are not affected in this case.

The AX, BX, CX, and DX parts of the above registers are subdivided into 8-bit (byte) parts. The lowest bytes are AL, BL, CL, and DL, and the higher bytes are AH, BH, CH, and DH (see Fig. 7.1). All the above 8 bytes can be individually accessed by data operations. Seven of the above correspond to the registers of the 8-bit Intel microprocessors 8080/8085, as follows:

	AL - A
BH - H	BL - L
CH - B	CL - C
DH - D	DL - E

The instruction pointer is a 32-bit register called EIP. It holds the offset (internal address within a segment; see Sec. 7.7) of the next instruction to be executed. The offset is always relative to the *base* (first byte of a segment) of the code segment currently executed. The lower 16 bits (bits 15 to 0) of the EIP contain the 16-bit instruction pointer named IP, which is used for 16-bit offset addressing. The IP corresponds to the IP of the 8086, 80186, and 80286 (see Chap. 10), and to the program counter of the 8080 and 8085.

The *flags register,* shown in Fig. 7.2, is a 32-bit register called EFLAGS. The specified bits and bit fields of EFLAGS control a number of operations and indicate the status of the processor. The lower 16 bits of EFLAGS, called FLAGS, are used when executing 8086 or 80286 code (see Sec. 7.9 and Chap. 10). After i386 there is an extra bit

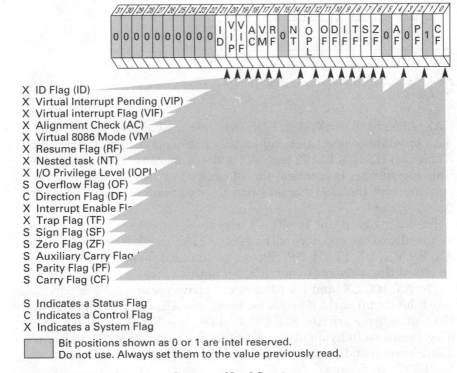

X ID Flag (ID)
X Virtual Interrupt Pending (VIP)
X Virtual interrupt Flag (VIF)
X Alignment Check (AC)
X Virtual 8086 Mode (VM)
X Resume Flag (RF)
X Nested task (NT)
X I/O Privilege Level (IOPL)
S Overflow Flag (OF)
C Direction Flag (DF)
X Interrupt Enable Fl
X Trap Flag (TF)
S Sign Flag (SF)
S Zero Flag (ZF)
S Auxiliary Carry Flag
S Parity Flag (PF)
S Carry Flag (CF)

S Indicates a Status Flag
C Indicates a Control Flag
X Indicates a System Flag

Bit positions shown as 0 or 1 are intel reserved.
Do not use. Always set them to the value previously read.

Figure 7.2 EFLAGS register. (*Courtesy of Intel Corp.*)

on the i486, and 3 more bits specified on the Pentium. Bits 1, 3, 5, 15, and 22 through 31 of EFLAGS are undefined. The description of the function of the specified bits follows.

- Bit 21, ID—Identification Flag: The ability of a program to set and clear the ID flag indicates that the processor supports the *CPU identification* (CPUID) instruction.

- Bit 20, VIP—Virtual Interrupt Pending Flag: The VIP flag together with the VIF (bit 19) enable each applications program in a multitasking environment to have virtualized versions of the system's IF flag (bit 9).

- Bit 19, VIF—Virtual Interrupt Flag: The VIF is a virtual image of the IF flag used with VIP.

- Bit 18, AC—Alignment Check: Setting the AC flag and the AM bit in the *control register 0* (CR0) enables alignment checking on memory references. An alignment check exception is generated when reference is made to an unaligned operand, such as a word at an odd byte address.

- Bit 17, VM—Virtual Mode: If VM is set (VM – 1), the processor will be placed in virtual 8086 mode (see Sec. 7.9), which is an emulation of the programming environment of the 8086 microprocessor.

- Bit 16, RF—Resume Flag: When RF is set (RF = 1), it temporarily disables debug faults so that an instruction can be restarted after a debug fault without immediately causing another debug fault.

- Bit 14, NT—Nested Task: If NT is set (NT = 1), it indicates that the currently executing task is nested within another task and has a valid link to the previous task (see Sec. 7.11).

- Bits 13,12, IOPL—Input/Output Privilege Level: The IOPL encoded values (0, 1, 2, 3) indicate the numerically maximum current privilege level (see Sec. 7.10) permitted to access I/O address space.

- Bit 11, OF—Overflow Flag: The OF is set (OF = 1) if the operation resulted in a signed overflow.

- Bit 10, DF—Direction Flag: DF defines whether ESI and/or EDI registers are incremented (postincrement) or decremented (postdecrement) during the execution of string instructions (see Sec. 7.5). Postincrement occurs if DF = 0; postdecrement occurs if DF = 1.

- Bit 9, IF—Interrupt Enable Flag: When IF is set (IF = 1), it allows recognition of external interrupts signaled on the INTR pin. When IF = 0, external interrupts on INTR are not recognized.

- Bit 8, TF—Trap Enable Flag: When TF is set (TF = 1), the processor is put into single-step mode for debugging. In this mode, the processor generates a debug exception after each instruction, which allows a program to be inspected as it executes each instruction.

- Bit 7, SF—Sign Flag: SF is set (SF = 1) if the MSB of the result is set (MSB = 1), or in other words, the result is negative. SF reflects the state of bit 7, 15, 31, for 8-, 16-, and 32-bit operations, respectively.

- Bit 6, ZF—Zero Flag: ZF is set (ZF = 1) if all bits of the result are zero. Otherwise, ZF = 0.

- Bit 4, AF—Auxiliary Carry Flag: The AF is used for BCD operations. AF is set (AF = 1) if the operation resulted in a carry out of bit 3. Otherwise, AF = 0.

- Bit 2, PF—Parity Flag: PF is set (PF = 1) if the low-order 8 bits of the operation contain an even number of 1s (even parity). PF is reset (PF = 0) if the low-order 8 bits have odd parity (odd number of 1s).

- Bit 0, CF—Carry Flag: CF is set (CF = 1) if the operation resulted in a carryout of the MSB (the sign bit). Otherwise, CF = 0. For 8-, 16-, or 32-bit operations, CF is set according to the carryout of bit 7, 15, or 31, respectively.

Segment registers

Six 16-bit *segment registers* CS, SS, DS, ES, FS, and GS (Fig. 7.1) hold segment selector values identifying the currently addressable memory segments (see Sec. 7.7). The selector in CS indicates the current code segment, the selector in SS indicates the current stack segment, and the selectors in DS, ES, FS, and GS indicate the current four data segments. Detailed discussion of the segment registers is deferred to Sec. 7.7 dealing with segmentation.

System registers

System registers consist of memory management and control groups of registers. Each of the two groups will be presented in the following.

Memory management registers

Four memory management registers are illustrated in Fig. 7.3. They specify the locations of the data structures which control segmented memory management. The *global descriptor table register* (GDTR) and *interrupt descriptor table register* (IDTR) can be loaded with instructions which get a 6-byte data item from memory. The *local descriptor table register* (LDTR) and *task register* (TR) can be loaded with instructions which take a 16-bit segment selector (see Sec. 7.7) as an operand. The remaining bytes of these registers are then loaded automatically

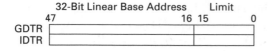

Figure 7.3 Memory management registers. (*Courtesy of Intel Corp.*)

by the processor from the descriptor referenced by the operand. More details about this group of registers will be given in the section on segmentation (Sec. 7.7).

Control registers

There are five control registers (CR0, CR1, CR2, CR3, CR4) illustrated in Fig. 7.4. Only four of them are used by the current implementation; register CR1 is reserved for future use.

The CR0 register contains system control flags, which control modes of operation or indicate states of the processor. Only bits 0 to 5, 16, 18, and 29 to 31 are currently used. The other bits are reserved for future implementation. The function of the CR0 bits is briefly explained in the following.

- Bit 31, PG—Paging Enable: When PG is set (PG = 1), paging is enabled. When PG = 0, paging (see Sec. 7.8) is disabled.

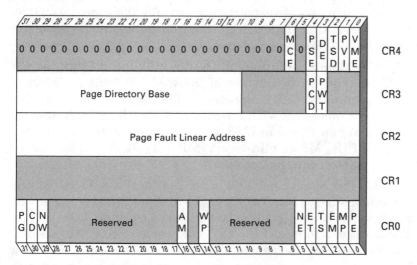

Figure 7.4 Control registers. (*Courtesy of Intel Corp.*)

- Bit 30, CD—Cache Disable: The CD bit is used to enable or disable the on-chip cache fill mechanism. When CD = 1, the cache will not be filled on cache misses. When CD = 0, cache fills may be performed on misses. Cache hits are not disabled by CD.

- Bit 29, NW—Not Write-through: When NW is cleared (NW = 0), it enables on-chip cache write-throughs and write-invalidate cycles. When NW = 0, all writes, including cache hits, are sent out to the pins. When NW = 1, write-throughs and write-invalidate cycles are disabled. The only write cycles that reach the external bus when NW = 1 are cache misses. Invalidate cycles are ignored. Write cycles with NW = 1 do not update main memory.

- Bit 18, AM—Alignment Mask: The AM bit allows alignment checking when set (AM = 1) and disables alignment checking when clear (AM = 0).

- Bit 16, WP—Write Protect: When WP is set (WP = 1), it offers write-protection to user-level pages against supervisor-level write operations. When WP is clear (WP = 0), read-only user-level pages can be written by a supervisor process.

- Bit 5, NE—Numeric Error: When NE is set (NE = 1), it enables the standard mechanism for reporting floating-point numeric errors.

- Bit 4, ET—Extension Type: The ET bit indicates support of the i387 mathematical coprocessor instructions.

- Bit 3, TS—Task Switched: The TS is set (TS = 1) whenever a task switch operation is performed.

- Bit 2, EM—Emulation: When EM is set (EM = 1), execution of a numeric floating-point instruction generates the coprocessor-not-available exception. The EM bit must be set when the processor does not have a floating-point unit.

- Bit 1, MP—Monitor coProcessor: On the i286 and i386 processors, the MP bit controls the function of the WAIT instruction, which is used to synchronize with a coprocessor. The WAIT instruction is not needed on processors with on-chip FPU, such as i486 and the Pentium. When running i286 and i386 programs on i486 and Pentium FPUs, MP should be set (MP = 1). It should be cleared (MP = 0) on i486 and the Pentium.

- Bit 0, PE—Protection Enable: When PE is set (PE = 1), the protection mechanism (see Sec. 7.10) is enabled. When PE = 0, the processor operates in unprotected real (8086) mode (see Sec. 7.9).

    The low-order 16 bits of the CR0 are also known as the machine status word (MSW), for compatibility with the i286. The MSW of the i286 has 4 bits that are used: bits 3 through 0, TS, EM, MP, and PE (see Chap. 10).

The CR2 register holds the 32-bit linear address that caused the last page fault detected.

The CR3 register contains in its upper 20 bits (bits 31 through 12) the 20 MSBs of the address of the page directory (see Chap. 4 and Sec. 7.8). The CR3 is also known as the page directory base register (PDBR). The page directory occupies a regular page frame, that is, it must be aligned to a page boundary, so the low 12 bits of the CR3 are not used as address bits. On the i486 and the Pentium (and possibly on future implementations) the state of bits 4 and 3 is driven on the outside pins PCD and PWT respectively, and they are used as follows:

- Bit 4, PCD—Page-level Cache Disable: When PCD = 1, the on-chip cache is disabled. When PCD = 0, on-chip caching is enabled, provided it is not disabled by other means (such as cache deactivation by a signal from an external pin).

- Bit 3, PWT—Page-level Write Transparent: The PWT bit can be used to control the write policy of an external second-level cache (see Chap. 4). When PWT = 1, it allows a write-through policy for the external cache. If PWT = 0, a write-back policy for the external cache is adopted.

Register CR4, new on the Pentium, contains bits that enable certain architectural extensions. Only bits 6 and 4 through 0 are currently used as follows:

- Bit 6, MCE—Machine Check Enable: Setting MCE (MCE = 1) enables the machine check exception.

- Bit 4, PSE—Page Size Extension: Setting PSE (PSE = 1) enables paging with large 4-Mbyte pages.

- Bit 3, DE—Debugging Extensions: Setting DE (DE = 1) enables I/O breakpoints.

- Bit 2, TSD—Time Stamp Disable: Setting TSD (TSD = 1) makes the *read from time stamp counter* (RDTSC) a privileged instruction.

- Bit 1, PVI—Protected-mode Virtual Interrupts: Setting PVI (PVI = 1) enables support for a virtual interrupt flag in protected mode. This feature can enable some programs designed for execution at privilege level 0 (see Sec. 7.10) to execute at privilege level 3 (applications level; least privileged).

- Bit 0, VME—Virtual-8086 Mode Extensions: Setting VME (VME = 1) enables support for a virtual interrupt flag in virtual-8086 mode (see Sec. 7.9). This feature may improve performance in this mode.

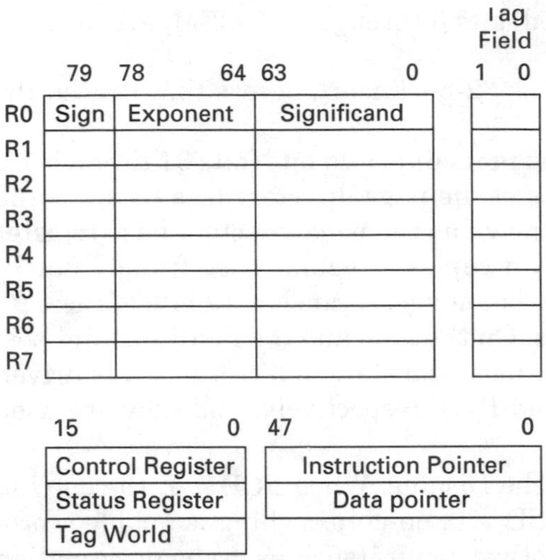

Tag
Field

	79 78	64 63	0	1 0

R0 | Sign | Exponent | Significand
R1
R2
R3
R4
R5
R6
R7

15 0	47 0
Control Register	Instruction Pointer
Status Register	Data pointer
Tag World	

Figure 7.5 Floating-point registers. (*Courtesy of Intel Corp.*)

Floating-point registers

The floating-point registers are shown in Fig. 7.5. The on-chip FPU includes eight 80-bit data registers R0 to R7, a 16-bit tag word, a 16-bit control register, a 16-bit status register, a 48-bit instruction pointer, and a 48-bit data pointer.

Data registers

The data registers R0 to R7 are used by floating-point computations. These registers can be accessed in two ways:

1. As a stack whose top is pointed to by bits 13 to 11 (TOP field) of the status register (or status word) with instructions operating on the top one or two stack elements.

2. As a fixed register set with instructions operating on explicitly designated registers.

A push operation decrements TOP by 1 and loads a value into the new top data register. A pop operation stores the value from the current top data register and then increments TOP by one. Like other x86 stacks in memory, the FPU data register stack grows down towards lower-addressed registers.

15   0

| TAG(7) | TAG(6) | TAG(5) | TAG(4) | TAG(3) | TAG(2) | TAG(1) | TAG(0) |

NOTE:
The index i of tag(1) is not top-relative. A program typically uses the "top"
field of Status World to determine which tag(1) field refers to logical top
of stack.
TAG VALUES
    00 = Valid
    01 = Zero
    10 = QNaN, SNaN, Infinity, Denormal and Unsupported Formats
    11 = Empty

Figure 7.6 FPU tag word. NaN is not a number; QNaN is quiet NaN; SNaN is signaling NaN. (*Courtesy of Intel Corp.*)

Tag word

The tag word marks the content of each data register, R0 to R7, as
shown in Fig. 7.6. Each 2-bit tag TAG(0) to TAG(7) represents one of
the R0 to R7 registers, respectively. The principal function of the tag
word is to characterize the content of the data registers according to
the 2-bit encoding listed in Fig. 7.6. The tag value of 10 should be
noted. It represents unusual, irregular possible values in the data registers R0 to R7. Floating-point operations that have no mathematical
interpretation, such as zero/zero, produce a value called "Not a
Number" (NaN). A NaN has an exponent 11..11 (binary), may have
either sign, and may have any mantissa except 1.00..00 (binary),
which is assigned to infinities. There are two classes of NaNs:
Signaling NaN (SNaN) and Quiet NaN (QNaN). A SNaN is a NaN that
has a zero as the MSB of its mantissa. The rest of the mantissa may be
set to any value. Arithmetic operations of the FPU on a SNaN cause an
invalid operation exception. A QNaN is a NaN that has a one as the
MSB of its mantissa. The FPU creates the QNaN as its default
response to certain exceptional conditions [Pnt3 93].

Status word

The 16-bit status word, located in the status register, reflects the overall state of the FPU. It is shown in Fig. 7.7.

Control word

The FPU 16-bit control word is shown in Fig. 7.8. The control word
provides the user with several programmable processing options, as
encoded in Fig. 7.8. The low-order 6 bits contain individual masks for
each of the six exceptions that the FPU recognizes. They fit the low-
order 6 bits of the status word.

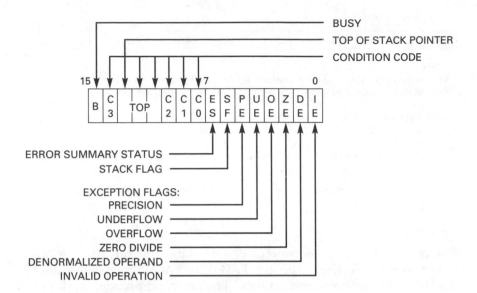

Figure 7.7 FPU status word. (*Courtesy of Intel Corp.*)

Instruction and data pointers

In case of an FPU error, the 48-bit instruction pointer contains the address of the failing instruction and the 48-bit data pointer contains the address of its numeric memory operand, if appropriate.

Debug registers

The x86 architecture features eight *debug registers,* DR0 to DR7. Only programs executing at the highest privilege level can access these registers. Registers DR0 to DR3 specify the four linear breakpoint addresses. The debug status register, DR6, displays the current state of the breakpoints. The debug control register, DR7, is used to set the breakpoints.

7.3   Data Formats

The Intel x86 architecture data formats are shown in Table 7.1. Since the x86 architecture evolved from the 16-bit 8086 and i286, the term

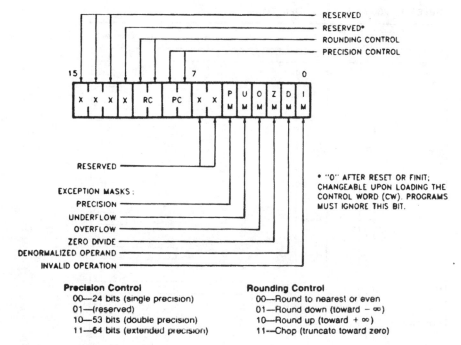

Figure 7.8   FPU control word. (*Courtesy of Intel Corp.*)

word is associated with a 16-bit data item. The 32-bit data item is called doubleword, or more briefly dword. The unsigned integer data are also called ordinals. The signed integer data are simply referred to as integers. The architecture also recognizes packed and unpacked *binary coded decimal* (BCD) data, and bit, byte, word, and dword strings. The floating-point data are the single-precision 32-bit, and double-precision 64-bit IEEE standard. In addition, the architecture recognizes the extended 80-bit format, as illustrated in Table 7.1 (see also Chap. 3).

There are two additional data formats, associated with the concept of segmentation, discussed in Sec. 7.7. The presentation of these data formats is therefore deferred to Sec. 7.7, where it will be better appreciated by the reader.

7.4   Addressing Modes

The addressing modes featured by the Intel x86 architecture use the following four basic components:

Displacement. The displacement indicates the offset in the address of an operand. It may be an 8-, 16-, or 32-bit immediate value at the end of an instruction.

TABLE 7.1   x86 Architecture Data Types

Data Format	Supported by Base Registers	Supported by FPU	Range	Precision	Bit Layout
Byte	X		0–255	8 Bits	7 0
Word	X		0–64K	16 Bits	15 0
Dword	X		0–4G	32 Bits	31 0
8-Bit Interger	X		10^2	8 Bits	Two's Complement; Sign Bit ↑; 7 0
16-Bit Interger	X	X	10^4	16 Bits	Two's Complement; Sign Bit ↑; 15 0
32-Bit Interger	X	X	10^9	32 Bits	Two's Complement; Sign Bit ↑; 31 0
64-Bit Interger		X	10^{19}	64 Bits	Two's Complement; Sign Bit ↑; 63 0
8-Bit Unpacked BCD	X		0–9	1 Digit	One BCD Digit per Byte; 0 0
8-Bit Packed BCD	X		0–9	2 Digits	Two BCD Digits per Byte; 0 0
80-Bit Packed BCD		X	$\pm10^{\pm18}$	18 Digits	Ignored; ↑ Sign Bit; 79 72 0
Single Precision Real		X	$\pm10^{\pm38}$	24 Bits	Sign Bit ↑; Bassed Exp. 31 23 0
Double Precision Real		X	$\pm10^{\pm308}$	53 Bits	Sign Bit ↑; Bassed Exp. 63 52; Significant 0
Extended Precision Real		X	$\pm10^{\pm4932}$	64 Bits	↑ Sign Bit; Bassed Exp. 79 63; Significant 0

Least Significant Byte

String Data Types

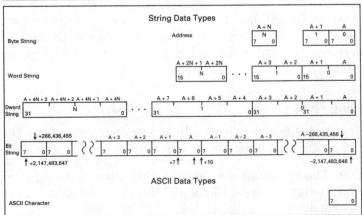

Byte Stnng

Word Stnng

Dword Stnng

Bit Stnng

ASCII Data Types

ASCII Character   7 0

Base. The base may be the content of any general-purpose register. A register containing the base is called a base register. The base registers are generally used by compilers to point to the start of a local variable area.

Index. The index may be the content of any general-purpose register, except ESP. A register containing the index is called an index register. Index registers are used to access elements of an array, or a string of characters. The main difference between the index and the base is that the index can be scaled.

Scale. The scale is a number which may be used to multiply the value of the index. The scale values may be 1, 2, 4, or 8. The scaling factor permits efficient indexing into an array when the array elements are 2, 4, or 8 bytes. For byte arrays the scale is 1 (or no scaling).

The above components are used to calculate an EA of an operand in memory. The EA is actually an offset address within a segment, if segmentation is used (see Sec. 7.7). The EA for the most general addressing mode, called based scaled index with displacement, is expressed as follows:

```
EA = (base register) + (index register)*scale + displacement
```

where (y) denotes the content of register y. The different memory access modes of the x86 architecture differ in the presence or absence of any of the above four components in the calculation of the EA. The addressing modes of the x86 architecture are summarized in Table 7.2.

Some x86 assemblers allow a double-square bracket notation, such as [EyX][EzX], instead of [EyX + EzX]. The displacement and the address may be expressed either by a number or by a symbolic address. A number of examples using the addressing modes are presented in the following. It should be noted that in the x86 architecture up to two operands are allowed. The destination is specified after the instruction mnemonic, followed by the source, separated by a comma:

```
INSTRUCTION dst,src
```

Examples

```
(1) MOV EAX, 11FF00H; EAX ← 11FF00H
```

The destination is in register mode and the source is immediate. A number is considered decimal by default. A hexadecimal, binary, or octal number must be followed by H, B, or O, respectively.

```
(2) MOV EAX, [05AA12H]; EAX ← (05AA12H)
```

The destination is in register mode, the source is in direct mode.

TABLE 7.2   Intel x86 Architecture Addressing Modes

Mode	Assembly notation	EA
Register	EyX	EyX
Immediate	Literal	—
Direct	[Address]	Address
Base	[EyX]	(EyX)
Base + disp	[EyX + disp]	(EyX) + disp
Index + disp	[EzX + disp]	(EzX) + disp
Scaled Index + disp	[EzX * SC + disp]	(EzX) * SC + disp
Based Index	[EyX + EzX]	(EyX) + (EzX)
Based Scaled Index	[EyX + EzX * SC]	(EyX) + (EzX) * SC
Based Index + disp	[EyX + disp + EzX]	(EyX) + disp + (EzX)
Based Scaled Index + disp	[EyX + disp + EzX * SC]	(EyX) + disp + (EzX) * SC
Relative	Address	(EIP) + address

Note: disp = a displacement; EyX = any general-purpose register; EzX = any general-pur-pose register except ESP; literal = an immediate number.

```
(3) MOV ECX, [EAX + 24]; ECX ← (memory location) at EA = (EAX) + 24
```

The source is at the base + displacement mode. The displacement is 24 (decimal), and the base register is EAX.

```
(4) MOV EAX, [ESI + EBX]; EAX ← (memory location) at EA = (ESI) +
(EBX)
```

The source is in the based index mode, with ESI as the base register, and EBX as the index register (actually it does not matter if the roles of ESI and EBX are reversed).

```
(5) MOV EAX, [EBP + FFF0H + EDI*4]; EAX ← (memory location) at ;EA =
(EBP) + FFF0H + (EDI)*4
```

```
(6) DISPL DW 0120H; define word, displacement = 0120H
```

```
MOV EAX, DISPL[EBX]; EAX ← (memory location) at
```

```
; EA = (EBX) + 0120H
```

The source is in the base + displacement mode, where the displacement is expressed in a previously defined symbolic notation.

In order to provide compatibility with the earlier 16-bit members of the x86 family, the 32-bit members (such as i386, i486, and Pentium) can run in an 8086 or i286 mode (see Sec. 7.9), where 16-bit registers are used, and only 16-bit offset addressing is permitted. In this case there are certain limitations imposed on the use of the EA components, summarized in Table 7.3.

TABLE 7.3   Properties of EA for 16- and 32-bit Addressing

EA component	16-bit addressing	32-bit addressing
Base register	BX, BP	Any 32-bit g.p. register
Index register	SI, DI	Any 32-bit g.p. reg., except ESP
Scale	None	1, 2, 4, or 8
Displacement	0, 8, or 16 bits	0, 8, or 32 bits

Note: g.p = general-purpose.

7.5   Instruction Set and Assembly Directives

The Intel x86 architecture instruction set can be subdivided into the following categories:

1. Data transfer
2. Arithmetic
3. Logic, shift, and rotate
4. String manipulation
5. Bit manipulation
6. Control transfer
7. HLL support
8. Protection support
9. Processor control
10. Floating-point

The x86 architecture instruction set is summarized in Appendix 2.A, by name, clock count, and machine language representation. A detailed description of the x86 instruction formats is included in Appendix 2.A. The x86 instructions operate on either 0, one, two, or three operands (very few three-operand instructions are allowed). An operand resides in a CPU register, in the instruction itself (immediate), or in memory. The following operation types are allowed:

Register-to-register

Memory-to-register

Register-to-memory

Memory-to-memory (for very few instructions)

Immediate-to-register

Immediate-to-memory

TABLE 7.4   Data Transfer Instructions

Assembly notation		Description
MOV	dst, src	; Move source to destination, dst ← src.
MOVSX	dst, src	; Move with sign extension.
MOVZX	dst, src	; Move with zero extension.
XCHG	src1, src2	; Exchange, src1 ⇄ src2.
PUSH	src	; Push operand on stack.
PUSHA(D)		; Push all general registers on stack.
POP	dst	; Pop data from stack into destination.
POPA(D)		; Pop all general registers.
CBW		; Convert byte AL, sign extended, into a word in AX.
CWDE		; Convert word AX, sign extended, into a doubleword in EAX.
CWD		; Convert word AX, sign extended, into a doubleword in DX, AX.
CDQ		; Convert doubleword EAX, sign extended, into a quadword in EDX, EAX.
IN	dst, port	; Input data from port into destination.
OUT	port, src	; Output data from source to port.
BSWAP	reg	; Byte swap in register.
LEA	reg, m	; Load effective address of memory location m into register.
LDS	reg, m	; Load DS and register with a full pointer from memory.
LES	reg, m	; Load ES and register with a full pointer from memory.
LFS	reg, m	; Load FS and register with a full pointer from memory.
LGS	reg, m	; Load GS and register with a full pointer from memory.
LSS	reg, m	; Load SS and register with a full pointer from memory.
		Flag Manipulation
LAHF		; Load low byte of EFLAGS into AH.
SAHF		; Store AH into the low byte of EFLAGS.
PUSHF		; Push FLAGS on stack.
PUSHFD		; Push EFLAGS on stack.
POPF		; Pop top of stack into FLAGS.
POPFD		; Pop top of stack into EFLAGS.
CLC		; Clear carry flag CF ← 0.
CLD		; Clear direction flag DF ← 0.
CLI		; Clear interrupt flag IF ← 0.
CMC		; Complement carry flag CF ← CF#.
STC		; Set carry flag CF ← 1.
STD		; Set direction flag DF ← 1.
STI		; Set interrupt flag IF ← 1.

Notes: ; indicates the beginning of a comment; src, src1, src2 are source operands; dst is a destination operand; reg is a CPU register; m is a memory location; port is an input or output device address; BSWAP is *new* on the 80486, not previously available on the 80386.

The operands can be either 8 (byte), 16 (word), or 32 bits (dword) long. The data transfer instructions are summarized in Table 7.4. The operands of the MOV (move) instruction can be 8, 16, or 32-bit for both source and destination. The source can also be an 8-, 16-, or 32-bit immediate operand. The source and destination can be either a CPU register or a location in memory. Only one memory operand is allowed. When the source or destination is either a control or debug register, the MOV instruction becomes privileged, executable at the highest privilege level. In case of a control or debug register, the other operand must be a 32-bit data register.

Examples

```
MOV CL, AH; CL ← (AH), byte transfer
MOV AX, BX; AX ← (BX), word transfer
MOV EDX, EAX; EDX ← (EAX), dword transfer
*MOV EAX, CR0; EAX ← (CR0)
MOV [2000H], EAX; move (EAX) to memory location 2000H
*MOV CR3, EBX; CR3 ← (EBX)
```

where (x) denotes the content of x, and "*" indicates a privileged instruction.

The *move with sign extension* instruction (MOVSX) transfers a byte or word from a register or memory location into a 16- or 32-bit register, extending the sign. Extending the sign means filling the upper bits of the destination register with the sign of the operand; 1s if negative, 0s if positive.

Examples

```
(AL) = A0H, a negative operand.
MOVSX BX, AL; BL ← (AL), BH filled with 1s.
(AX) = 4000H, a positive word.
MOVSX EDX, AX; DX ← (AX), upper 16 bits od EDX filled with 0s.
```

The *move with zero extension* instruction (MOVSZ) works on the same type of operands as the MOVSX, filling the upper half of the destination register with zeros.

Example

```
MOVZX BX, AL; BL ← (AL), BH filled with zeros.
```

The *exchange* (XCHG) instruction exchanges (swaps) the contents of two registers or a memory location with a register. Byte, word, and dword operands are permitted.

Examples

```
XCHG DI, AX; (DI) ↔ (AX)
XCHG EBX, ESI; (EBX) ↔ (ESI)
```

The PUSH instruction stores on top of the stack, pointed to by the ESP, a word or a dword contents of a general-purpose register or a memory location, the content of a 16-bit segment register, or an immediate byte, word, or dword operand. The content of the ESP is decremented by a number equal to the size of the pushed operand in bytes. Thus, pushing a word causes the ESP to be decremented by 2, and pushing a dword causes it to be decremented by 4.

Examples

```
PUSH CX; (CX) → top of stack, (ESP) - 2 → ESP
PUSH ES; (ES) → top of stack, (ESP) - 2 → ESP
PUSH FFH; byte FFH → top of stack, (ESP) - 1 → ESP
```

The push all (PUSHA) instruction pushes all 16-bit general-purpose registers on stack in the following order:

```
AX, CX, DX, BX, SP, BP, SI, DI
```

The SP value pushed is the initial value, before AX was pushed. SP is subsequently decremented by 2 for each register pushed on stack (by 16 total). The PUSHAD instruction does the same for the 32-bit registers:

```
EAX, ECX, EDX, EBX, ESP, EBP, ESI, EDI
```

The POP instruction retrieves from the top of the stack 16- or 32-bit values and places them in CPU registers or memory locations. The POP CS combination is not allowed. Popping the stack into the CS register is accomplished automatically with a RET (return from subroutine) instruction. The ESP is incremented according to the byte size of the popped operand.

Examples

```
POP ES; ES ← top of stack, (ESP) + 2 → ESP
POP EAX; EAX ← top of stack, (ESP) + 4 → ESP
```

The POPA(D) instruction restores the 16(32)-bit general-purpose registers from the stack in an order reversed to that of PUSHA(D):

```
(E)DI, (E)SI, (E)BP, (E)SP, (E)BX, (E)DX, (E)CX, (E)AX
```

The ESP value, stored previously on stack, is discarded. The ESP is incremented appropriately (by 2 for each 16-bit pop, by 4 for each 32-bit pop) for each register popped.

The *convert byte to word* (CBW) instruction sign extends the byte in AL into a word in AX. Similarly, the *convert word to doubleword* (CWDE) instruction sign extends the word in AX into a doubleword in EAX. In a similar manner, the other *convert word to doubleword* (CWD) instruction sign extends the word in AX into a doubleword stored in DX,AX (most-significant word in DX, least-significant word in AX). The *convert doubleword to quadword* (CDQ) instruction sign extends the 32-bit doubleword in EAX into a 64-bit quadword in EDX,EAX (most-significant 32 bits in EDX, least-significant 32 bits in EAX).

The input instruction, IN dst,port, transfers a byte, word, or doubleword into AL, AX, or EAX, respectively, from an input device at the "port" address. The port address can be specified either as an immediate number, or indirectly as the content of register DX.

Examples

```
IN EAX, 6; transfer 32 bits from input port 6 into EAX
IN AL, DX; transfer a byte from an input port, whose address is in DX
```

The output instruction, OUT port,src, works in a similar manner to IN. The source can be AL, AX, or EAX, depending on the size of the data transferred out. The port address can be either an immediate number, or a 16-bit address in DX.

The *byte swap* (BSWAP) instruction converts a 32-bit data item in a 32-bit register from *little-endian* byte ordering into *big-endian* (see Chap. 3), and the other way around.

Instructions LEA, LDS, LES, LFS, LGS, and LSS deal with moving address information and implementation of the segmentation mechanism. They will be discussed in Sec. 7.7. The operation of the flag manipulation instructions is clarified by the comments in Table 7.4.

The arithmetic instructions of the x86 architecture are summarized in Table 7.5. The ADD (add) and SUB (subtract) instructions are standard. The operands can be both registers, one register and one memory location, and either a memory location or a register with an immediate value for a source operand. Memory-to-memory operation is not featured. The operands can be bytes, words, or dwords, identified by the registers used or by the definition of the memory locations, which can be defined by assembly directives, as explained at the end of this section.

Examples

```
ADD AL,8; AL ← (AL) + 8, byte operation
ADD BX,CX; BX ← (BX) + (CX), word operation
SUB EAX,EBX; EAX ← (EAX) - (EBX), dword operation
```

The add with carry (ADC) instruction adds the two specified operands and the content of the carry flag (CF). If (CF) = 0, the result of ADC is identical to that of ADD. The ADC is useful in adding values longer than 32 bits.

TABLE 7.5   Arithmetic Instructions

Assembly notation		Description
ADD	dst, src	; dst ← (dst) + (src).
ADC	dst, src	; Add with carry, dst ← (dst) + (src) + (CF)
SUB	dst, src	; dst ← (dst) – (src).
SBB	dst, src	; Subtract with borrow, dst ← (dst) – (src + CF).
INC	dst	; dst ← (dst) + 1.
DEC	dst	; dst ← (dst) – 1.
NEG	dst	; Two's complement negation of (dst).
AAA		; ASCII adjust AL after addition.
AAD		; ASCII adjust AX before division.
AAM		; ACSII adjust AX after multiply.
AAS		; ASCII adjust AL after subtraction.
DAA		; Decimal adjust AL after addition.
DAS		; Decimal adjust AL after subtraction
CMP	dst, src	; Compare dst with src by subtraction (dst) – (src).
MUL	dst, src	; Unsigned multiply, dst ← (dst)*(src).
IMUL	dst, src	; Signed multiply, dst ← (dst)*(src).
IMUL	dst, src, imm	; Signed multiply, dst ← (src)* imm.
DIV	dst, src	; Unsigned divide, dst ← (dst)/(src).
IDIV	dst, src	; Signed divide, dst ← (dst)/(src).
XADD	dst, src	; Atomic exchange and add.
CMPXCHG	dst, src	; Atomic compare end exchange.

Notes: imm is an immediate value; XADD and CMPXCHG are new on the 80486, not featured on the 80386; the multiply and divide instructions require further clarification, given in the text.

Example   Perform a 64-bit addition of (EBX,ECX) to (EAX,EDX) with the sum in (EAX,EDX).

```
ADD EDX,ECX; EDX ← (EDX) + (ECX), add lower 32 bits
ADC EAX,EBX; EAX ← (EAX) + (EBX), add upper 32 bits and carry from
the lower 32 bits.
```

The *subtract with borrow* (SBB) instruction subtracts the two specified operands and also subtracts the content of the CF. If (CF) = 0, the SBB effect is identical to that of SUB. Similarly to the ADC instruction, the SBB may be used to subtract numbers larger than 32 bits.

The *increment* (INC) and *decrement* (DEC) instructions increment by 1 and decrement by 1, respectively, an operand (8, 16, or 32 bits) in a register or in memory. The INC and DEC instructions do not affect the CF. The *negate* (NEG) instruction produces a 2's complement value (in the same location) of an 8-, 16-, or 32-bit register or memory operand.

The *ASCII adjust after addition* (AAA) instruction changes the content of AL to a valid unpacked decimal number (see Chap. 3) and clears the top 4 bits. The AAA must always follow the addition of two unpacked decimal operands in AL. The CF is set and AH is incremented if a decimal carry out from AL is generated.

The *ASCII adjust AX before division* (AAD) instruction modifies the dividend (numerator) in AH and AL, to prepare for the division of two valid unpacked decimal operands. After the execution of AAD, AH will be cleared, and AL will contain the binary equivalent of the original unpacked two-digit number. Initially, AH contains the most-significant unpacked digit, and AL the least-significant.

The *ASCII adjust AX after multiplication* (AAM) instruction corrects the result of a multiplication of two valid unpacked decimal numbers. The high-order digit is placed in AH and the low-order in AL. The *ASCII adjust AL after subtraction* (AAS) instruction changes the content of AL to a valid unpacked decimal number and clears the top 4 bits of AL. The AAS must always follow the subtraction of one unpacked decimal operand from another in AL. The CF is set and AH is decremented if a decimal carry occurred.

The *decimal adjust AL after addition* (DAA) instruction adjusts the decimal packed result of adding two valid packed decimal operands in AL. The DAA adjusts AL to contain the correct two-digit packed decimal result. The CF is set if a carry was generated. The *decimal adjust AL after subtraction* (DAS) instruction similarly adjusts AL after the subtraction of two valid packed decimal operands, setting CF if a borrow was generated.

The *compare* (CMP) instruction works in a similar way and on the same type of operands as the SUB instruction, except that the operands remain unchanged and only the condition flags are affected.

Example   Assume AL contains the number 8, (AL) = 8. Then the instruction

```
CMP AL,8; (AL) - 8 = 0
```

will set the zero flag, ZF = 1. The content of AL will remain 8. Comparing (AL) with 9,

```
CMP AL,9; (AL) - 9 = -1 < 0.
```

Since the result is negative, the sign flag is set, SF = 1. The zero flag will be clear in this case, ZF = 0. The content of AL will remain unchanged.

The *unsigned multiply* (MUL) instruction is actually specified by a single operand:

```
MUL src
```

where src is an 8-, 16-, or 32-bit register or memory operand. The destination in this case is implicit and depends on the size of the src operand:

Source size (bits)	Multiplicand (bits)	Destination (bits)
8	AL (8)	AX (16)
16	AX (16)	DX,AX (32)
32	EAX (32)	EDX,EAX (64)

The DX or EDX will contain the most-significant half of the product, and AX or EAX the least-significant.

The *signed multiply* (IMUL) instruction can work in the same way as the MUL; however, the destination on IMUL can also be any other (other than EAX or part of it) 16- or 32-bit register. The source can be as in MUL and also an 8-, 16-, or 32-bit immediate operand. There is also a three-operand format for IMUL, where the third operand (multiplier) can be an 8-, 16-, or 32-bit immediate constant.

Examples

```
IMUL CX; DX,AX ← (AX)*(CX)
IMUL CX,3; CX ← (CX)*3
IMUL DX,BX,300; DX ← (BX)*300
IMUL EDX,EBP; EDX ← (EDX)*(EBP)
```

The *unsigned divide* (DIV) instruction operates with implied operands as the MUL instruction. Only the divisor, src, is specified.

```
DIV src
```

The source (src) may be an 8-, 16-, or 32-bit register or memory operand. The dividend depends on the size of the divisor src:

Dividend (bits)	Divisor (bits)	Quotient (bits)	Remainder (bits)
AX (16)	src (8)	AL (8)	AH (8)
DX,AX (32)	src (16)	AX (16)	DX (16)
EDX,EAX (64)	src (32)	EAX (32)	EDX (32)

The EAX and EDX (or parts of the above) contain the dividend before the operation and the quotient and the remainder after the operation, as shown above.

The *signed division* (IDIV) instruction works as the DIV, with the same rules for all operands.

Examples

```
DIV  CL;  (AX)/(CL),  unsigned,  quotient  →  AL,  remainder  →  AH
IDIV BX;  (DX,AX)/(BX),  signed,  quotient  →  AX,  remainder  →  DX
IDIV ECX;  (EDX,EAX)/(ECX),  signed,  quotient  →  EAX,  remainder  →  EDX
```

The *exchange and add* (XADD) and the *compare and exchange* (CMPXCHG) instructions are both atomic, uninterruptible instructions (see Chap. 3). They are particularly used as semaphore handling instructions in multiprocessors [AlGo 89, Hwan 93, Ston 93, Tabk 90a]. Both instructions perform an atomic read/modify/write (RMW) bus cycle. Both instruction operands can be either two registers or a register and a memory location. The src can only be a register, while the dst (see Table 7.5) can be either a register or a memory location.

The Intel x86 architecture logical, shift, and rotate instructions are shown in Table 7.6. In the two-operand logical instructions AND, OR, and XOR (exclusive OR) the destination (dst) can be any 8-, 16-, or 32-bit register or memory location. The src can be any of the above or an 8-, 16-, or 32-bit immediate value. Only one memory operand per instruction is permitted.

Example

```
AND AX,FFAAH; AX ← (AX) AND FFAAH
OR CL,DL; CL ← (CL) OR (DL)
XOR EAX,EDX; EAX ← (EAX) XOR (EDX)
```

The single-operand NOT instruction converts the dst operand into its 1's complement, by complementing all its bits. The dst can be an 8-, 16-, or 32-bit register or memory location.

The TEST instruction, in analogy to the CMP, performs an AND operation on the source and destination operands, without changing their contents. Only the flags are affected. The dst can be an 8-, 16-, or 32-bit register or memory location. The src can be an 8-, 16-, or 32-bit register or an immediate value.

TABLE 7.6 Logical, Shift, and Rotate Instructions

Assembly notation		Description
AND	dst, src	; dst ← (dst) AND (src), and
OR	dst, src	; dst ← (dst) OR (src), or
XOR	dst, src	; dst ← (dst) XOR (src), exclusive or
NOT	dst	; dst ← (dst) #, one's complement negation.
TEST	dst, src	; (dst) AND (src), flags affected, operands unchanged.
SAL	dst, count	; Shift dst arithmetic left by "count" bits.
SAR	dst, count	; Shift dst arithmetic right by "count" bits.
SHL	dst, count	; Shift dst logical left by "count" bits.
SHR	dst, count	; Shift dst logical right by "count" bits.
SHLD	dst, reg, count	; Double-precision shift left dst, reg by "count" bits.
SHRD	dst, reg, count	; Double-precision shift right dst, reg by "count" bits.
RCL	dst, count	; Rotate dst left through carry by "count" bits.
RCR	dst, count	; Rotate dst right through carry by "count" bits.
ROL	dst, count	; Rotate dst left by "count" bits.
ROR	dst, count	; Rotate dst right by "count" bits.

Note: "Count" can be either the contents of CL or an immediate 8-bit value.

The SAL, SAR, SHL, and SHR instructions perform arithmetic and logical shifts (see Chap. 3) left or right by a number of bits specified by the count operand. The shifted operand, located in dst, can be an 8-, 16-, or 32-bit register or memory location. The count operand can be either the content of CL or an immediate 8-bit value.

The SHLD and SHRD instructions shift left or right, respectively, two concate-nated operands dst, reg. The dst operand can be a 16- or 32-bit register or mem-ory location. The reg operand can be a 16- or 32-bit register. The number of bits to be shifted, count, can be either the content of CL or an immediate 8-bit value. The above instructions perform either (1) shift two word (16-bit) operands and pro-duce a single-word output in dst. The reg operand remains unchanged, or (2) shift two doubleword (32-bit) operands and produce a doubleword output in dst. The reg operand remains unchanged.

In a left shift, the reg operand is the least significant. In a right shift the reg operand is the most significant.

Example

```
MOV AX,0000H; clear AX
MOV BX,1234H; BX ← 1234H
SHLD AX,BX,8; shift left 2 hexadecimal digits (8 bits)
```

After SHLD, (AX) = 0012H, (BX) = 1234H (unchanged). If instead of SHLD, an SHRD is executed for the same operands, (AX) = 3400H, (BX) = 1234H.

In the rotate instructions RCL, RCR, ROL, and ROR, the dst (destination operand being rotated) can be an 8-, 16-, or 32-bit register or memory location. The count can be the content of CL or an immediate 8-bit value. The RCL and RCR instructions include the carry bit CF along with the dst in the rotation (see Chap. 3). In the ROL and ROR instructions only the dst is rotated.

Examples

```
MOV AL,81H; AL ← 81H
MOV CL,02H; set up count for shift by 2 bits
RCL AL,CL
After RCL, (AL) = 05H, (CF) = 0. If the next instruction is RCR AL,CL ,
AL returns to its original value (AL) = 81H and (CF) = 0. If the next
instruction is:
ROL AL,CL
```

then after execution (AL) = 06H, (CF) = 0. CF was not included in the rotation but it receives the MSB of AL. If the next instruction is ROR AL,CL , AL returns to its original value (AL) = 81H, but CF is set this time, (CF) = 1.

The x86 string manipulation instructions are summarized in Table 7.7. The string instructions operate on elements of strings of bytes, words, or doublewords. Registers ESI and EDI contain the offset (address within a segment) of an element in the source string and the destination string, respectively. The source string is inherently in the data segment pointed to by DS and the destination string is in the data segment pointed to by ES (see Sec. 7.7).

The MOVS instruction moves a byte, word, or dword element of a source string, pointed to by ESI, into a location within the destination string, pointed to by EDI. For 8- or 16-bit elements, SI and DI are used. After each string operation, ESI and EDI are automatically updated. If (DF) = 0 (DF bit of EFLAGS, see Sec. 7.2), they are incremented; if (DF) = 1, they are decremented to point to the next element in the string. The amount of increment or decrement is 1, 2, or 4, for a byte, word, or dword string element, respectively. The MOVS instruction can be expressed synonymously for byte, word, or dword string elements, by MOVSB, MOVSW, or MOVSD, respectively. The synonymous notation is used for all other string manipulation instructions, listed in Table 7.7. The sizes of the source string elements must be equal to the sizes of the destination string elements for all two-operand instructions. These instructions are among the few memory-to-memory instructions featured by the Intel x86 architecture.

If MOVS is preceded by a REP prefix, it operates as a memory-to-memory string transfer. The ECX must be preloaded with the total number of elements to be transferred. If (DF) = 0, ESI and EDI must point to the first string elements, since both will be automatically incremented after each element transfer. If (DF) = 1, ESI and EDI must point to the last string elements, since both will be decremented. The repeated execution of MOVS will stop when the content of ECX (automatically decremented at each element transfer) reaches the zero value.

The *compare string elements* (CMPS) instruction works in a similar manner to CMP, comparing elements in the source and destination strings, by subtracting them, without changing their values and affecting the condition flags appropriately. Contrary to CMP, the destination string element is subtracted from the source string element. The ESI and EDI are automatically updated after each CMPS operation.

The *load string* (LODS) instruction loads a source string element into EAX. The ESI is updated after the operation. The *store string* (STOS) instruction stores the content of EAX into an element location in the destination string. The EDI is updated after the operation. The *compare string* (SCAS) instruction compares an element from the destination string with the content of EAX by subtracting them without changing their contents and affecting the condition flags. The EDI is updated after the operation.

The *repeat* (REP) prefix instruction causes the next string instruction to be repeated. Similarly, the *conditional repeat while equal* (REPE) and *repeat while zero* (REPZ) cause the next string instruction to be repeated while (ZF) = 1, and *repeat while not equal* (REPNE) and *repeat while not zero* (REPNZ) cause a repeat while (ZF) = 0. The REP and the following string instruction must be within a looping construct. The looping construct is controlled by the value in ECX, preset to be equal to the desired number of loop iterations. If conditional REP prefixes are used, the loop will be exited as soon as the appropriate ZF

TABLE 7.7   String Manipulation Instructions

Assembly notation		Description
MOVS	sdst, ssrc	; Move string element from ssrc to sdst.
MOVSB		; Move string byte from DS:[(E)SI] to ES:[(E)DI].
MOVSW		; Move string word from DS:[(E)SI] to ES:[(E)DI].
MOVSD		; Move string doubleword from DS:[(E)SI] to ES:[(E)DI].
CMPS	sdst, ssrc	; Compare string operands (ssrc)-(sdst), flags affected.
CMPSB		; Compare string byte, DS:[(E)SI]-ES:[(E)DI].
CMPSW		; Compare string word, DS:[(E)SI]-ES:[(E)DI].
CMPSD		; Compare string doubleword, DS:[(E)SI]-ES:[(E)DI].
LODS	ssrc	; Load string operand into EAX.
LODSB		; Load string byte, DS:[(E)SI] into AL.
LODSW		; Load string word, DS:[(E)SI] into AX.
LODSD		; Load string doubleword, DS:[(E)SI] into EAX.
STOS	sdst	; Store string operand, from EAX.
STOSB		; Store string byte, (AL) into ES:[(E)DI].
STOSW		; Store string word, (AX) into ES:[(E)DI].
STOSD		; Store string doubleword, (EAX) into ES:[(E)DI].
SCAS	sdst	; Compare string data, (EAX)-ES:[(E)DI].
SCASB		; Compare string bytes, (AL)-ES:[(E)DI].
SCASW		; Compare string words, (AX)-ES:[(E)DI].
SCASD		; Compare string doublewords, (EAX)- ES:[(E)DI].
REP		; Repeat following string operation until (ECX) = 0.
REPE or REPZ		; Repeat while ZF = 1.
REPNE or REPNZ		; Repeat while ZF = 0.
INS	sdst, DX	; Input string ES:[(E)DI] from port pointed to by DX.
INSB		; Input string byte.
INSW		; Input string word.
INSD		; Input string doubleword.
OUTS	DX, ssrc	; Output string DS:[(E)SI] to port pointed to by DX.
OUTSB		; Output string byte.
OUTSW		; Output string word.
OUTSD		; Output string doubleword.

Notes: DS:[(E)SI] denotes an element from a source string in a segment (see Sec. 5.7) pointed to by DS. The element's offset within the segment is in ESI for 32-bit elements and in SI for 16- or 8-bit elements.

ES:[(E)DI] denotes, in a similar manner, an element from a destination string in a segment pointed to by ES. The element's offset is given by EDI (for 32-bit elements) or DI for 16- or 8-bit elements.

ssrc—string source operand.

sdst—string destination operand.

condition is satisfied. For instance, if REPE (repeat if ZF is set) is used, the loop will be exited when ZF is cleared, (ZF) = 0.

The *input string* (INS) and *output string* (OUTS) instructions permit the input or output of string elements. The address of the I/O port is preloaded into the DX register.

The bit manipulation instructions are summarized in Table 7.8. These instructions can reach any single bit in a CPU register or in memory and just test it, by storing it in the CF bit of EFLAGS, or even modify it, by clearing it, setting it, or complementing it in its original

TABLE 7.8   Bit Manipulation Instructions

Assembly notation		Description
BT	dst, bit	; Bit test, transfer bit of dst into CF.
BTC	dst, bit	; Bit test and complement, transfer bit of dst into CF, then complement bit in dst.
BTR	dst, bit	; Bit test and reset, transfer bit of dst into CF, then clear bit in dst.
BTS	dst, bit	; Bit test and set, transfer bit of dst into CF, then set bit = 1 in dst.
BSF	dst, src	; Bit scan forward, scan src from bit 0 to the left, store the bit number of the first bit set in dst.
BSR	dst, src	; Bit scan reverse, scan src from most-significant bit (31 or 15) to the right, store the bit number of the first bit set in dst.

Notes: bit is the bit number ($0 \leq$ bit ≤ 31), either stored in a 16- or 32-bit register, or given as an 8-bit immediate value; *dst* in BT, BTC, BTR, and BTS is a 16- or 32-bit register or memory location; *dst* in BSF and BSR is a 16- or 32-bit register; *src* is a 16- or 32-bit register or memory location.

location. The test of the copy of the bit in CF can be accomplished by performing a conditional jump instruction (discussed next in this section) subsequent to the bit manipulation instruction. Two more instructions, the BSF and the BSR, can locate the least significant "1" and the most significant "1," respectively, within a 16- or 32-bit register or memory location. Such a capability might become useful in resource allocation. For instance, a disk OS might determine which of its disk sectors are free by maintaining a Boolean array of bits. The *n*th bit of the array is set if sector *n* is free, and it is cleared if the sector is in use. Thus, in order to locate the first free sector, the OS can execute either the BSF or BSR instruction.

The control transfer instructions of the x86 architecture are summarized in Table 7.9. The jump instructions of the x86 architecture can be classified in two ways. First, by target address specification:

1. Direct: The jump target address is a part of the instruction.

2. Indirect: The jump target address is stored in a register or in memory and the instruction contains a pointer to the target address location.

Thus, the "addr" operand in Table 7.9 can be either the target address, specified explicitly or symbolically, or an address pointer, specified by any memory addressing mode (see Sec. 7.4), pointing to the target address location in memory.

Examples

```
JMP TARGET; a direct jump to a symbolically expressed target address
CALL SUB1; a direct call to subroutine SUB1
```

TABLE 7.9   Control Transfer Instructions

		Unconditional Transfers
JMP	addr	; Jump unconditionally to addr.
CALL	addr	; Call procedure starting at addr.
RET		; Return from procedure.
RET	imm 16	; Return from procedure and add the 16-bit immediate constant to the new top-of-stack pointer.

		Conditional Transfers
JA	addr	; Jump if above to addr, if CF = ZF = 0.
JAE	addr	; Jump if above or equal to addr, if CF = 0.
JB	addr	; Jump if below to addr, if CF = 1.
JBE	addr	; Jump if below or equal to addr, if CF = 1 or ZF = 1.
JC	addr	; Jump if carry to addr, if CF = 1.
JCXZ	addr	; Jump if (CX) = 0.
JECXZ	addr	; Jump if (ECX) = 0.
JE	addr	; Jump if equal, if ZF = 1.
JG	addr	; Jump if greater, if ZF = 0 and SF = OF.
JGE	addr	; Jump if greater or equal, if SF = OF.
JL	addr	; Jump if less, if SF $\neq$ OF.
JLE	addr	; Jump if less or equal, if ZF = 1 or SF $\neq$ OF.
JNA	addr	; Jump if not above, if CF = 1 or ZF = 1.
JNAE	addr	; Jump if not above or equal, if CF = 1.
JNB	addr	; Jump if not below, if CF = 0.
JNBE	addr	; Jump if not below or equal, if CF = ZF = 0.
JNC	addr	; Jump if not carry, if CF = 0.
JNE	addr	; Jump if not equal, if ZF = 0.
JNG	addr	; Jump if not greater, if ZF = 1 or SF $\neq$ OF.
JNGE	addr	; Jump if not greater or equal, if SF $\neq$ OF.
JNL	addr	; Jump if not less, if SF = OF.
JNLE	addr	; Jump if not less or equal, if ZF = O and SF = OF.
JNO	addr	; Jump if not overflow, if OF = 0.
JNP	addr	; Jump if not parity, if PF = 0.
JNS	addr	; Jump if not sign, if SF = 0.
JNZ	addr	; Jump if not zero, if ZF = 0.
JO	addr	; Jump if overflow, if OF = 1.
JP	addr	; Jump if parity, if PF = 1.
JPE	addr	; Jump if parity even, if PF = 1.
JPO	addr	; Jump if parity odd, if PF = 0.
JS	addr	; Jump if sign, if SF = 1.
JZ	addr	; Jump if zero, if ZF = 1.

		Loop Instructions
LOOP	addr	; Loop to addr if (ECX) $\neq$ 0.
LOOPE	addr	; Loop while equal, while (ECX) $\neq$ 0 and ZF = 1.
LOOPZ	addr	; Loop while zero, while (ECX) $\neq$ 0 and ZF = 1.
LOOPNE	addr	; Loop while not equal, while (ECX) $\neq$ 0 and ZF = 0.
LOOPNZ	addr	; Loop while not zero, while (ECX) $\neq$ 0 and ZF = 0.
Interrupts		
INT	n	; Interrupt type n.
INTO		; Interrupt on overflow.
IRET		; Return from interrupt.
IRETD		; Return from interrupt—32-bit mode.

Note: addr is a specification of an address in memory, discussed in the text.

```
JMP EBX; an indirect jump, EBX contains the target address
JMP [EBX]; double indirect jump, EBX contains the address of the
target address location in memory.
```

Second, the jump instructions can be classified by the segment allocation of the target address (see Sec. 7.7):

1. Intrasegment jump: The target address is in the same segment as the jump instruction. It is also called a NEAR jump.

2. Intersegment jump: The target address is in another segment than the jump instruction. It is also called a FAR jump.

In an assembly language program, the above jumps can be distinguished by using assembly directives NEAR, FAR, and PTR. Thus, for an intrasegment jump to target address ADDR, the instruction expression is

```
JMP NEAR PTR ADDR
```

For an intersegment jump to ADDR:

```
JMP FAR PTR ADDR
```

The conditional jump target address is specified relative to the content of EIP (i.e., by the relative addressing mode). After the jump instruction was fetched, the EIP is incremented to point to the next instruction. This is the EIP value taken into account to compute the target address. In this case the "addr" operand in Table 7.9 can be either an 8-, 16-, or 32-bit relative value, specified either explicitly or symbolically. When the relative value is 8-bit, the jump is considered to be SHORT (the relative value, a signed, 2's complement byte, ranges from -128 to $+127$). Otherwise, the jump is NEAR:

A 16-bit relative value ranges from -32768 to $+32767$.

A 32-bit relative value ranges from -2^{31} to $+2^{31} - 1$.

An unconditional jump or call also has the EIP relative target address option. The relative value for a call instruction must be 16- or 32-bit. The CALL instruction works basically in the same way as the JMP as far as target address is concerned. The basic difference is that the address of the next instruction, following the CALL, is pushed on stack (in the stack segment; see Sec. 7.7). It is popped back from the stack into the EIP during the execution of the return from procedure instruction RET.

There is also a provision for automatic removal from the stack of any arguments that the calling procedure may have pushed, before executing the CALL. One can specify the RET instruction with an immediate

16-bit operand (see Table 7.9). This will cause the specified operand to be added to the value in the stack pointer as part of the execution of RET.

Some of the conditional jump mnemonics have an identical effect. The operationally identical pairs are listed below.

JE	JZ
JNE	JNZ
JL	JNGE
JNL	JGE
JG	JNLE
JNG	JLE
JB	JNAE
JNB	JAE
JA	JNBE
JNA	JBE
JP	JPE
JNP	JPO

The LOOP instructions are basically conditional jumps, dependent on the value stored in ECX. The ECX is to be loaded with the desired number of loop iterations, prior to entering the loop. The LOOP first decrements ECX by 1 and then tests ECX. If ECX is not zero, the program jumps to the target address. If (ECX) = 0, control transfers to the instruction immediately following the LOOP instruction. The four conditional LOOP instructions (actually only two, since LOOPE is identical to LOOPZ and LOOPNE is identical to LOOPNZ) introduce an extra jump condition by testing the ZF flag, as specified in Table 7.9.

Example

```
    MOV ECX,10; load ECX for 10 iterations
L: MOV EAX,[EBP]; move a dword from memory, pointed to by EBP into
    EAX
    OUT port,EAX; output of (EAX)
    ADD EBP,4; EBP ← (EBP) + 4
    LOOP L; ECX ← (ECX) - 1, if (ECX) non-zero, go to L,
    ; otherwise, if (ECX) = 0, proceed to next instruction
```

The above program takes 10 consecutive dwords from memory (the address of the first dword pointed to by EBP) and transfers them to the output port.

The INT n instruction permits the user to cause an interrupt of a specified type n. The INTO causes an interrupt on overflow. The IRET instruction, placed at the end of an interrupt handling routine, returns control to the interrupted procedure. The IRET differs from the RET in that it also pops the flags from the stack into the FLAGS register (IRETD, into EFLAGS). The EFLAGS is stored on the stack by the interrupt mechanism (see Sec. 7.7).

TABLE 7.10   HLL Support Instructions

Assembly notation		Description
BOUND	reg, addr	; Check if (addr) ≤ (reg) ≤ (addr + s), ; s = 2 for word operands, ; s = 4 for doubleword operands, ; if not, interrupt 5 is caused.
ENTER	imm16, imm8	; Make stack frame of imm16 bytes ; at nesting level imm8.
LEAVE		; High-level procedure exit.
SETcc byte		; Set byte on condition, reset byte ; to zero if condition not met.

Notes: reg can be a 16- or 32-bit general register; addr is a 16- or 32-bit (corresponding to reg) memory location; imm16 is a 16-bit immediate value; imm8 is an 8-bit immediate value; byte is an 8-bit register or memory location; cc represents a condition same as the conditions for the conditional jump Jcc, as listed in Table 7.9.

The HLL support instructions are summarized in Table 7.10. The *check array index against bounds* (BOUND) instruction verifies that the signed value, contained in the specified register reg (16 or 32 bits), lies within specified limits, stored at memory location "addr," the lower limit first. The upper and lower limit size fits the size of reg. If the value checked is below the lower bound or above the upper bound, interrupt 5 will occur. The BOUND instruction checks whether an array index, stored in reg, is within the limits specified for the array. Since arrays are structures used very often by HLLs [Myrs 82], the inclusion of an instruction such as BOUND can be considered as a machine-based, hardware-supported, general support of HLLs (not particularly associated with any specific HLL).

Procedures are also very widely used features of HLLs. The ENTER and LEAVE instructions directly support the handling of procedures and thus generally support HLLs. The ENTER instruction creates a stack frame that can be used to implement block-structured HLLs. A LEAVE instruction at the end of the procedure complements the ENTER. The first parameter of ENTER (imm16, a 16-bit immediate value) specifies the number of bytes of dynamic storage to be allocated on the stack for the procedure being entered.

The second parameter (imm8, a byte immediate value) corresponds to the lexical nesting level of the procedure: 0 to 31. This level determines how many sets of stack frame pointers the CPU copies into the new stack frame from the preceding frame. This list of stack frames is often called the display. ENTER creates the new display for a procedure. Then it allocates the dynamic storage space for that procedure by decrementing ESP by the number of bytes specified by imm16. This new value of ESP serves as the starting point for all stack-related operations within that procedure. The EBP register is the current stack frame pointer. It is set to point to the beginning (highest stack address) of the stack frame.

The LEAVE instruction reverses the action of ENTER. It copies EBP to ESP to release all stack space allocated to the procedure by the most recent ENTER. Then LEAVE pops the old value of EBP from stack. A subsequent RET can then remove any arguments that were pushed on the stack by the calling procedure for use by the called procedure.

Example Consider the following part of a program involving a procedure call from a main procedure (at lexical nesting level 0):

```
CALL NEAR PTR PROC1 ; the called procedure PROC1 (at lexical nesting
level 1)
; is in the same segment with the CALL instruction
...............
PROC1: ENTER 000CH, 01H; the stack frame consists of 12 bytes, level 1
........
LEAVE
RET
```

For the sake of simplicity, 16-bit execution mode is assumed. It is also assumed that the stack is in the same segment as the CALL, and other PROC1 instructions are assumed not to involve the stack. The display will have one 2-byte item. The values in the SP and BP registers at selected places in the program (in hexadecimal) are

Place in the program	(SP)	(BP)
Before CALL	FA12	0000
After CALL	FA10	0000
After ENTER	FA00	FA0E
After LEAVE	FA10	0000
After RET	FA12	0000

The memory map corresponding to this example is illustrated in Fig. 7.9.

The protection support instructions are listed in Table 7.11. They will be discussed, however, in Secs. 7.7 and 7.10.

The processor control instructions are listed in Table 7.12. The halt (HLT) instruction stops the execution of all instructions and places the processor in a halt state. An interrupt or a reset signal will cause the processor to resume execution. The LOCK instruction asserts for the processor an exclusive hold on the use of the system bus. It asserts an external locking signal and places a special lock prefix (see Appendix 2.A) in front of the instruction for which a lock is to be asserted. The LOCK will only function with the following instructions:

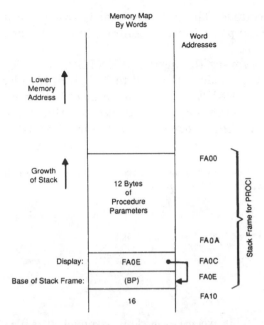

Figure 7.9 Illustration of a stack frame.

TABLE 7.11 Protection Support Instructions

Assembly notation		Description
CLTS		; Clear task switched (TS) flag in CR0.
SGDT	addr	; Store GDTR in addr in memory.
SIDT	addr	; Store IDTR in addr in memory.
STR	dst	; Store TR in dst (16-bit register or memory location).
SLDT	dst	; Store LDTR in dst (16-bit register or memory location).
LGDT	addr	; Load GDTR from addr in memory.
LIDT	addr	; Load IDTR from addr in memory.
LTR	src	; Load TR from src (16-bit register or memory location).
LLDT	src	; Load LDTR from src (16-bit register or memory location).
ARPL	sel, reg	; Adjust RPL of selector (sel, 16-bit) to no less than ; RPL of reg (16-bit).
LAR	dst, src	; Load access rights byte.
LSL	dst, src	; Load segment limit.
VERR	dst	; Verify a segment for reading, selector in dst (16-bit register or memory location), set ZF = 1 if segment can be ; read.
VERW	dst	; Verify a segment for writing, set ZF = 1 if ; segment can be written, dst as in VERR.
LMSW	src	; Load MSW from src (16-bit register or memory location).
SMSW	dst	; Store MSW into dst (16-bit register or memory location).

Notes: GDTR = global descriptor table register; IDTR = interrupt descriptor table register; TR = task register; LDTR = local descriptor table register; RPL = requestor privilege level; MSW = machine status word.

TABLE 7.12   Processor Control Instructions

Assembly notation	Description
HLT	; Enter halt state.
LOCK	; Assert LOCK# signal prefix.
NOP	; No operation.
XLAT	; Table lookup translation,
	; Replace AL by a byte in a segment
	; pointed to by DS, at offset address
	; [EBX + unsigned (AL)]
XLATB	; same as XLAT.
INVD	; Invalidate on-chip cache.
WBINVD	; Write-back and invalidate on-chip cache.
INVLPG addr	; Invalidate TLB entry for a
	; specified memory address "addr,"
	; if present in the TLB.

Note: Instructions INVD, WBINVD, and INVLPG are *new* on the 80486, not previously available on the 80386. The following instruction was featured on the 80386, but it is not needed on the 80486 because it has an on-chip FPU:
```
ESC ; Escape to coprocessor execution.
```

```
BT, BTS, BTR, BTC, XCHG, ADD, OR, ADC, SBB, AND, SUB, XOR, NOT, NEG,
INC, DEC
```

when they involve a memory operand. An undefined opcode trap will be generated if a LOCK prefix is used with any instruction not listed above.

The table lookup translation (XLAT) instruction is useful for translating from one coding system to another, such as from Extended Binary Coded Decimal Interchange (EBCDIC) to ASCII. The INVD instruction invalidates (flushes) the entire on-chip cache. The WBINVD instruction also invalidates the entire on-chip cache. In contrast to the INVD instruction, the processor must also write back into memory all modified lines before flushing the cache. The INVLPG instruction invalidates a single entry in the TLB.

The details of the x86 floating-point instructions are given in Appendix to Part 2. As mentioned in Sec. 7.2, bits 13 to 11 (TOP field) of the FPU status word point to the FPU data register, which is currently the top of stack. The top of stack register is denoted ST(0) or just ST. Other registers are indexed with respect to the top of stack register.

Example   Assume the TOP field of the status word contains 011 (3). Thus, R3 is the top of stack register ST(0). The other registers are then denoted:

R4 ST(1)
R5 ST(2)
R6 ST(3)
R7 ST(4)
R0 ST(5)
R1 ST(6)
R2 ST(7)

The stack annotation of the FPU data registers proceeds in a circular manner around the 8 data registers from R3 through R7 to R2.

The top of stack register ST(0) is used as an implied destination operand in many floating-point instructions.

Examples   Assume ST(0) = ST = R3.

```
FADD TEMP; floating add (TEMP) + ST → ST = R3, TEMP is a memory
location
FADD ST(3); (R6) + (R3) → R3 = ST
FADD ST; (ST) + (ST) = 2*ST → ST = R3
FABS ; replace the content of ST = R3 by its absolute value.
```

There are special floating load and store instructions that push and pop variables from the FPU data register stack, while affecting the TOP field in the status word.

Examples   Assume ST = R3, (TOP) = 011, before the execution of any of the following instructions.

```
FLD TEMP; push (TEMP) from memory on stack into R2, which now becomes
ST(0),
; decrement (TOP) by 1, which becomes 010, former ST (or R3) becomes
ST(1).
```

Entry	Before	After
TOP	011	010
R0	ST(5)	ST(6)
R1	ST(6)	ST(7)
R2	ST(7)	ST(0)
R3	ST(0)	ST(1)
R4	ST(1)	ST(2)
R5	ST(2)	ST(3)
R6	ST(3)	ST(4)
R7	ST(4)	ST(5)

```
FLD ST(3); push ST(3) = R6 on top of stack into R2, which becomes
ST(0),
; former ST(3) becomes ST(4). Now (ST(0)) = (ST(4)).
FSTP ST(3); transfer (pop) ST(0) into ST(3), increment TOP. Now ST
(0) = R4,
; and former ST(3) becomes ST(2), and (R6) = (R3).
```

Entry	Before	After
TOP	011	100
R0	ST(5)	ST(4)
R1	ST(6)	ST(5)
R2	ST(7)	ST(6)
R3	ST(0)	ST(7)
R4	ST(1)	ST(0)
R5	ST(2)	ST(1)
R6	ST(3)	ST(2)
R7	ST(4)	ST(3)

There is also a floating-point store instruction that does not change the stack structure and does not affect the TOP field in the status word.

Example

```
FST ST(3); copy the content of ST(0) into ST(3), TOP not affected.
```

Assembly directives

When we use a register as an instruction operand, the size of the operand is obvious; using AL, AX, or EAX, for instance, establishes the operand to be 8-bit (byte), 16-bit (word), or 32-bit (dword), respectively. When we use symbolic memory addresses, the size of the operands is not clearly obvious. Symbolic names of operands must be declared for their size and possibly initial values by using assembly directives, to be discussed next.

Assembly directives constitute commands to the assembler. They are not executable instructions. The first type of assembly directives to be discussed are the data definition directives. An example follows each directive name.

Define byte, DB (8-bit data)

```
DATAB DB A0H,150,?,?
```

Define word, DW (16-bit data)

```
DATAW DW FF00H,2000,0,?,?
```

Define doubleword, DD (32-bit data, single-precision floating-point)

```
DATAD DD 0FFAA12H,0,?,4.53e5
```

Define quadword, DQ (64-bit data, double-precision floating-point)

```
DATAQ DQ 6.84e6
```

Define 10 bytes, DT (80-bit data, extended-precision floating-point)

```
DATAT DT 8.83e10
```

DATAB, DATAW, DATAD, DATAQ, and DATAT are symbolic labels for the data defined in the examples. Note that no colon (:) is used with data definitions. The symbol "?" indicates a storage allocation without any value assigned. Exponent notation, such as

```
4.53e5 = 4.53*10⁵
```

is acceptable. The define byte (DB) example above defines 5 bytes; the first 3 with numerical values, the last two just storage allocation of 2 bytes. If we want to allocate a whole array of bytes, say 100, without value assignment, we can use:

```
ARRAY1 DB 100(?)
```

If we want all of them cleared:

```
ARRAY1 DB 100(0)
```

In the DB statement the symbolic address statement for "A0H" is DATAB, for "150": DATAB + 1, and so on. Similarly, for the DW, the address of "FF00H" is DATAW, for "2000": DATAW + 2, and so on. Doubleword arrays, DD, have a step of 4, DQ a step of 8, and DT a step of 10 bytes, or addresses.

There is also a possibility of data definition by using the EQU expression as follows:

```
DATAB EQU 8
```

which assigns the numerical value of 8 to DATAB. It can also be used to establish equivalence between symbolic names:

```
DATAB EQU DATA1
```

It is suggested to put data definition statements (such as DB, DW) first and EQU statements next, in order to avoid possible errors.

In order to ensure the alignment of a word array, we can place the directive EVEN before the DW directive:

```
EVEN
DATAW DW 50 DUP(1)
```

This will ensure that the DATAW array of 50 words, loaded with the value one, will start at an even address.

The statement

```
ORG 2000H
```

will cause the next byte to start at the address (offset within a segment) 2000H.

The PTR directive is used to identify a memory-addressed operand that has the same address of some other memory-addressed operand, of a different type. The PTR does not allocate any memory space.

Example Define a word

```
DBYTE DW ?; 2 bytes allocated as one word.
```

We can now assign alternate names to each one of the above 2 bytes. The first byte will be called FBYTE.

```
FBYTE EQU BYTE PTR DBYTE
```

The second byte will be called SBYTE.

```
SBYTE EQU BYTE PTR (DBYTE + 1)
```

The last statement is equivalent to

```
SBYTE EQU FBYTE + 1
```

7.6 Interrupt

An *interrupt* is usually understood to be an event when an external signal stops the execution of a program by a processor. While the program is interrupted, the processor runs an *interrupt service routine* (ISR), or an interrupt handling routine. A possible cause for an interrupt could be an I/O interface device requesting the CPU to perform a certain function, such as transmit outside specified information. A more general concept is that of an exception, which also interrupts the processor and causes the CPU to run an exception handling routine. In most cases exceptions are caused by internal processor faults, errors, or other exceptional events. The treatment of interrupts and exceptions is basically the same in most microprocessors. In some cases, interrupts are considered to be a particular case of exceptions. The Intel x86 architecture uses both terms.

The x86 architecture handles up to 256 interrupts and exceptions. The starting addresses of the interrupt handling routines are called interrupt vectors. An interrupt vector is 8 bytes long. The interrupt vectors are stored in a special interrupt vector table in memory. Of the 256 possible interrupts, 32 are reserved by Intel, and the remaining 224 are free to be designed by the user. The interrupt vector assignments are summarized in Table 7.13.

TABLE 7.13   Interrupt Vector Assignments

Function	Interrupt Number	Instruction Which Can Cause Exception	Return Address Points to Faulting Instruction	Type
Divide Error	0	DIV, IDIV	YES	FAULT
Debug Exception	1	Any Instruction	YES	TRAP*
NMI Interrupt	2	INT 2 or NMI	NO	NMI
One Byte Interrupt	3	INT	NO	TRAP
Interrupt on Overflow	4	INTO	NO	TRAP
Array Bounds Check	5	BOUND	YES	FAULT
Invalid OP-Code	6	Any Illegal Instruction	YES	FAULT
Device Not Available	7	ESC, WAIT	YES	FAULT
Double Fault	8	Any Instruction That Can Generate an Exception		ABORT
Intel Reserved	9			
Invalid TSS	10	JMP, CALL, IRET, INT	YES	FAULT
Segment Not Present	11	Segment Register Instructions	YES	FAULT
Stack Fault	12	Stack References	YES	FAULT
General Protection Fault	13	Any Memory Reference	YES	FAULT
Page Fault	14	Any Memory Access or Code Fetch	YES	FAULT
Intel Reserved	15			
Floating Point Error	16	Floating Point, WAIT	YES	FAULT
Alignment Check Interrupt	17	Unaligned Memory Access	YES	FAULT
Intel Reserved	18–31			
Two Byte Interrupt	32–255	INT n	NO	TRAP

*Some debug exceptions may report both traps on the previous instruction, and faults on the next instruction.

SOURCE: Courtesy of Intel Corp.

When an interrupt is acknowledged by the CPU, the following sequence of actions occurs:

1. The EFLAGS register and the address of the next instruction to be executed are pushed onto the stack to allow resumption of the interrupted program.

2. An 8-bit interrupt number (0 to 255) is supplied to the CPU, which identifies the appropriate entry in the interrupt vector table.

3. The starting address of the appropriate ISR is extracted from the vector table, loading the EIP and the CS registers (see Sec. 7.7 on segmentation).

4. The ISR is executed. It is terminated by the IRET instruction. This restores the processor state, and program execution resumes at the instruction that was to be executed next, when the interrupt occurred.

The 8-bit interrupt number is supplied to the processor in several different ways: exceptions supply the interrupt number internally, software INT instructions contain or imply the interrupt number, and maskable hardware interrupts supply the 8-bit interrupt number via the interrupt acknowledge bus sequence. Nonmaskable hardware interrupts are assigned to interrupt number 2 (see Table 7.13).

The IF bit in the EFLAGS register is reset when an interrupt is being serviced. This effectively disables servicing additional interrupts during the running of an ISR. However, the IF may be set explicitly by the ISR to allow nesting of interrupts. When an IRET instruction is executed, the original state of the IF is restored.

7.7 Segmentation

The concept of segmentation in general was discussed in Chap. 4. The option of using segmentation is featured by the Intel x86 family of microprocessors. The x86 architecture segmentation mechanism will be discussed in this section.

The memory is subdivided into parts called segments. A segment can be from 1 byte to 4 Gbytes long. Segments can start at any base address in memory, and storage overlapping between segments is allowed.

A virtual (logical) address in the x86 architecture is formed out of two components:

1. A 16-bit selector, used to determine the linear base address (the address of the first byte of the segment) of the segment.

2. A 32-bit offset, the internal address within a segment. The offset of a given memory location address is its distance in bytes from the segment base address.

The addressing of a memory operand within a segment is illustrated in Fig. 7.10. When a 32-bit x86 processor is reset or powered up, it is initialized in real mode (see Sec. 7.9). Real mode has the same architecture as the 8086 (see Chap. 10) but allows access to the 32-bit register set. The default operand size in real mode is 16 bits. However, the regular mode of operation of a 32-bit x86 architecture processor is in *protected virtual address mode* (PVAM), or simply, protected mode. In protected mode the 16-bit selector is used to specify an index in an OS-defined table (this mechanism will be elaborated later in this section). The table contains the 32-bit base address of a given segment. The physical address is formed by adding the base address obtained from the table to the offset.

As any other modern microprocessor, the x86 architecture supports paging, discussed in Chap. 4. Paging on x86 systems is provided in the

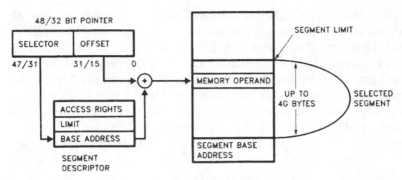

Figure 7.10 Protected mode addressing. (*Courtesy of Intel Corp.*)

protected mode. It provides an efficient mechanism for the handling of virtual memory. The paging mechanism is optional on the x86 systems; it can be enabled or disabled by software [enabled by setting bit 31, (PG) = 1, in CR0, as described in Sec. 7.2]. The addressing sequence in the x86 architecture, featuring both segmentation and paging, is illustrated in Fig. 7.11. The address at the output of the segmentation mechanism is called the linear address. It is the input to the paging mechanism (described in Sec. 7.8). The 32-bit address output from the paging mechanism is the physical address. When paging is disabled, the linear address and the physical address are identical. When both segmentation and paging are disabled, the virtual and the physical addresses are identical. Segmentation can be essentially disabled by choosing the segment size equal to the size of the whole physical memory, up to 4 Gbytes.

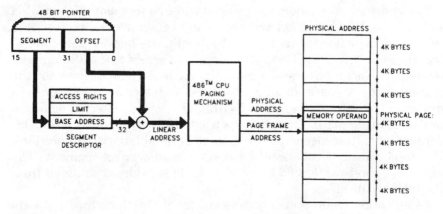

Figure 7.11 Paging and segmentation. (*Courtesy of Intel Corp.*)

Each segment has a segment descriptor associated with it. The segment descriptor is 8 bytes long and contains the following information about the segment:

1. A 32-bit segment base linear address
2. A 20-bit segment limit, specifying the size of the segment
3. Access rights byte, containing protection mechanism information (see Sec. 7.10)
4. Control bits

The segment limit field of the segment descriptor has only 20 bits (and not 32) because the segment size does not have a byte granularity for all segment sizes. Segments have a byte granularity (i.e., segments may differ in size by a single byte) for segment sizes up to 1 Mbyte (2^{20} bytes). For segments above 1 Mbyte and up to 4 Gbytes, there is a page granularity, that is, segment sizes may differ by a page size, which is 4 kbytes (2^{12} bytes).

A general form of a segment descriptor is illustrated in Fig. 7.12. The descriptor fields and control bits are defined in the figure. The access rights byte will be discussed in more detail in Sec. 7.10 on the protection mechanism. As can be seen, the segment base and the segment limit fields are not contiguous, but distributed in a number of subfields.

Segment descriptors are stored in descriptor tables in memory. The descriptor tables define all the segments which are used in the system. There are three types of descriptor tables:

1. The *global descriptor table* (GDT) contains descriptors that are possibly available to all tasks in the system.

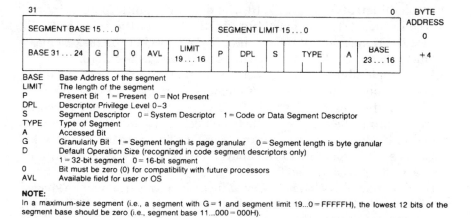

Figure 7.12   Segment descriptors. (*Courtesy of Intel Corp.*)

2. The *local descriptor table* (LDT) contains descriptors associated with a given task. Each task may have a separate LDT. A segment cannot be accessed by a task if its segment descriptor does not exist in either the current LDT or the GDT.

3. The *interrupt descriptor table* (IDT) contains descriptors that point to the location of up to 256 ISRs (see Sec. 7.6). The IDT is basically the interrupt vector table. Each interrupt vector is a descriptor.

All of the above descriptor tables are variable-length memory arrays. They can range in size between 8 bytes (a single descriptor) and 64 kbytes (upper limit: $2^{13} = 8192$ descriptors, 8 bytes each). The upper 13 bits of a selector are used as an index into the descriptor table. Each of the above tables has a register, located in the CPU (see Sec. 7.2), associated with it and pointing to it:

1. GDT register (GDTR), 48 bits, associated with the GDT

2. LDT register (LDTR), 16 bits, associated with the LDT

3. IDT register (IDTR), 48 bits, associated with the IDT

The LGDT, LLDT, and LIDT instructions (see Table 7.11) load the base and the limit of the GDT, LDT, and IDT, respectively, into the appropriate register: GDTR, LDTR, and IDTR, respectively. The SGDT, SLDT, and SIDT instructions store the contents of GDTR, LDTR, and IDTR, respectively, into a specified destination address.

A 16-bit selector, which can be stored in a 16-bit segment register, points to a segment descriptor in GDT or in LDT, as illustrated in Fig. 7.13. The upper 13 bits of a selector constitute a pointer to one of the $8192 = 2^{13}$ descriptors in the descriptor table. Bit TI (bit 2) of the selector differentiates between the LDT (TI = 1) and GDT (TI = 0). In the example illustrated in Fig. 7.13, the selector points to descriptor 3 (011B) in the LDT. The least significant 2 bits of the selector (RPL, requestor privilege level) reflect the privilege level of the task requesting memory access. This aspect will be discussed in Sec. 7.10.

An example illustrating the access of an operand in memory using the GDT is shown in Fig. 7.14. A memory data operand is accessed in the DS-pointed segment. The EIP contains the offset of the accessed operand within the segment. The base address of the data segment is stored in its descriptor, located in the GDT. The segment register DS contains a 13-bit pointer to the above descriptor. The GDTR contains the base address of the GDT.

The LDT usually belongs to a specific task; however, it can be shared between tasks. It is optional; a task can operate using the GDT only. A descriptor, pointing to the base of LDT, is stored in the GDT. The LDT descriptor within the GDT is pointed to by the LDTR, which has the

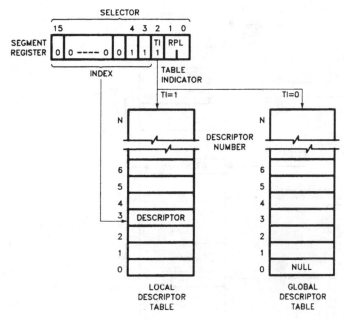

Figure 7.13 Example descriptor selection. (*Courtesy of Intel Corp.*)

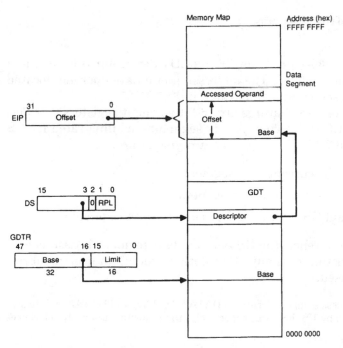

Figure 7.14 Example of an operand addressing.

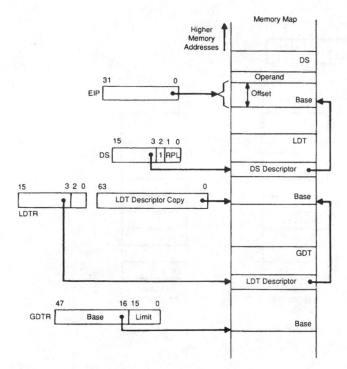

Figure 7.15 Operand addressing using LDT.

format of a 16-bit selector. A copy of the LDT descriptor is located in a 64-bit register in the CPU. The addressing of a data operand, located in a DS-pointed segment, is illustrated in Fig. 7.15.

The x86 architecture features six 16-bit segment registers (see Sec. 7.2) which constitute selectors for six segments, as illustrated for DS in Figs. 7.14 and 7.15. The six segment registers are:

CS indicating the current code segment

SS indicating the current stack segment

DS, ES, FS, and GS indicating four current data segments

On any data reference, the DS-pointed data segment is assumed by default. In order to access any other data segment, an override directive should be used.

Example To access symbolic locations DATA1, DATA2, and DATA3 in data segments pointed to by ES, FS, GS, respectively, the following assembly references should be used:

```
ES:DATA1, FS:DATA2, GS:DATA3
```

If the above addresses are referred to without prefixes, the assembler will assume that they are in the DS-pointed data segment.

Each segment register has a segment descriptor cache register associated with it, shown in Fig. 7.16. These registers, located in the CPU, contain the same information as the segment descriptors, located in GDT or LDT, associated with the segments pointed to by the segment registers. This permits them to obtain the base address of any active segment (currently pointed to by a segment register) in the fastest way, without incurring an extra memory access into the GDT or LDT. When a selector value is loaded into a segment register, the associated descriptor cache register is automatically updated with the current descriptor information. Whenever a memory reference occurs, the segment descriptor cache register associated with the segment being used is automatically utilized. The 32-bit segment base address becomes a component of the linear address calculation, the limit field is used for the limit check operation, and the attributes are checked against the type of memory reference requested (see Sec. 7.10).

It is appropriate to introduce at this point the x86 architecture pointer data types, not specified in Sec. 7.3. The pointer data types, shown in Fig. 7.17, are:

48-bit pointer: 16-bit selector and 32-bit offset

32-bit pointer: 32-bit offset

The above pointers can be stored in memory and from there loaded into the appropriate CPU registers: the 32-bit offset into any 32-bit general register and the 16-bit selector into any 16-bit segment register.

For instance, such a loading would occur in an intersegment JMP or CALL (see Sec. 7.5). The 48-bit pointer would then be a part of the instruction, following the opcode. As part of the execution of the instruction, the 32-bit offset (stored first) will be loaded into the EIP, and the 16-bit selector (contains a pointer to the new target code segment) will be loaded into the CS.

A more complete illustration of memory addressing for various addressing modes (see Sec. 7.4) is given in Fig. 7.18.

Figure 7.16 x86 architecture segment registers and associated descriptor cache registers. (*Courtesy of Intel Corp.*)

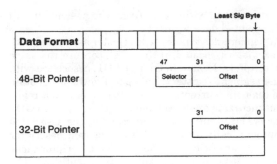

Figure 7.17 Pointer data types. (*Courtesy of Intel Corp.*)

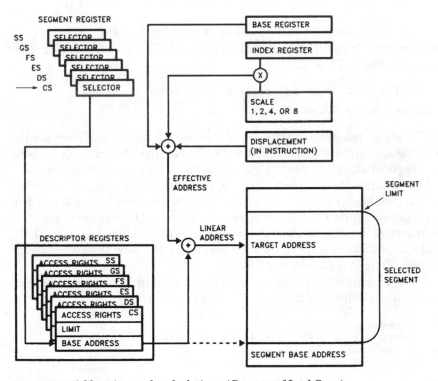

Figure 7.18 Addressing mode calculations. (*Courtesy of Intel Corp.*)

There is a set of six data transfer instructions that supports the handling of effective addresses and address pointers (see Table 7.4). The load effective address (LEA) instruction transfers the offset (which is the EA) of a memory-stored operand, rather than its value, to the destination operand, which is a 16- or a 32-bit general register.

The instructions LDS, LES, LFS, LGS, and LSS load a full 48-bit pointer from memory in the following way:

The 32-bit offset into a specified general register (reg)

The 16-bit selector into the DS, ES, FS, GS, and SS segment register, respectively

When the system is running in an 8086 or 80286 mode (see Sec. 7.9), the offset is 16-bit and it is transferred into a 16-bit general register.

Segmentation-related assembly directives

The assembly directives to generate a segment are as follows [LiGi 86]:

```
segment name SEGMENT
[contents of the segment]
segment name ENDS
```

The assembly directive reserved names are italicized.

Example   Definition of two data segments:

```
DATA1 SEGMENT
[contents of DATA1]
DATA1 ENDS
DATA2 SEGMENT
[contents of DATA2]
DATA2 ENDS
```

A segment can be started at a user-specified address as follows:

```
DATA1 SEGMENT AT FFF000H
```

The definition of a code segment requires additional assembly directives, as illustrated in the following example.

Example   Definition of a code segment called PROG:

```
PROG SEGMENT
ASSUME CS:PROG, DS:DATA1, ES:DATA2
START: MOV AX,DATA1; first executable statement
       MOV DS,AX
       MOV AX,DATA2
       MOV ES,AX; segment registers loaded
[rest of the code of PROG]
PROG ENDS
END START; end of assembly
```

Note: On some assemblers the *ASSUME* directive may be sufficient, and then there is no need in the MOV to segment register instructions.

7.8   Paging

A general discussion of paging was given in Chap. 4. The Intel x86 architecture paging mechanism will be discussed in this section. The

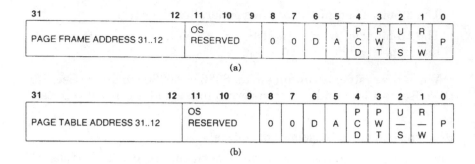

31		12	11	10	9	8	7	6	5	4	3	2	1	0
PAGE FRAME ADDRESS 31..12			OS RESERVED			0	0	D	A	P C D	P W T	U — S	R — W	P

(a)

31		12	11	10	9	8	7	6	5	4	3	2	1	0
PAGE TABLE ADDRESS 31..12			OS RESERVED			0	0	D	A	P C D	P W T	U — S	R — W	P

(b)

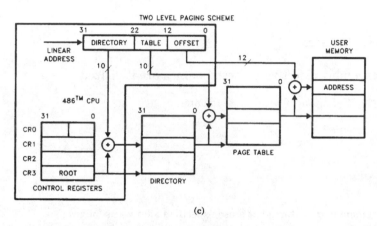

(c)

Figure 7.19 Paging mechanisms. (a) Page table entry (points to page); (b) page directory entry (points to page table). (*Courtesy of Intel Corp.*)

standard page size of the x86 is 4 kbytes = 2^{12} bytes. The x86 uses two levels of tables to translate the linear address into a physical address. There are three components to the paging mechanism: the page directory, the page tables, and the page frame. A uniform size for all the elements simplifies memory allocation and reallocation schemes, since there is no problem with memory fragmentation. Figure 7.19 illustrates the paging mechanism.

The control register CR2 is the page fault linear address register. It holds the 32-bit linear address which caused the last page fault detected. Register CR3 points to the base of the page directory (see Sec. 7.2).

The page directory is 4 kbytes long and allows up to 1024 PDEs. Each PDE, shown in Fig. 7.19(b), contains the address of the next level tables, the page tables, and information about the page table pointed to. The upper 10 bits of the linear address (the directory field; bits 31 to 22) are used as an index to select the correct PDE.

Each page table is 4 kbytes and holds up to 1024 PTEs. A PTE, shown in Fig. 7.19(a), contains the starting address of the page frame and access information about the page. Address bits 31 to 22 (table field) are used as an index to select one of the 1024 PTEs. Bits 31 to 12 of the PTE contain the upper 20 bits of the base of the page frame pointed to by the PTE. The lower 12 bits of the PTE are identical to those of the PDE. The function of the bits currently in use is as follows:

Bit 6, D—dirty. Bit D is set before a write to an address covered by the PTE occurs. It is undefined for the PDE.

Bit 5, A—accessed. Bit A is set before a read or write access occurs to an address covered by the entry.

Bit 4, PCD—page cache disable. The PCD bit controls the page on-chip cacheability. When (PCD) = 0, the on-chip cache is enabled. When (PCD) = 1, on-chip caching is disabled.

Bit 3, PWT—page write-through. The PWT bit controls page write policy. (PWT) = 1 defines a write-through policy for the current page. (PWT) = 0 allows the possibility of write-back (see Chap. 4). Bits PCD and PWT are also bits 4 and 3 on the CR3 (see Sec. 7.2). The state of the PCD and PWT bits is driven out on the PCD and PWT pins during a memory access.

Bit 2, U/S—user/supervisor. Bit U/S differentiates between lower-privilege user mode and higher-privilege supervisor mode (see also Sec. 7.10).

Bit 1, R/W—read/write. Bit R/W establishes read and write protection privileges for the page (see Sec. 7.10).

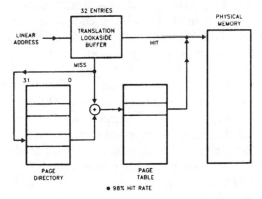

Figure 7.20 Translation lookaside buffer. (*Courtesy of Intel Corp.*)

Bit 0, P—present. Bit P indicates whether the PDE or PTE can be used in address translation. If (P) = 1, the entry can be used. If (P) = 0, it cannot be used.

As in most modern systems, there is a TLB (Chap. 4), shown in Fig. 7.20, for faster address translation.

If there is a miss in the TLB, and (P) = 1 for the missing PTE, the CPU will read the missing PTE into the TLB and set the A bit. If (P) = 0, a page fault exception will be generated.

7.9   Real and Virtual Mode Execution

The x86 architecture allows the 32-bit processors (i386, i486, Pentium) to execute 8086 application programs in two modes:

1. Real mode: This has the same base architecture as the 8086 but allows access to the 32-bit register set. When the processor is reset or powered up, it is initialized in real mode.

2. Virtual mode (or virtual 8086 mode): This allows the execution of 8086 applications, while still allowing the system designer to take full advantage of the protection mechanism (see Sec. 7.10).

All 32-bit architecture instructions are available in the real or virtual modes, except the following:

```
ARPL, LAR, LLDT, LIDT, LSL, LTR, SLDT, LGDT, STR, VERR, VERW, LMSW,
CLTS, HLT, MOV to or from control or debug registers.
```

The default operand size in real mode is 16 bits. The segment size in real mode is fixed at 64 kbytes. Segments may overlap. The maximum memory size in real mode is 1 Mbyte. Since paging is not allowed in real mode, the linear addresses are the same as physical addresses. Physical addresses are formed in real mode by adding the content of the appropriate segment register (16-bit), which is shifted left by 4 bits, to the offset (the EA). Real-mode segments always start on 16-byte boundaries. The real-mode addressing is illustrated in Fig. 7.21

In virtual mode the segment registers are used in an identical fashion to real mode. The address calculation is also done as in the real mode, as shown in Fig. 7.21. Through the use of paging, the 1-Mbyte address space of the virtual mode task can be mapped to anywhere in the 4-Gbyte linear address space of the 32-bit processor. Virtual mode is entered by executing an IRET instruction or by a task switch (see Sec. 7.11). The transition out of the virtual to the protected mode occurs only by an interrupt or exception.

The transition from real mode to protected mode can be done by setting the PE bit in CR0 (see Sec. 7.2), using a MOV to CR0 or a LMSW

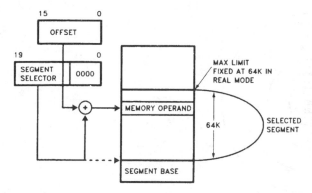

Figure 7.21   Real address mode addressing. (*Courtesy of Intel Corp.*)

instruction. A subsequent return to real mode can be done by resetting (clearing) the above PE bit in a similar manner.

7.10 Protection Mechanism

The x86 architecture has a very elaborate protection mechanism, compared to other systems. It was started with the introduction of the protected mode of operation on the 16-bit i286 and subsequently expanded for a 32-bit system on the i386. It is continued on the subsequent 32-bit systems (i486, Pentium).

Unlike most other systems which have only two protection levels (user and supervisor), the x86 architecture features four levels of protection, called *privilege levels* (PL). They are designed to support the needs of a multitasking OS to isolate and protect user programs from each other and the OS from unauthorized access. The privilege levels control the use of privileged instructions, I/O instructions, and access to segments and segment descriptors. The x86 architecture offers an additional type of protection on a page basis, when paging is enabled (see Sec. 7.8). This feature will be elaborated later in this section.

The four-level hierarchical privilege system is illustrated in Fig. 7.22. The PLs are numbered 0, 1, 2, 3. Level 0 is the most privileged level. Level 3 is the least privileged, used for regular user applications. Level 2 is used for OS extensions, level 1 for system services, and the most privileged level 0 is used for the OS kernel [BiSh 88].

The x86 architecture controls access to both data and code between levels of a task, according to the following rules of privilege:

1. Data stored in a segment with PL = p can be accessed only by code executing at a PL at least as privileged as p.

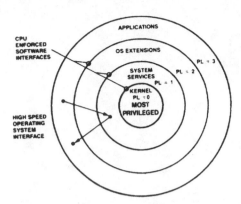

Figure 7.22 Four-level hierarchical protection. (*Courtesy of Intel Corp.*)

2. A code segment (a procedure) with PL = p can be called only by a task executing at the same or a lower PL than p.

The PLs are classified into four categories:

1. Requestor PL, RPL, the PL of the original supplier of the selector (see Sec. 7.7). RPL is determined by the two LSBs of the selector.
2. Descriptor PL, DPL, the least PL at which a task may access that descriptor and the segment associated with that descriptor (see Sec. 7.7). The DPL is determined by bits 6 and 5 in the access rights byte of a descriptor.
3. Current PL, CPL, the PL at which a task is currently executing, which equals the PL of the code segment being executed. The CPL is stored in the two LSBs of the CS register, except for conforming code segments (explained later in this section).
4. Effective PL, EPL, the least privileged of the RPL and CPL. Since smaller PL values indicate greater privilege, EPL is the numerical maximum of RPL and CPL.

```
EPL = max (RPL, CPL)
```

At this point, the access rights byte of a descriptor can be discussed in detail. Table 7.14 lists the bit interpretations of the access rights byte.

Code and data segments (bit 4, S = 1 for both, bit 3, E = 0 for data, E = 1 for code) have several descriptor fields in common. The accessed bit (A; bit 0) is set (A = 1) whenever the processor accesses the descriptor. The A bit is used by the OS to keep usage statistics on a given segment. The executable bit (E; bit 3) indicates whether a segment is a code (E = 1) or a data (E = 0) segment. A code segment may be execute

TABLE 7.14   Access Rights Byte Definition for Code and Data Descriptions

	Bit Position	Name	Function	
	7	Present (P)	P = 1 Segment is mapped into physical memory.	
			P = 0 No mapping to physical memory exits, base and limit are not used.	
	6–5	Descriptor Privilege Level (DPL)	Segment privilege attribute used in privilege tests.	
	4	Segment Descriptor (S)	S = 1 Code or Data (includes stacks) segment descriptor.	
			S = 0 System Segment Descriptor or Gate Descriptor.	
Type Field Definition	3	Executable (E)	E = 0 Descriptor type is data segment:	If
	2	Expansion Direction (ED)	ED = 0 Expand up segment, offsets must be ≤ limit.	Data
			ED = 1 Expand down segment, offsets must be > limit.	Segment
	1	Writeable (W)	W = 0 Data segment may not be written into.	(S = 1,
			W = 1 Data segment may be written into.	E = 0)
	3	Executable (E)	E = 1 Descriptor type is code segment:	If
	2	Conforming (C)	C = 1 Code segment may only be executed when CPL ≥ DPL and CPL remains unchanged.	Code Segment (S = 1,
	1	Readable (R)	R = 0 Code segment may not be read.	E = 1)
			R = 1 Code segment may be read.	
	0	Accessed (A)	A = 0 Segment has not been accessed.	
			A = 1 Segment selector has been loaded into segment register or used by selector test instructions.	

SOURCE: Courtesy of Intel Corp.

only or execute/read as determined by the read bit (R; bit 1). If R = 0, code segments are execute only, and execute/read if R = 1. Code segments may never be written into.

The conforming bit (C; bit 1, when E = 1) determines if the code segment, pointed to by the descriptor, is of a conforming type (conforming when C = 1, nonconforming when C = 0). A conforming segment has no inherent fixed PL; it conforms to the PL that CALLs it or JMPs to it. For instance, if a program in a PL = 3 segment [Turl 88] transfers control to a conforming code segment, then the conforming code runs with CPL = 3. A conforming segment can be executed and shared by programs at different PLs. The RPL bits of the CS are not changed when execution starts, reflecting the previous code segment executed. The following inequality is to be satisfied:

```
(DPL of conforming code segment descriptor) ≤ (current CPL)
```

This means that control can be transferred only across (at the same PL) or up to a more privileged segment. Control can never be transferred to a segment whose DPL is greater in value (less privileged) than the current segment.

Data segments (E = 0, S = 1) can be data or stack segments. The expansion direction bit (ED; 2 when E = 0) specifies whether a segment expands downward (stack, ED = 1) or upward (data, ED = 0). If a segment is a stack segment, all offsets must be greater than the segment limit. On a data segment, all offsets must be less than or equal to the limit. In other words, stack segments start at the base linear address plus the maximum segment limit and grow down to the base linear address. On the other hand, data segments start at the base linear address and expand to the base linear address plus limit.

The write bit (W; bit 1 when E = 0) controls the ability to write into a segment. Data segments are read only if W = 0. Writing into a segment is permitted if W = 1. The stack segment must be writable; it must have W = 1.

System segments (bit S = 0) describe information about OS tasks and data structures. The general format of a system segment descriptor is shown in Fig. 7.23 along with the 4-bit encoding of various types of system segments in the TYPE field of the access rights byte (bits 3 to 0). Types defined for the 486 in Fig.7.23 are the same for the other 32-bit processors (i386, Pentium) as well. The system descriptor types will be discussed next.

Types 1 (0001), 3 (0011), 9 (1001), task state segment (TSS) descriptors: will be discussed in Sec. 7.11.

Type 2 (0010), LDT descriptor: located in the GDT and points to the base of the LDT (see Sec. 7.7).

Types 4 (0100), 7 (0111), C (1100), E (1110), F, gate descriptors: gates are used to control access to entry points within the target code segment. Gates provide a level of indirection between the source and destination of the control transfer. This indirection allows the processor to perform protection checks automatically. It also allows system designers to control entry points to the OS. There are four types of gates:

1. Call gates serve as an intermediary between code segments at different PLs. Call gates are used to change PLs.

2. Task gates are used to perform a task switch (see Sec. 7.11).

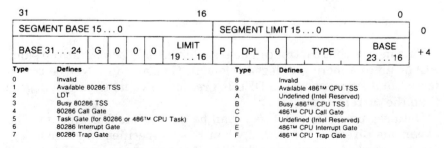

Type	Defines	Type	Defines
0	Invalid	8	Invalid
1	Available 80286 TSS	9	Available 486™ CPU TSS
2	LDT	A	Undefined (Intel Reserved)
3	Busy 80286 TSS	B	Busy 486™ CPU TSS
4	80286 Call Gate	C	486™ CPU Call Gate
5	Task Gate (for 80286 or 486™ CPU Task)	D	Undefined (Intel Reserved)
6	80286 Interrupt Gate	E	486™ CPU Interrupt Gate
7	80286 Trap Gate	F	486™ CPU Trap Gate

Figure 7.23 System segments descriptors. (*Courtesy of Intel Corp.*

3. Interrupt gates are used to specify interrupt service routines (ISRs).

4. Trap gates are used to specify trap handling routines.

The format of the four types of gate descriptors is shown in Fig. 7.24. Wherever there is an entry of "486," it is also valid for all 32-bit systems (i386, Pentium).

Call gates are primarily used to transfer program control to a more privileged level. The call gate descriptor consists of three fields:

1. The access rights byte.

2. The long pointer (selector and offset), which points to the start of a target routine.

3. The word count (5 bits), which specifies how many parameters are to be copied from the caller's stack to the stack of the called routine. This field is used only be call gates when there is a change in the PL. Other types of gates ignore this field.

The difference between interrupt and trap gates is that the interrupt gate disables interrupts (resets the IF bit in EFLAGS; see Sec. 7.2), while the trap gate does not disable interrupts.

Call gates are accessed via a CALL instruction and are syntactically identical to calling a normal procedure. When an interlevel (to a program with a different PL) call gate is activated, the following actions occur:

1. Load CS:EIP from the selector and offset fields of the gate, respectively, and check for validity.

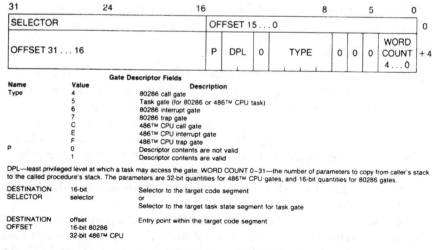

Figure 7.24 Gate descriptor formats. (*Courtesy of Intel Corp.*)

2. SS is pushed on stack zero-extended to 32 bits.

3. ESP is pushed on stack.

4. Copy word count 32-bit parameters from the old stack to the new stack.

5. Push return address on stack.

Gate descriptors follow the data access rules of privilege; that is, gates can be accessed by a task if its EPL is equal to, or more privileged than the gate descriptor's DPL (i.e., its value is less than or equal to that of the DPL). The PL check can be expressed by the following inequality [Turl 88]:

```
(target DPL) ≤ EPL = max (RPL, CPL) ≤ (gate DPL)
```

Example A call gate with a DPL = 2 references a code segment with a DPL = 0 (highest PL), as shown in Fig. 7.25. The call gate at PL = 2 can be called from levels 0, 1, and 2. It cannot be called from level 3, since in this case

```
EPL = max (RPL, CPL) > (gate DPL)
```

contradicting the privilege check inequality. Naturally, the target code segment at PL = 0 can be called directly from level 0.

The x86 architecture provides a set of protection attributes for paging. The paging mechanism (see Sec. 7.8) distinguishes between two levels of protection:

User—corresponding to PL = 3,

Supervisor—corresponding to PL = 0, 1, or 2.

The combined page protection effect of bits U/S, R/W from PTEs or PDEs, and bit WP (bit 16 of CR0) is summarized in Table 7.15.

7.11 Task Management

The x86 architecture was particularly designed for efficient handling of tasks in a multitasking environment. A *task* can be defined as an instance of the execution of a program. A very important attribute of any multitasking, multiuser OS is the ability to switch rapidly between tasks. The x86 supports the task switching operation in hardware. The task switch operation saves the entire state of the machine (all the registers, the address space, and a link to the previous task), loads a new execution state, performs protection checks, and begins execution of the new task.

The task switch operation is invoked by executing an intersegment JMP or CALL instruction (see Secs. 7.5 and 7.7), which refers to a task state segment (TSS) or a task gate descriptor in the GDT or LDT. An

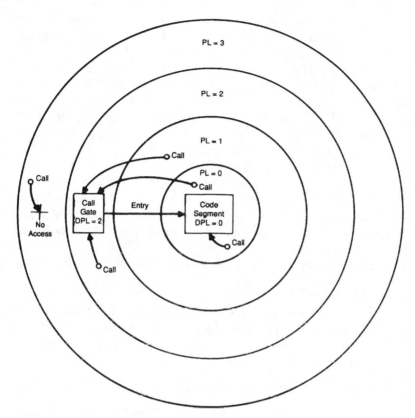

Figure 7.25   Example of a call gate access.

TABLE 7.15   Page Level Protection Attributes

U/S	R/W	WP	User Access	Supervisor Access
0	0	0	None	Read/Write/Execute
0	1	0	None	Read/Write/Execute
1	0	0	Read/Execute	Read/Write/Execute
1	1	0	Read/Write/Execute	Read/Write/Execute
0	0	1	None	Read/Execute
0	1	1	None	Read/Write/Execute
1	0	1	Read/Execute	Read/Execute
1	1	1	Read/Write/Execute	Read/Write/Execute

Source: Courtesy of Intel Corp.

INT n instruction, exception, trap, or external interrupt may also invoke the task switch operation if there is a task gate descriptor in the associated IDT descriptor slot. The TSS descriptor points to the TSS that contains the entire processor execution state, as illustrated in Fig. 7.26. A task gate descriptor contains a TSS selector. The limit of a TSS must be greater than 64H (100 decimal) bytes and can be as large as 4 Gbytes. In the additional TSS space, the OS is free to store additional information such as the reason the task is inactive, time the task spent running, and open files belonging to the task.

Each task has a TSS associated with it. The current TSS is identified by a special CPU register called the TSS register (TR), shown in Fig. 7.26. The 16-bit TR contains a selector pointing to the TSS descriptor (in the GDT) that defines (and points to) the current TSS. The CPU also contains a user-invisible base, limit, and access rights register, associated with the TR, as shown in Fig. 7.26. This register is appropriately loaded (from the TSS descriptor in GDT) whenever TR is loaded with a new selector. Returning from a task is accomplished by the IRET instruction. When IRET is executed, control is returned to the task which was interrupted. The current executing task's state is saved in its TSS and the old task's state is restored from its TSS.

The task switch operation proceeds according to the following steps:

1. The entire task state is saved in its TSS, pointed to by the TR.

2. The CPU registers are loaded from the new TSS. The TR is loaded with a selector for the new TSS, and the user-invisible register is loaded with the contents of the new TSS descriptor from the GDT.

3. Protection checks are performed.

4. The execution of the new task begins.

A task gate descriptor [Turl 88] acts as an interface point between the user code and the TSS. The task gate descriptor contains a selector, pointing to the TSS and an access rights byte. The current program with its CPL and RPL must be privileged enough to invoke the task gate. The privilege rule is

```
max (CPL, RPL) ≤ (task gate DPL)
```

That is, the calling program must be at least as privileged as the task gate.

The T bit (debug trap bit, Fig. 7.26) in the TSS indicates that the processor should generate a debug exception when switching to a task. If $T = 1$, then upon entry to a new task a debug exception 1 will be generated.

The I/O permission bitmap (Fig. 7.26) grants selective access to I/O ports in the system. If the nth bit is zero, then the task can access port n. If the nth bit is 1, then any attempt by the task to access port n will cause a general protection exception. Each bit in the bitmap corresponds to a bytewide port. For a wordwide (2-byte) port, 2 consecutive bitmap bits must be cleared to 0, and for a dword-wide (4-byte) port, 4 consecutive bits must be cleared. In the completely general case, the

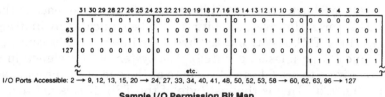

I/O Ports Accessible: 2 → 9, 12, 13, 15, 20 → 24, 27, 33, 34, 40, 41, 48, 50, 52, 53, 58 → 60, 62, 63, 96 → 127

Sample I/O Permission Bit Map

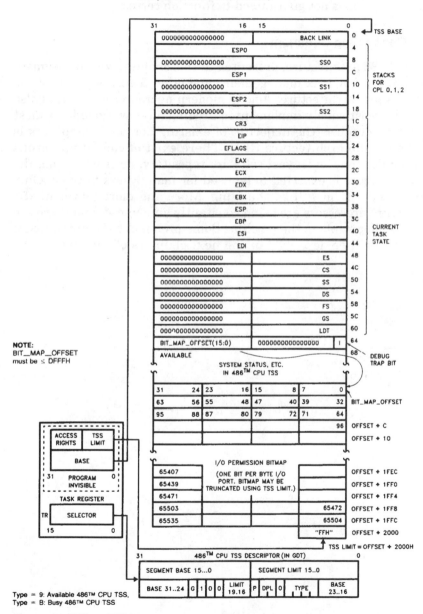

NOTE:
BIT__MAP__OFFSET
must be ≤ DFFFH

Type = 9: Available 486™ CPU TSS,
Type = B: Busy 486™ CPU TSS

Figure 7.26   TSS and TSS registers.

bitmap must be 8 kbytes long (64 kbits) to specify access for each of the possible 64-k ports. However, the bitmap can be truncated by limiting the length of the TSS. Any truncated bytes are interpreted as containing all 1s (no I/O permission granted). The system was designed in a manner requiring to provide an all 1s byte at the end of the bitmap, as shown in Fig. 7.26. This is the last byte of the TSS. If it is not available, the bitmap is not guaranteed to function correctly.

7.12   Concluding Comment

The Intel x86 architecture, presented in this chapter, with its segmentation, protection, task management, and other features, is one of the most versatile architectures among modern microprocessors. It is also one of the most widely implemented architectures worldwide. A most significant number (the majority) of existing personal computers is based on a x86 architecture CPU. There exist of course numerous other applications and a vast software repository. For this reason, the x86 architecture is expected to be used for many years to come. Only its implementation will be changing. More and more powerful x86 architecture processors are being constantly announced, and there are future developments in the pipeline. Some prominent x86 architecture processors will be described in the next chapters of this part of the text.

8

The Pentium

The Pentium is the third among Intel's 32-bit microprocessors, following the i386 and i486. It has the distinction of being the first CISC-type processor implementing ILP; it is a two-issue superscalar (see Chap. 5). Its performance is comparable to some of the most advanced RISC-type systems (see Chap. 6).

The Pentium is a 0.8-micron, BiCMOS technology, three-layer metal, 273-pin grid array package microprocessor [AlAv 93, Pnt1 93]. It is a 32-bit system with a double 64-bit data bus inside and outside of the chip. The gradual development of the Intel 32-bit microprocessors is illustrated in Table 8.1 [AlAv 93].

A block diagram of the Pentium processor is shown in Fig. 8.1 [Pnt1 93]. The internal and external data buses are 64-bit, and the internal and external address buses are 32-bit. There are two separate 8-kbyte caches: one for code and one for data. Each cache has a separate address translation TLB (see Chap. 4) associated with it. The availability of a dual cache and a dual TLB permits the CPU to handle simultaneous instruction and data operand access, thus facilitating efficient handling of the pipeline (see Chap. 5). There are 256 lines between the code cache and the prefetch buffers, permitting the prefetching of 32 bytes of instructions. Since the Pentium is a two-issue superscalar, two instructions are fetched and decoded simultaneously. There are two parallel integer instruction pipelines: the U pipeline and the V pipeline. The U pipeline has a barrel shifter in addi-

TABLE 8.1   Intel 32-bit Microprocessor Evolution

CPU	Year	Technology	Layers	Trans. * 10^6	Frequency (MHz)
i386	1985	1.5-micron CMOS	2	0.275	16–20
i486	1989	1.0-micron CMOS	2	1.2	25–66
Pentium	1993	0.8-micron BiCMOS	3	3.1	60–66

Note: BiCMOS = Bipolar CMOS.

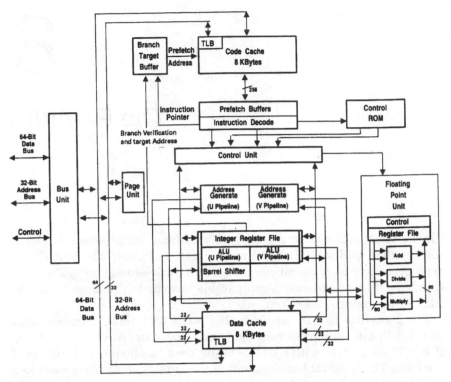

Figure 8.1 Pentium processor block diagram. (*Courtesy of Intel Corp.*)

tion to the regular ALU. There is also a separate FPU pipeline with individual floating-point add, multiply, and divide operational units. The integer and the floating-point register files are as described in Chap. 7. The data cache is dual-ported, accessible by the U and V pipelines simultaneously. The cache tags are triple-ported, allowing parallel snooping. The cache itself is interleaved, allowing two parallel accesses. There is also a *branch target buffer* (BTB), supplying jump target prefetch addresses to the code cache.

The Pentium has a five-stage integer pipeline, branching out into two paths U and V in the last three stages, as illustrated in Fig. 8.2 (*a*) and (*b*). The Pentium pipeline stages are as follows [AlAv 93, Pnt1 93]:

1. PF—prefetch. The CPU prefetches code from the code cache and aligns the code to the initial byte of the next instruction to be decoded.

2. D1—first decode. The CPU decodes the instruction to generate a control word. A single control word causes direct execution of an instruction. More complex instructions require microcoded control sequencing [HaVZ 90, Hays 88] in stage D1.

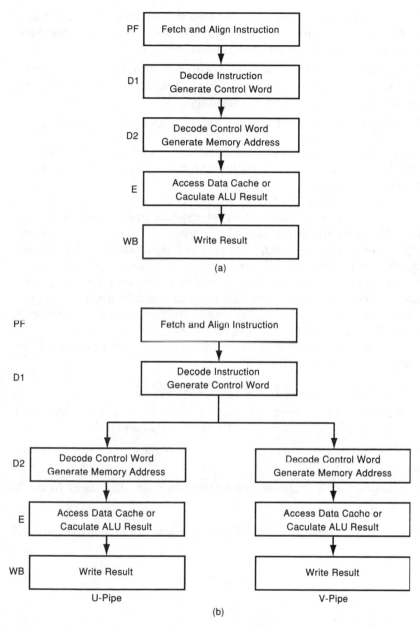

Figure 8.2 The Pentium pipeline. (*a*) Interior pipeline. (*b*) Superscalar execution. (*Courtesy of Intel Corp.*)

3. D2—second decode. The CPU decodes the control word, generated in stage D1, for subsequent use in the next E stage. In addition, addresses for data memory references are generated.

4. E—execute. The instruction is executed in the ALU. If necessary, the barrel shifter or other operational units are used. If necessary, the data cache is accessed at this stage.

5. WB—write back. The CPU stores the results and updates the flags.

In order to fully understand the handling of instructions by the two pipelines U and V, we have to distinguish between simple and complex instructions [Pnt1 93].

Simple instructions of the Pentium are entirely hardwired (as opposed to microcoded), and in general, execute in one clock cycle as in a RISC. The exceptions are the ALU register-to-memory or memory-to-register instructions which take two and three clock cycles, respectively. The instructions considered simple are: move register or memory or immediate into register, move register or immediate into memory, integer arithmetic instructions, increment, decrement, push register or memory, pop register, load effective address, jump, call, jump conditional near, and no operation.

The following are the conditions for simultaneous issue of two instructions on the Pentium [Pnt1 93]:

1. Both instructions must be simple, as defined above.

2. There must be no *read-after-write* (RAW) or *write-after-write* (WAW) data dependencies (see Chap. 5) between them.

3. Neither instruction may contain both a displacement and an immediate.

4. Instructions with prefixes (other than Jcc) can only occur in the U pipeline.

We can also formulate the following instruction issue algorithm [AlAv 93]:

```
Decode two consecutive instructions I1 and I2.
If the following are all true:
I1 and I2 are simple instructions.
I1 is not a jump instruction.
Destination of I1 is not a source of I2.
Destination of I1 is not a destination of I2.
Then issue I1 to pipeline U and I2 to pipeline V.
Else issue I1 to pipeline U.
```

If the first of the two decoded instructions is a jump, it is forwarded to pipeline U for execution, and no instruction is forwarded to pipeline

V. In case of a mispredicted jump the pipeline is delayed by three or four clock cycles (the jump destination is fetched during the E cycle). When a jump instruction is first taken [AlAv 93], the CPU allocates an entry in the 256-entry BTB to associate the jump instruction's address with its target address and to initialize the history used in the prediction algorithm. As instructions are decoded, the CPU searches the BTB to determine whether it holds an entry for a corresponding jump instruction. When there is a hit, the CPU uses the history to determine whether the jump should be taken. If it should, the CPU uses the target address to begin fetching and decoding instructions from the target path. The jump is resolved early in the WB stage, and if the prediction was incorrect, the CPU flushes the pipeline and resumes fetching instructions along the correct path. The CPU updates the dual-ported history in the WB stage. Correctly predicted jumps execute without any delay.

The Pentium has an eight-stage floating-point pipeline, illustrated in Fig. 8.3. The floating-point pipeline stages are [AlAv 93]:

1. PF—prefetch. Prefetch instructions from the code cache.

2. D1—first decode. Same as in the integer pipeline.

3. D2—second decode. Same as in the integer pipeline.

4. E—operand fetch. Operands are fetched either from the floating-point register file or the cache.

5. X1—first execute. First step in the floating-point execution by the FPU.

6. X2—second execute. Second step in the floating-point execution by the FPU.

7. WF—write float. The FPU completes the floating-point computation and writes the result into the floating-point register file.

8. ER—error reporting. The FPU reports internal special situations that might require additional processing to complete execution and updates the floating-point status word.

• Three Dedicated Arithmetic Units
• Eight Stage Pipeline

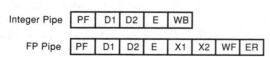

• Three Execution Stages

Figure 8.3 Floating-point pipeline. (*Courtesy of Intel Corp.*)

The initial three stages of a floating-point instruction are executed in the U pipeline. A floating-point instruction cannot be paired with any other instruction, integer or floating-point. The dual port of the data cache is used for access of 64-bit double-precision operands.

The code and the data caches are each 8-kbyte, two-way set-associative, with 32 bytes/line (see Chap. 4). In a two-way set-associative cache, with only two lines/set, we need only a single byte per set (a single flip-flop) to realize the LRU replacement algorithm. In this case, the designers' decision was in favor of simplicity of the logic circuitry, without giving up too much on the hit ratio (see Chap. 4). The data cache is write-back, and uses a *modified/exclusive/shared/invalid* (MESI) cache coherency protocol.

The Pentium uses the regular Intel 4-kbyte page. However, it also features an optional large 4-mbyte page intended for large software packages. Each cache has its own TLB (a new feature on the Pentium). The data cache TLB is 4-way set-associative, 64-entry, for 4-kbyte pages. There is a separate data cache TLB for 4-Mbyte pages. It has eight entries and the same parameters as the TLB for 4-kbyte pages. There is only one 32-entry code cache TLB for 4-kbyte pages. This TLB is also four-way set-associative. Replacement in the TLBs is handled by a pseudo-LRU algorithm, similar to the one implemented on the i486 (see Chap. 9), requiring 3 bits per set.

The functional grouping of Pentium pins is listed in Table 8.2. The addressing interconnections to address 8-, 16-, 32-, and 64-bit memories is illustrated in Fig. 8.4.

It should be noted that the Pentium transmits only address bits A31 to A3, pointing to a 64-bit quadword (the internal and external data buses are 64-bit). The 8 bytes of a 64-bit quadword are activated outside by the eight byte enable signals BE7# to BE0# (# denotes a low-asserted signal). Signal BE0# enables the low byte D7 to D0, and so on.

Intel provides a set of auxiliary chips for a 256- or 512-kbyte external second-level cache. The chips are 280-pin 82496 cache controller and 84-pin 82491, 256-kbit SRAM [Pnt2 93]. The 82496 cache controller features a choice of write-back or write-through policies and implements the MESI cache coherency protocol. The 82491 cache SRAM supports 32-, 64-, and 128-byte line sizes. The tightly coupled 82496 cache controller/82491 cache SRAM separates the Pentium CPU bus from the memory bus. The 82496 cache controller and memory can exchange handshake signals synchronously, asynchronously, or with a strobed protocol. Concurrent CPU bus and memory bus operation is possible.

TABLE 8.2   Pin Functional Grouping

Function	Pins
Clock	CLK
Initialization	RESET, INIT
Address Bus	A31-A3, BE7# - BE0#
Address Mask	A20M#
Data Bus	D63-D0
Address Parity	AP, APCHK#
Data Parity	DP7-DP0, PCHK#, PEN#
Internal Parity Error	IERR#
System Error	BUSCHK#
Bus Cycle Definition	M/IO#, D/C#, W/R#, CACHE#, SCYC, LOCK#
Bus Control	ADS#, BRDY#, NA#
Page Cacheability	PCD, PWT
Cache Control	KEN#, WB/WT#
Cache Snooping/Consistency	AHOLD, EADS#, HIT#, HITM#, INV
Cache Flush	FLUSH#
Write Ordering	EWBE#
Bus Arbitration	BOFF#, BREQ, HOLD, HLDA
Interrupts	INTR, NMI
Floating-Point Error Reporting	FERR#, IGNNE#
System Management Mode	SMI#, SMIACT#
Functional Redundancy Checking	FRCMC# (IERR#)
TAP Port	TCK, TMS, TDI, TDO, TRST#
Breakpoint/Performance Monitoring	PM0/BP0, PM1/BP1, BP3-2
Execution Tracing	BT3-BT0, IU, IV, IBT
Probe Mode	R/S#, PRDY

SOURCE: Courtesy of Intel Corp.

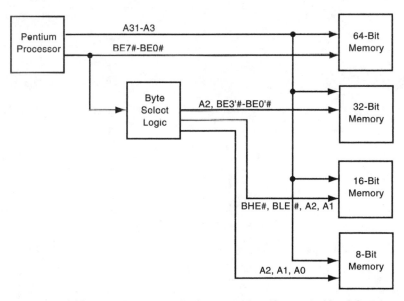

Figure 8.4   Addressing 32-, 16-, and 8-bit memories. (*Courtesy of Intel Corp.*)

The i486 and i386 Microprocessors

The Intel i486, also denoted as 80486 or just 486, is a CHMOS IV technology, 1.2 million transistors, 168-pin, 32-bit microprocessor. Its block diagram is shown in Fig. 9.1 [i486 90]. The i486 consists of the following main subsystems:

1. The bus interface, connected to the external system bus and to the on-chip cache and prefetcher units.

2. The prefetcher, which includes a 32-byte queue of prefetched instructions and is connected to the bus interface, cache, instruction decoder, and segmentation unit (see Sec. 7.7).

3. The cache unit, which includes an 8-kbyte cache, storing both code and data, and cache management logic. It is connected through a 64-bit interunit transfer (data) bus to the segmentation unit, ALU, and FPU. The cache unit is also directly connected to the paging unit, bus interface, and prefetcher (through 128 lines, permitting the prefetching of 16 bytes of instructions simultaneously). The cache is 4-way set-associative, write-through, with 16 bytes/line (see Chap. 4).

4. The instruction decode unit, which receives 3 bytes (24 bits) of undecoded instructions from the prefetcher queue and transmits decoded instructions to the control and protection test unit.

5. The control and protection test unit, which generates microinstructions transmitted to other units and performs protection testing (see Sec. 7.10).

6. The ALU, which includes the general-purpose register file (see Sec. 7.2), a barrel shifter, and registers for microcode use.

7. The FPU, which includes the floating-point registers, an adder, a multiplier, and a shifter.

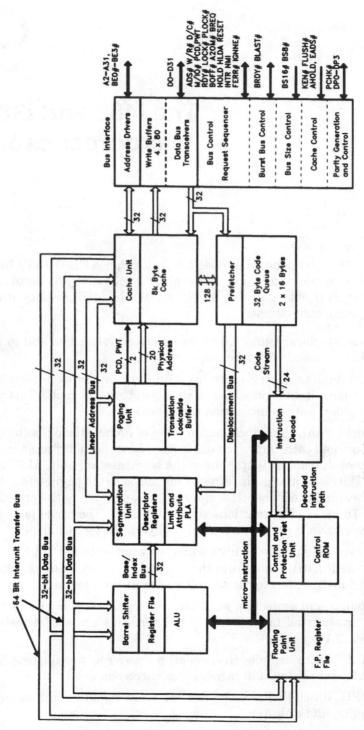

Figure 9.1 The 80486 block diagram. (*Courtesy of Intel Corp.*)

8. The segmentation unit, which includes segmentation management logic, descriptor registers, and breakpoint logic (see Sec. 7.7).

9. The paging unit, which includes paging management logic and a 32-entry TLB (see Chap. 4 and Sec. 7.8).

The pinout diagram of the i486 is shown in Fig. 9.2 [i486 90]. Some of the pin labels are also shown at the external interconnections of the bus interface unit in Fig. 9.1. The subdivision of 99 of the i486 signals into functional groups is listed in Table 9.1. In addition, there are 24 power pins (Vcc, + 5V), 28 ground pins (Vss), and 17 unconnected pins (NC), for a total of 168 pins.

Only address lines A31 to A2 are transmitted outside to point to dword addresses. Bytes within a 32-bit dword are selected by the four byte enable signals BE3# to BE0# (the # symbol denotes a low-asserted signal) as follows:

BE3# D31–D24

BE2# D23–D16

BE1# D15–D8

BE0# D7–D0

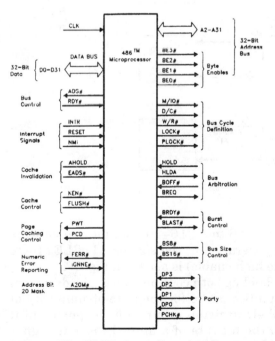

Figure 9.2 The 80486 pinout diagram. (*Courtesy of Intel Corp.*)

TABLE 9.1   i486 Signal Groups

Signal groups	Pins
Clock	1
Address bus (A2–A31, BE0#–BE3#)	34
Data bus (D0–D31)	32
Data parity	5
Bus cycle definition	5
Bus control	2
Burst control	2
Interrupts	3
Bus arbitration	4
Cache invalidation	2
Cache control	2
Page cacheability	2
Numeric error reporting	2
Bus size control	2
Address mask	1
Total	99

SOURCE: Courtesy of Intel Corp.

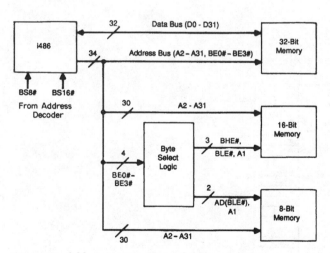

Figure 9.3   Addressing 32-, 16-, and 8-bit memories.

The interconnection of the i486 to 8-, 16-, and 32-bit memories is illustrated in Fig. 9.3. The generation of the signals A1, BLE# (byte low enable), and BHE# (byte high enable) is shown in Fig. 9.4.

The i486 cache is unified, holding both code and data. The designers of i486 decided to adopt a unified cache for the sake of simplicity of design. They also based their decision on previous experimental results, according to which the hit rate of a unified cache is higher than that of separate (dual) caches of the same total size [Craw 90,

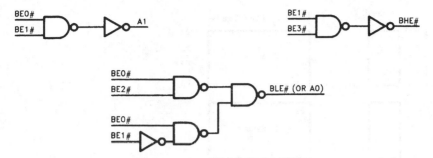

Figure 9.4 Logic to generate A1, BHE#, and BLE# for 16-bit buses. (*Courtesy of Intel Corp.*)

Smit 82, Smit 87]. However, as argued in Chaps. 5 and 6, having a dual cache facilitates better handling of pipelines and enhances the overall performance. This is particularly true for superscalar systems. This is why the Pentium designers chose the dual cache feature, in addition to doubling its overall size (see Chap. 8). The competitive performance of the Pentium, compared to some leading RISC-type systems, indicates that the Pentium designers made the correct decision.

The i486 designers decided on a four-way set-associative organization in order to increase the hit ratio, notwithstanding the higher hardware complexity involved (see Chap. 4). Here again, the Pentium designers changed the parameters. They adopted a two-way set-associative mapping, thus simplifying the cache management hardware, while giving up somewhat on the hit ratio (the letdown of the hit ratio is not significant, though). The line size on the Pentium caches was doubled to 32 bytes/line (16 bytes/line on the i486). Both parameters are within established and widely accepted practical line sizes [Smit 87]; it is not completely clear which one yields a better performance. Some experimental results on some benchmarks did indicate that better performance results are obtained for 16-byte lines, as opposed to the 32-byte lines [GrSh 89]. General line size design considerations were discussed in Chap. 4.

The organization of the i486 cache is shown in Fig. 9.5. It contains 128 sets. The cache is logically organized into three major parts. The largest part is the data storage part on the right in Fig. 9.5. Addressing into the data part consists of two components: the set select and the line select. The set select is 7 bits wide since there are 2^7 = 128 sets. The line select is 2 bits wide to select one of the four lines in the set. The second part is the tag array, shown on the left side of Fig. 9.5. The tag array is physically organized as 128 sets, each containing four 21-bit-wide tag entries (see Chap. 4 for cache addressing and address mapping). The third part consists of 128 seven-bit struc-

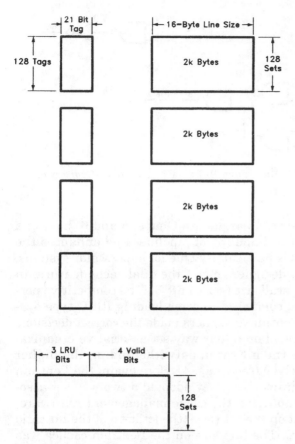

Figure 9.5 On-chip cache physical organization.
(*Courtesy of Intel Corp.*)

tures that contain 4 valid bits for each set (1 bit per line), and 3 bits per set used in the pseudo-LRU replacement algorithm (see Chap. 4). These 3 bits are called the LRU bits (see Fig. 9.5).

The four lines in each set are labeled I0, I1, I2, and I3. All valid bits are cleared when the processor is reset or when the cache is flushed. Replacement in the cache is handled by a pseudo-LRU mechanism when all four lines in a set are valid. Three bits, B0, B1, and B2, are defined for each of the 128 sets in the cache (Fig. 9.5). These bits are called the LRU bits. The LRU bits are updated for every hit or replacement in the cache. If the most recent access to the set was to line I0 or I1, B0 is set to 1. If the most recent access was to line I2 or I3, B0 is cleared to 0. If the most recent access to I0:I1 was to I0, B1 is set to 1; otherwise B1 is cleared to 0. If the most recent access to I2:I3 was to I2, B2 is set to 1; otherwise, B2 is cleared to 0. The decision tree for the pseudo LRU mechanism is shown in Fig. 9.6.

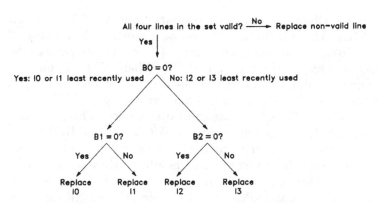

Figure 9.6   On-chip cache replacement strategy. (*Courtesy of Intel Corp.*)

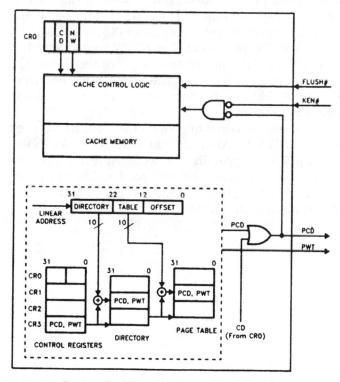

Figure 9.7   Page cacheability. (*Courtesy of Intel Corp.*)

As presented in Chap. 7, there are two control bits in CR3 and in all PTEs and PDEs, which influence the cache operation. The value of these bits is also driven out on outside pins. These bits are: *page cache disable* (PCD) and *page write through* (PWT). When PCD = 0, on-chip caching of a page is enabled. Bit PCD alone does not enable caching; it depends on the activation of the input signal cache enable (KEN#), and on the status of the CD (cache disable) bit in CR0. Thus, for the caching to be enabled we must have: PCD = 0, CD = 0, and KEN# = 0. When PWT = 1, we have a write-through policy, when PWT = 0, we have write-back. Since the internal cache is inherently write-through, PWT is intended for an external second level cache. The interrelationship between bits and signals PCD and PWT, and the KEN# signal, is illustrated in Fig. 9.7.

The predecessor of the i486 in the Intel x86 family, and its first 32-bit microprocessor, is the i386 (or:80386, or simply 386). The i386 is a CHMOS III technology, 132-pin microprocessor. It has the same architecture as the i486, described in Chap. 7. It has no on-chip cache and no on-chip FPU. Floating-point computations in an i386-based system are performed on the 80387 coprocessor. The 80387 coprocessor has essentially the same register and instruction set architecture as the FPU on the i486 or the Pentium. Of course, the organization and technology of the FPUs on the i486 and the Pentium are different, yielding better floating-point performance.

There are only six instructions unavailable on the i386, which were introduced on the i486: BSWAP, XADD, CMPXCHG, INVD, WBINVD, and INVLPG (see Chap. 7 and Appendix to Part 2). Naturally, all control bits related to cache management are new on the i486 (not previously available on the i386). The prefetch queue is only 16 bytes on the i386. It was doubled to 32 bytes on the i486.

10

Earlier Systems: 8086, 80186, and 80286

10.1 The 8086 and 8088

The 8086 microprocessor, announced by Intel in 1978, was Intel's first 16-bit microprocessor and the first in the x86 family. A block diagram of the 8086 is shown in Fig. 10.1. It is subdivided into two principal units [LiGi 86]:

1. The execution unit, EU, including the ALU, eight 16-bit general registers, a 16-bit FLAGS register (also called status word), and a control unit.

2. The bus interface unit, BIU, including an adder for address calculations, four 16-bit segment registers, a 16-bit instruction pointer (IP), a 6-byte instruction queue, and bus control logic.

The 8086 chip has 40 pins, including 16 data pins and 20 address pins, for direct memory addressing of up to 2^{20} bytes = 1 Mbyte. The 16 data lines are multiplexed with the lower 16 address lines (pins AD0 to AD15). Subsequent to the 8086, Intel featured the 8088, which has essentially the same architecture as the 8086 but has an 8-bit external data bus. A block diagram of the 8088 is almost identical to the one shown in Fig. 10.1, with the following differences:

1. The data bus connected to the adder and the segment registers in the BIU, is 8-bit, instead of 16 for the 8086. The ALU data bus is 16-bit in the 8088, as in the 8086.

2. The instruction queue in the BIU of the 8088 is 4 bytes long (6 in the 8086).

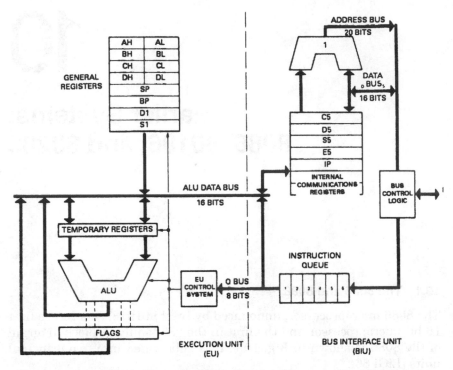

Figure 10.1   8086 simplified functional block diagram. (*Courtesy of Intel Corp.*)

The motivation for featuring the 8088 is hardware cost reduction, while maintaining the same 8086 16-bit architecture. With an 8-bit, instead of a 16-bit data bus, we can save about 50 percent on interface devices. The 8088 was the CPU implemented in the original IBM PC and its numerous clones. From this point on, whenever an 8086 feature is mentioned, it will be understood that it applies to the 8088 as well, unless stated otherwise.

The register structure of the 8086 is shown in Fig. 10.2, and the FLAGS or staus register, in Fig. 10.3. The general registers, the IP and the FLAGS, exist also on the other members of the x86 family (see Sec. 7.2) and have essentially the same function. The four segment registers CS, DS, SS, and ES also exist on the other members of the x86 family. However, they are used on the 16-bit systems in a different way than on the 32-bit systems of the same family. The segment registers on the 8086 contain the upper, nonzero, 16 bits of the segment base address. The segment size on the 8086 is fixed at 2^{16} bytes = 64 kbytes (the maximum size of the memory for an 8080/8085-based 8-bit microcomputer). The base address of a segment must be divisible by 16; that is, its 4 LSB must be zero. Thus, a physical address on the

DATA REGISTERS

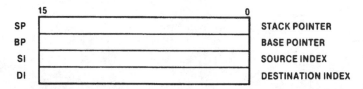

	7	0 7	0
AX	AH	AL	
BX	BH	BL	
CX	CH	CL	
DX	DH	DL	

POINTER AND INDEX REGISTERS

	15	0	
SP			STACK POINTER
BP			BASE POINTER
SI			SOURCE INDEX
DI			DESTINATION INDEX

SEGMENT REGISTERS

	15	0	
CS			CODE
DS			DATA
SS			STACK
ES			EXTRA

INSTRUCTION POINTER AND FLAGS

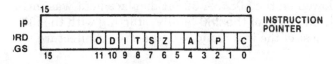

IP — INSTRUCTION POINTER

WORD/FLAGS: O D I T S Z A P C
15   11 10 9 8 7 6 5 4 3 2 1 0

Figure 10.2 Register structure. (*Courtesy of Intel Corp.*)

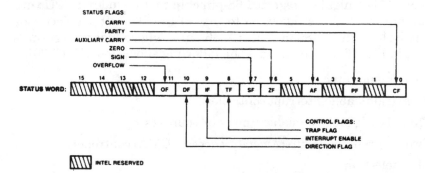

Figure 10.3 Status word or flags format. (*Courtesy of Intel Corp.*)

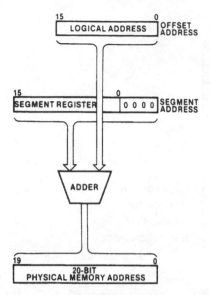

Figure 10.4   8086 physical address.

8086 is calculated by shifting the content of a segment register 4 bits to the left and adding to it the offset within the segment, as shown in Fig. 10.4 (see also Chap. 7).

The offset is the EA, calculated by one of the x86 addressing modes (see Sec. 7.4), and stored in IP. The addressing modes on the 8086 are the same, except that the index is not scaled. Only 8- or 16-bit displacements are allowed on the 8086. A 32-bit displacement was introduced on the 32-bit members of the x86 family, starting with the i386. The 8086 features most of the instructions of the x86 architecture.

10.2   The 80186 and 80188

The 80186 is a highly integrated 68-pin chip that includes a CPU with an architecture almost identical to that of the 8086. It is object code compatible with the 8086. The 80186 chip, whose block diagram is shown in Fig. 10.5, also includes the following subsystems [LiGi 86]:

A clock generator

A programmable interrupt controller

Three 16-bit programmable timers or counters

Two programmable *direct memory access* (DMA) controllers

Chip select unit

Programmable control registers

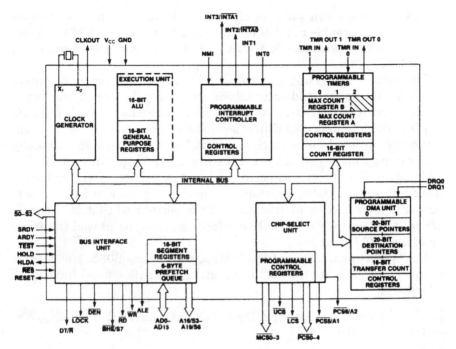

Figure 10.5 80186 block diagram. (*Courtesy of Intel Corp.*)

Bus interface unit

Six-byte prefetch queue

Similarly to the 8088, there is also an 80188. The 80188 is identical to the 80186 with the following differences:

1. The 80188 has only an 8-bit external data bus (16-bit on 80186).

2. The 80188 has only a 4-byte prefetch queue (6-byte on 80186).

From this point on, any feature described for the 80186 will be understood to apply to the 80188 as well, unless stated otherwise.

The two-channel DMA unit of the 80186 performs transfers to or from any combination of I/O space and memory space in either byte or word units. Each DMA channel maintains independent source and destination pointers which are used to access the source and destination of transferred data. The 80186 timer unit contains three independent 16-bit timer-counters. Two of these timers can be used to count external events, to provide waveforms derived from either the CPU or an external clock, or to interrupt the CPU after a specified number of timer events. The third timer counts only CPU clock cycles and can be used to interrupt the CPU after a programmable number of CPU clocks to give a count pulse to either or both of the other two timers after a programmable number of CPU clocks, or to give a DMA request pulse to the integrated DMA unit after a programmable number of CPU clock cycles.

The 80186 interrupt controller arbitrates interrupt requests between all internal and external sources. It can be directly cascaded as the master to two external 8259A interrupt controller chips [Uffe 91]. The 80186 integrated chip select logic can be used to enable memory or peripheral devices. Six output lines are used for memory addressing and seven output lines are used for peripheral device addressing. The integrated peripheral and chip select circuitry is controlled by sets of 16-bit registers accessed using standard input, output, or memory access instructions. These peripheral control registers are all located within a 256-byte block that can be placed in either memory or I/O space.

The 80186 includes all of the instructions of the 8086. However, a number of new instructions (and instruction extensions) have been added:

1. New instructions on the 80186: BOUND, ENTER, LEAVE, INS, OUTS (see Sec. 7.5)

2. Instruction extensions on the 80186:

    PUSH immediate

    PUSHA

    POPA

    IMUL by an immediate value

    Shifts/rotates by an immediate value

All the 80186 instructions and their extensions are available on the subsequent x86 systems. The register set and the addressing modes of the 80186 are exactly the same as those of the 8086.

10.3 The 80286

The 80286 [LiGi 86, Strs 86] is a 68-pin microprocessor, constituting a milestone in the development of the Intel x86 family. The protected mode operation (see Chap. 7), or the protected virtual address mode (PVAM), was introduced with the 80286. A block diagram of the 80286 is shown in Fig. 10.6. Another innovation on the 80286 is the extension of the directly addressed memory space to 2^{24} bytes = 16 Mbytes. Moreover, the 24 address lines A23 to A0 are separate and are not multiplexed with the 16 data lines D15 to D0 (as on the 8086 and 80186). The 80286 is subdivided into four main subunits:

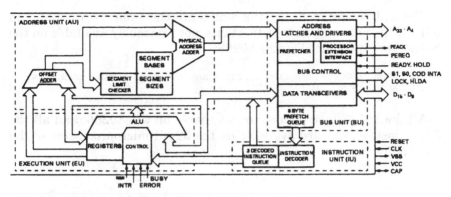

Figure 10.6 80286 internal block diagram. (*Courtesy of Intel Corp.*)

1. The EU, including the ALU, general registers (same as on 8086 and 80186), and the CU

2. The address unit (AU), including the segment registers (same as on 8086 and 80186), an offset adder, and a physical address adder

3. The bus unit (BU), including address latches and data transceivers (previously implemented on separate chips for the 8086), bus interface and control circuitry, instruction prefetcher, and a 6-byte instruction queue

4. The instruction unit, IU, including an instruction decoder and a three decoded instructions queue

The concepts of the four-level protection mechanism (see Sec. 7.10), task management (see Sec. 7.11), descriptors and descriptor tables (see Sec. 7.7), the use of selectors (see Sec. 7.7), and call gates (see Sec. 7.10) were first introduced on the 80286 and later expanded for a 32-bit architecture on the i386.

The 80286 segment can be 1 byte and up to 64 kbytes long. For this reason, the limit field in an 80286 descriptor is 16 bits wide. Since the 80286 has 24 address lines, the descriptor base field is 24 bits wide. The structure of the 80286 access rights byte is the same as on the 32-bit x86 systems. An 80286 descriptor is shown in Fig. 10.7.

31						0
SEGMENT BASE 15 . . . 0		SEGMENT LIMIT 15 . . . 0				0
Intel Reserved Set to 0	P	DPL	S	TYPE	BASE 23 . . . 16	+4

BASE	Base Address of the segment	DPL	Descriptor Privilege Level 0–3	
LIMIT	The length of the segment	S	System Descriptor	0 = System	1 = User
P	Present Bit 1 = Present 0 = Not Present	TYPE	Type of Segment	

Figure 10.7 80286 code and data segment descriptors. (*Courtesy of Intel Corp.*)

The addressing modes of the 80286 are the same as on 8086 and 80186. The following new instructions, not previously available on the 80186, were introduced on the 80286:

```
CTS, LGDT, SGDT, LIDT, SIDT, LLDT, SLDT, LTR, STR, LMSW, SMSW, LAR,
LSL, ARPL, VERR, VERW.
```

All the 80186 instructions are available on the 80286, and all the 80286 instructions are available on the 32-bit microprocessors of the x86 family.

2

Intel x86 Architecture
Instruction Set

x86 Architecture Instruction Set

Instruction	Format	Clocks	Notes
AAA – ASCII Adjust after Addition	0011 0111	3	
AAD – ASCII Adjust AX before Division	1101 0101 : 0000 1010	10	
AAM – ASCII Adjust AX after Multiply	1101 0100 : 0000 1010	18	
AAS – ASCII Adjust AL after Subtraction	0011 1111	3	
ADC – ADD with Carry			
reg1 to reg2	0001 000w : 11 reg1 reg2	1	
reg2 to reg1	0001 001w : 11 reg1 reg2	1	
memory to register	0001 001w : mod reg r/m	2	
register to memory	0001 000w : mod reg r/m	3	U/L
immediate to register	1000 00sw : 11 010 reg : immediate data	1	
immediate to accumulator	0001 010w : immediate data	1	
immediate to memory	1000 00sw : mod 010 r/m : immediate data	3	U/L
ADD – Add			
reg1 to reg2	0000 000w : 11 reg1 reg2	1	
reg2 to reg1	0000 001w : 11 reg1 reg2	1	
memory to register	0000 001w : mod reg r/m	2	
register to memory	0000 000w : mod reg r/m	3	U/L
immediate to register	1000 00sw : 11 000 reg : immediate data	1	
immediate to accumulator	0000 010w : immediate data	1	
immediate to memory	1000 00sw : mod 000 r/m : immediate data	3	U/L
AND – Logical AND			
reg1 to reg2	0010 000w : 11 reg1 reg2	1	
reg2 to reg1	0010 001w : 11 reg1 reg2	1	
memory to register	0010 001w : mod reg r/m	2	
register to memory	0010 000w : mod reg r/m	3	U/L
immediate to register	1000 00sw : 11 100 reg : immediate data	1	
immediate to accumulator	0010 010w : immediate data	1	
immediate to memory	1000 00sw : mod 100 r/m : immediate data	3	U/L
ARPL – Adjust RPL Field of Selector			
from register	0110 0011 : 11 reg1 reg2	7	
from memory	0110 0011 : mod reg r/m	7	
BOUND – Check Array Against Bounds	0110 0010 : mod reg r/m		
if within bounds		8	
if out of bounds		INT + 32	21

x86 Architecture Instruction Set *(Cont.)*

Instruction	Format	Clocks	Notes
BSF – Bit Scan Forward			
reg1, reg2	0000 1111 : 1011 1100 : 11 reg2 reg1		
word		6–34	MN/MX, 12
doubleword		6–42	MN/MX, 12
memory, reg	0000 1111 : 1011 1100 : mod reg r/m		
word		6–35	MN/MX, 13
doubleword		6–43	MN/MX, 13
BSR – Bit Scan Reverse			
reg1, reg2	0000 1111 : 1011 1101 : 11 reg2 reg1		
word		7–39	MN/MX,14
doubleword		7–71	MN/MX,14
memory, reg	0000 1111 : 1011 1101 : mod reg r/m		
word		7–40	MN/MX,15
doubleword		7–72	MN/MX,15
BSWAP – Byte Swap	0000 1111 : 1100 1 reg	1	
BT – Bit Test			
register, immediate data	0000 1111 : 1011 1010 : 11 100 reg: imm8	4	
memory, immediate imm8 data	0000 1111 : 1011 1010 : mod 100 r/m :	4	
reg1, reg2	0000 1111 : 1010 0011 : 11 reg2 reg1	4	
memory, reg	0000 1111 : 1010 0011 : mod reg r/m	9	
BTC – Bit Test and Complement			
register, immediate data	0000 1111 : 1011 1010 : 11 111 reg: imm8	7	
memory, immediate imm8 data	0000 1111 : 1011 1010 : mod 111 r/m :	8	U/L
reg1, reg2	0000 1111 : 1011 1011 : 11 reg2 reg1	7	
memory, reg	0000 1111 : 1011 1011 : mod reg r/m	13	U/L
BTR – Bit Test and Reset			
register, immediate data	0000 1111 : 1011 1010 : 11 110 reg: imm8	7	
memory, immediate imm8 data	0000 1111 : 1011 1010 : mod 110 r/m :	8	U/L
reg1, reg2	0000 1111 : 1011 0011 : 11 reg2 reg1	7	
memory, reg	0000 1111 : 1011 0011 : mod reg r/m	13	U/L
BTS – Bit Test and Set			
register, immediate data	0000 1111 : 1011 1010 : 11 101 reg: imm8	7	
memory, immediate imm8 data	0000 1111 : 1011 1010 : mod 101 r/m :	8	U/L
reg1, reg2	0000 1111 : 1010 1011 : 11 reg2 reg1	7	
memory, reg	0000 1111 : 1010 1011 : mod reg r/m	13	U/L
CALL – Call Procedure (In same segment)			
direct	1110 1000 : full displacement	1	23
register indirect	1111 1111 : 11 010 reg	2	23
memory indirect	1111 1111 : mod 010 r/m	2	23

x86 Architecture Instruction Set (Cont.)

Instruction	Format	Clocks	Notes
CALL – Call Procedure (in other segment)			
direct	1001 1010 : unsigned full offset, selector	4	R,23
to same level		4–13	P,9,23,24
thru gate to same level		22	P,9,25
to inner level, no parameters		44	P,9,25
to inner level, x parameters (d)words		45+2x	P,9,25
to TSS		21+TS	P,10,9,25
thru task gate		22+TS	P,10,9,25
indirect	1111 1111 : mod 011 r/m	5	R,23
to same level		5–14	P,9,23,24
thru gate to same level		22	P,9,25
to inner level, no parameters		44	P,9,25
to inner level, x parameters (d)words		45+2x	P,9,25
to TSS		21+TS	P,10,9,25
thru task gate		22+TS	P,10,9,25
CBW – Convert Byte to Word **CWDE – Convert Word to Doubleword**	1001 1000	3	
CLC – Clear Carry Flag	1111 1000	2	
CLD – Clear Direction Flag	1111 1100	2	
CLI – Clear Interrupt Flag	1111 1010	7	
CLTS – Clear Task-Switched Flag in CR0	0000 1111 : 0000 0110	10	
CMC – Complement Carry Flag	1111 0101	2	
CMP – Compare Two Operands			
reg1 with reg2	0011 100w : 11 reg1 reg2	1	
reg2 with reg1	0011 101w : 11 reg1 reg2	1	
memory with register	0011 100w : mod reg r/m	2	
register with memory	0011 101w : mod reg r/m	2	
immediate with register	1000 00sw : 11 111 reg : immediate data	1	
immediate with accumulator	0011 110w : immediate data	1	
immediate with memory	1000 00sw : mod 111 r/m	2	
CMPS/CMPSB/CMPSW/CMPSD – Compare String Operands	1010 011w	5	16
CMPXCHG – Compare and Exchange			
reg1, reg2	0000 1111 : 1011 000w : 11 reg2 reg1	5	
memory, reg	0000 1111 : 1011 000w : mod reg r/m	6	U/L
CMPXCHG8B – Compare and Exchange 8 Bytes			
memory, reg	0000 1111 : 1100 0111 : mod reg r/m	10	U/L
CWD – Convert Word to Dword **CDQ – Convert Dword to Qword**	1001 1001	2	
DAA – Decimal Adjust AL after Addition	0010 0111	3	

x86 Architecture Instruction Set *(Cont.)*

Instruction	Format	Clocks	Notes
DAS – Decimal Adjust AL after Subtraction	0010 1111	3	
DEC – Decrement by 1			
reg	1111 111w : 11 001 reg	1	
or	0100 1 reg	1	
memory	1111 111w : mod 001 r/m	3	U/L
DIV – Unsigned Divide			
accumulator by register	1111 011w : 11 110 reg		
divisor —	byte	17	
	word	25	
	doubleword	41	
accumulator by memory	1111 011w : mod 110 r/m		
divisor —	byte	17	
	word	25	
	doubleword	41	
ENTER – Make Stack Frame level (L) **for Procedure Parameters**	1100 1000 : 16-bit displacement : 8-bit		
L = 0		11	
L = 1		15	
L > 1		15 + 2L	8
HLT – Halt	1111 0100		
IDIV – Signed Divide			
accumulator by register	1111 011w : 11 111 reg		
divisor —	byte	22	
	word	30	
	doubleword	46	
accumulator by memory	1111 011w : mod 111 r/m		
divisor —	byte	22	
	word	30	
	doubleword	46	
IMUL – Signed Multiply			
accumulator with register	1111 011w : 11 101 reg		
multiplier —	byte	11	
	word	11	
	doubleword	10	
accumulator with memory	1111 011w : mod 101 reg		
multiplier —	byte	11	
	word	11	
	doubleword	10	

x86 Architecture Instruction Set *(Cont.)*

Instruction	Format	Clocks	Notes
reg1 with reg2	0000 1111 : 1010 1111 : 11 : reg1 reg2		
multiplier —	byte	10	
	word	10	
	doubleword	10	
register with memory	0000 1111 : 1010 1111 : mod reg r/m		
multiplier —	byte	10	
	word	10	
	doubleword	10	
reg1 with imm. to reg2	0110 10s1 : 11 reg1 reg2 : immediate data		
multiplier —	byte	10	
	word	10	
	doubleword	10	
mem. with imm. to reg	0110 10s1 : mod reg r/m : immediate data		
multiplier —	byte	10	
	word	10	
	doubleword	10	
INC – Increment by 1			
reg	1111 111w : 11 000 reg	1	
or	0100 0 reg	1	
memory	1111 111w : mod 000 r/m	3	U/L
INT n – Interrupt Type n	1100 1101 : type	INT + 6	21,25
INT – Single-Step Interrupt 3	1100 1100	INT + 5	21,25
INTO – Interrupt 4 on Overflow	1100 1110		
taken		INT + 5	21,25
not taken		4	21,25
INVD – Invalidate Cache	0000 1111 : 0000 1000	15	
INVLPG – Invalidate TLB Entry	0000 1111 : 0000 0001 : mod 111 r/m	29	
IRET/IRETD – Interrupt Return	1100 1111		
real mode or virtual 8086 mode		7	R,23
protected mode			
to same level		10–19	P,9,23,24
to outer level		27	P,9,25
to nested task		10 + TS	P,9,10,25
Jcc – Jump if Condition is Met			
8-bit displacement	0111 tttn : 8-bit displacement	1	23
full displacement	0000 1111 : 1000 tttn : full displacement	1	23
JCXZ/JECXZ – Jump on CX/ECX Zero	1110 0011 : 8-bit displacement	6/5	T/NT,23
address size prefix differentiates JCXZ from JECXZ			

x86 Architecture Instruction Set *(Cont.)*

Instruction	Format	Clocks	Notes
JMP – Unconditional Jump (to same segment)			
short	1110 1011 : 8-bit displacement	1	23
direct	1110 1001 : full displacement	1	23
register indirect	1111 1111 : 11 100 reg	2	23
memory indirect	1111 1111 : mod 100 r/m	2	23
JMP – Unconditional Jump (to other segment)			
direct intersegment	1110 1010 : unsigned full offset, selector	3	R,23
to same level		3–12	P,9,23,24
thru call gate ro same level		18	P,9,25
thru TSS		19 + TS	P,10,9,25
thru task gate		20 + TS	P,10,9,25
indirect intersegment	1111 1111 : mod 101 r/m	4	R,23
to same level		4–13	P,9,23,24
thru call gate ro same level		18	P,9,25
thru TSS		19 + TS	P,10,9,25
thru task gate		20 + TS	P,10,9,25
LAHF – Load Flags into AH Register	1001 1111	2	
LAR – Load Access Rights Byte			
from register	0000 1111 : 0000 0010 : 11 reg1 reg2	8	
from memory	0000 1111 : 0000 0010 : mod reg r/m	8	
LDS – Load Pointer to DS	1100 0101 : mod reg r/m	4–13	9,24
LEA – Load Effective Address	1000 1101 : mod reg r/m	1	
LEAVE – High Level Procedure Exit	1100 1001	3	
LES – Load Pointer to ES	1100 0100 : mod reg r/m	4–13	9,24
LFS – Load Pointer to FS	0000 1111 : 1011 0100 : mod reg r/m	4–13	9,24
LGDT – Load Global Descriptor Table Register	0000 1111 : 0000 0001 : mod 010 r/m	6	
LGS – Load Pointer to GS	0000 1111 : 1011 0101 : mod reg r/m	4–13	9,24
LIDT – Load Interrupt Descriptor Table Register	0000 1111 : 0000 0001 : mod 011 r/m	6	
LLDT – Load Local Descriptor Table Register			
LDTR from register	0000 1111 : 0000 0000 : 11 010 reg	9	
LDTR from memory	0000 1111 : 0000 0000 : mod 010 r/m	9	
LMSW – Load Machine Status Word			
from register	0000 1111 : 0000 0001 : 11 110 reg	8	
from memory	0000 1111 : 0000 0001 : mod 110 r/m	8	
LOCK – Assert LOCK# Signal Prefix	1111 0000	1	
LODS/LODSB/LODSW/LODSD – Load String Operand	1010 110w	2	
LOOP – Loop Count	1110 0010 : 8-bit displacement	5/6	L/NL,23

x86 Architecture Instruction Set *(Cont.)*

Instruction	Format	Clocks	Notes
LOOPZ/LOOPE – Loop Count while Zero/Equal	1110 0001 : 8-bit displacement	7/8	L/NL,23
LOOPNZ/LOOPNE – Loop Count while not Zero/Equal	1110 0000 : 8-bit displacement	7/8	L/NL,23
LSL – Load Segment Limit			
from register	0000 1111 : 0000 0011 : 11 reg1 reg2	8	
from memory	0000 1111 : 0000 0011 : mod reg r/m	8	
LSS – Load Pointer to SS	0000 1111 : 1011 0010 : mod reg r/m	4–13/ 8–17	RV/P,9,24
LTR – Load Task Register			
from register	0000 1111 : 0000 0000 : 11 011 reg	10	
from memory	0000 1111 : 0000 0000 : mod 011 r/m	10	
MOV – Move Data			
reg1 to reg2	1000 100w : 11 reg1 reg2	1	
reg2 to reg1	1000 101w : 11 reg1 reg2	1	
memory to reg	1000 101w : mod reg r/m	1	
reg to memory	1000 100w : mod reg r/m	1	
immediate to reg	1100 011w : 11 000 reg : immediate data	1	
or	1011 w reg : immediate data	1	
immediate to memory	1100 011w : mod 000 r/m : immediate data	1	
memory to accumulator	1010 000w : full displacement	1	
accumulator to memory	1010 001w : full displacement	1	
MOV – Move to/from Control Registers			
CR0 from register	0000 1111 : 0010 0010 : 11 000 reg	22	
CR2 from register	0000 1111 : 0010 0010 : 11 010reg	12	
CR3 from register	0000 1111 : 0010 0010 : 11 011 reg	21	
CR4 from register	0000 1111 : 0010 0010 : 11 100 reg	14	
register from CR0-4	0000 1111 : 0010 0000 : 11 eee reg	4	
MOV – Move to/from Debug Registers			
DR0-3 from register	0000 1111 : 0010 0011 : 11 eee reg	11	
DR4-5 from register	0000 1111 : 0010 0011 : 11 eee reg	12	
DR6-7 from register	0000 1111 : 0010 0011 : 11 eee reg	11	
register from DR6-7	0000 1111 : 0010 0001 : 11 eee reg	11	
register from DR4-5	0000 1111 : 0010 0001 : 11 eee reg	12	
register from DR0-3	0000 1111 : 0010 0001 : 11 eee reg	2	

x86 Architecture Instruction Set *(Cont.)*

Instruction	Format	Clocks	Notes
MOV – Move to/from Segment Registers			
reg to segment reg	1000 1110 : 11 sreg3 reg	2–11	9,24
reg to SS	1000 1110 : 11 sreg3 reg	2–11/ 8–17	RV/P,9,24
memory to segment reg	1000 1110 : mod sreg3 r/m	3	9,24
memory to SS	1000 1110 : mod sreg3 r/m	3–12/ 8–17	RV/P,9,24
segment reg to reg	1000 1100 : 11 sreg3 reg	1	
segment reg to memory	1000 1100 : mod sreg3 r/m	1	
MOVS/MOVSB/MOVSW/ MOVSD – Move Data from String to String	1010 010w	4	16
MOVSX – Move with Sign-Extend			
reg2 to reg1	0000 1111 : 1011 111w : 11 reg1 reg2	3	
memory to reg	0000 1111 : 1011 111w : mod reg r/m	3	
MOVZX – Move with Zero-Extend			
reg2 to reg1	0000 1111 : 1011 011w : 11 reg1 reg2	3	
memory to reg	0000 1111 : 1011 011w : mod reg r/m	3	
MUL – Unsigned Multiplication of AL or AX			
accumulator with register	1111 011w : 11 100 reg		
multiplier —	byte	11	
	word	11	
	doubleword	10	
accumulator with memory	1111 011w : mod 100 reg		
multiplier —	byte	11	
	word	11	
	doubleword	10	
NEG – Two's Complement Negation			
reg	1111 011w : 11 011 reg	1	
memory	1111 011w : mod 011 r/m	3	U/L
NOP – No Operation	1001 0000	1	
NOT – One's Complement Negation			
reg	1111 011w : 11 010 reg	1	
memory	1111 011w : mod 010 r/m	3	U/L
OR – Logical Inclusive OR			
reg1 to reg2	0000 100w : 11 reg1 reg2	1	
reg2 to reg1	0000 101w : 11 reg1 reg2	1	
memory to register	0000 101w : mod reg r/m	2	
register to memory	0000 100w : mod reg r/m	3	U/L
immediate to register	1000 00sw : 11 001 reg : immediate data	1	
immediate to accumulator	0000 110w : immediate data	1	
immediate to memory	1000 00sw : mod 001 r/m : immediate data	3	U/L

x86 Architecture Instruction Set *(Cont.)*

Instruction	Format	Clocks	Notes
POP – Pop a Word from the Stack			
reg	1000 1111 : 11 000 reg	1	
or	0101 1 reg	1	
memory	1000 1111 : mod 000 r/m	3	1
POP – Pop a Segment Register from the Stack			
segment reg CS, DS, ES	000 sreg2 111	3–12	9,24
segment reg SS	000 sreg2 111	3–12/ 8–17	RV/P,9,24
segment reg FS, GS	0000 1111: 10 sreg3 001	3–12	9,24
POPA/POPAD – Pop All General Registers	0110 0001	5	
POPF/POPFD – Pop Stack into FLAGS or EFLAGS Register	1001 1101	4/14	RV/P
PUSH – Push Operand onto the Stack			
reg	1111 1111 : 11 110 reg	1	
or	0101 0 reg	1	
memory	1111 1111 : mod 110 r/m	2	1
immediate	0110 10s0 : immediate data	1	
PUSH – Push Segment Register onto the Stack			
segment reg CS,DS,ES,SS	000 sreg2 110	1	
segment reg FS,GS	0000 1111: 10 sreg3 000	1	
PUSHA/PUSHAD – Push All General Registers	0110 0000	5	
PUSHF/PUSHFD – Push Flags Register onto the Stack	1001 1100	3/9	RV/P
RCL – Rotate thru Carry Left			
reg by 1	1101 000w : 11 010 reg	1	
memory by 1	1101 000w : mod 010 r/m	3	
reg by CL	1101 001w : 11 010 reg	7–24	MN/MX,4
memory by CL	1101 001w : mod 010 r/m	9–26	MN/MX,5
reg by immediate count	1100 000w : 11 010 reg : imm8 data	8–25	MN/MX,4
memory by immediate count	1100 000w : mod 010 r/m : imm8 data	10–27	MN/MX,5
RCR – Rotate thru Carry Right			
reg by 1	1101 000w : 11 011 reg	1	
memory by 1	1101 000w : mod 011 r/m	3	
reg by CL	1101 001w : 11 011 reg	7–24	MN/MX,4
memory by CL	1101 001w : mod 011 r/m	9–26	MN/MX,5
reg by immediate count	1100 000w : 11 011 reg : imm8 data	8–25	MN/MX,4
memory by immediate count	1100 000w : mod 011 r/m : imm8 data	10–27	MN/MX,5
RDMSR – Read from Model- Specific Register	0000 1111 : 0011 0010	20–24	MN/MX

x86 Architecture Instruction Set *(Cont.)*

Instruction	Format	Clocks	Notes
REP LODS – Load String	1111 0011 : 1010 110w		
C = 0		7	
C > 0		7 + 3c	16
REP MOVS – Move String	1111 0011 : 1010 010w		
C = 0		6	
C = 1		13	16
C > 1		13 + c	16
REP STOS – Store String	1111 0011 : 1010 101w		
C = 0		6	
C > 0		9 + c	
REPE CMPS – Compare String (Find Non-Match)	1111 0011 : 1010 011w		
C = 0		7	
C > 0		8 + 4c	16
REPE SCAS – Scan String (Find Non-AL/AX/EAX)	1111 0011 : 1010 111w		
C = 0		7	
C > 0		8 + 4c	16
REPNE CMPS – Compare String (Find Match)	1111 0010 : 1010 011w		
C = 0		7	
C > 0		9 + 4c	16
REPNE SCAS – Scan String (Find AL/AX/EAX)	1111 0010 : 1010 111w		
C = 0		7	
C > 0		8 + 4c	16
RET – Return from Procedure (to same segment)			
	1100 0011	2	
adding immediate to SP	1100 0010 : 16-bit displacement	3	
RET – Return from Procedure (to other segment)			
intersegment	1100 1011	4	R,23
to same level		4–13	P,9,23,24
to outer level		23	P,9,25
adding immediate to SP	1100 1010 : 16-bit displacement	4	R,23
to same level		4–13	P,9,23,24
to outer level		23	P,9,25
ROL – Rotate (not thru Carry) Left			
reg by 1	1101 000w : 11 000 reg	1	
memory by 1	1101 000w : mod 000 r/m	3	
reg by CL	1101 001w : 11 000 reg	4	
memory by CL	1101 001w : mod 000 r/m	4	
reg by immediate count	1100 000w : 11 000 reg : imm8 data	1	
memory by immediate count	1100 000w : mod 000 r/m : imm8 data	3	

x86 Architecture Instruction Set *(Cont.)*

Instruction	Format	Clocks	Notes
ROR – Rotate (not thru Carry) Right			
reg by 1	1101 000w : 11 001 reg	1	
memory by 1	1101 000w : mod 001 r/m	3	
reg by CL	1101 001w : 11 001 reg	4	
memory by CL	1101 001w : mod 001 r/m	4	
reg by immediate count	1100 000w : 11 001 reg : imm8 data	1	
memory by immediate count	1100 000w : mod 001 r/m : imm8 data	3	
RSM – Resume from System Management Mode	0000 1111 : 1010 1010		
SAHF – Store AH into Flags	1001 1110	2	
SAL – Shift Arithmetic Left	same instruction as SHL		
SAR – Shift Arithmetic Right			
reg by 1	1101 000w : 11 111 reg	1	
memory by 1	1101 000w : mod 111 r/m	3	
reg by CL	1101 001w : 11 111 reg	4	
memory by CL	1101 001w : mod 111 r/m	4	
reg by immediate count	1100 000w : 11 111 reg : imm8 data	1	
memory by immediate count	1100 000w : mod 111 r/m : imm8 data	3	
SBB – Integer Subtraction with Borrow			
reg1 to reg2	0001 100w : 11 reg1 reg2	1	
reg2 to reg1	0001 101w : 11 reg1 reg2	1	
memory to register	0001 101w : mod reg r/m	2	
register to memory	0001 100w : mod reg r/m	3	U/L
immediate to register	1000 00sw : 11 011 reg : immediate data	1	
immediate to accumulator	0001 110w : immediate data	1	
immediate to memory	1000 00sw : mod 011 r/m : immediate data	3	U/L
SCAS/SCASB/SCASW/SCASD – Scan String	1101 111w	4	
SETcc – Byte Set on Condition			
reg	0000 1111 : 1001 tttn : 11 000 reg	1	
memory	0000 1111 : 1001 tttn : mod 000 r/m	2	
SGDT – Store Global Descriptor Table Register	0000 1111 : 0000 0001 : mod 000 r/m	4	
SHL – Shift Left			
reg by 1	1101 000w : 11 100 reg	1	
memory by 1	1101 000w : mod 100 r/m	3	
reg by CL	1101 001w : 11 100 reg	4	
memory by CL	1101 001w : mod 100 r/m	4	
reg by immediate count	1100 000w : 11 100 reg : imm8 data	1	
memory by immediate count	1100 000w : mod 100 r/m : imm8 data	3	

x86 Architecture Instruction Set *(Cont.)*

Instruction	Format	Clocks	Notes
SHLD – Double Precision Shift Left			
register by immediate count imm8	0000 1111 : 1010 0100 : 11 reg2 reg1 :	4	
memory by immediate count imm8	0000 1111 : 1010 0100 : mod reg r/m :	4	
register by CL	0000 1111 : 1010 0101 : 11 reg2 reg1	4	
memory by CL	0000 1111 : 1010 0101 : mod reg r/m	5	
SHR – Shift Right			
reg by 1	1101 000w : 11 101 reg	1	
memory by 1	1101 000w : mod 101 r/m	3	
reg by CL	1101 001w : 11 101 reg	4	
memory by CL	1101 001w : mod 101 r/m	4	
reg by immediate count	1100 000w : 11 101 reg : imm8 data	1	
memory by immediate count	1100 000w : mod 101 r/m : imm8 data	3	
SHRD – Double Precision Shift Right			
register by immediate count imm8	0000 1111 : 1010 1100 : 11 reg2 reg1 :	4	
memory by immediate count imm8	0000 1111 : 1010 1100 : mod reg r/m :	4	
register by CL	0000 1111 : 1010 1101 : 11 reg2 reg1	4	
memory by CL	0000 1111 : 1010 1101 : mod reg r/m	5	
SIDT – Store Interrupt Descriptor Table Register	0000 1111 : 0000 0001 : mod 001 r/m	4	
SLDT – Store Local Descriptor Table Register			
to register	0000 1111 : 0000 0000 : 11 000 reg	2	
to memory	0000 1111 : 0000 0000 : mod 000 r/m	2	
SMSW – Store Machine Status Word			
to register	0000 1111 : 0000 0001 : 11 100 reg	4	
to memory	0000 1111 : 0000 0001 : mod 100 r/m	4	
STC – Set Carry Flag	1111 1001	2	
STD – Set Direction Flag	1111 1101	2	
STI – Set Interrupt Flag	1111 1011	7	
STOS/STOSB/STOSW/STOSD – Store String Data	1010 101w	3	
STR – Store Task Register			
to register	0000 1111 : 0000 0000 : 11 001 reg	2	
to memory	0000 1111 : 0000 0000 : mod 001 r/m	2	
SUB – Integer Subtraction			
reg1 to reg2	0010 100w : 11 reg1 reg2	1	
reg2 to reg1	0010 101w : 11 reg1 reg2	1	
memory to register	0010 101w : mod reg r/m	2	
register to memory	0010 100w : mod reg r/m	3	U/L
immediate to register	1000 00sw : 11 101 reg : immediate data	1	
immediate to accumulator	0010 110w : immediate data	1	
immediate to memory	1000 00sw : mod 101 r/m : immediate data	3	U/L

x86 Architecture Instruction Set (Cont.)

Instruction	Format	Clocks	Notes
TEST – Logical Compare			
reg1 and reg2	1000 010w : 11 reg1 reg2	2	
memory and register	1000 010w : mod reg r/m	1	
immediate and register	1111 011w : 11 000 reg : immediate data	1	
immediate and accumulator	1010 100w : immediate data	1	
immediate and memory	1111 011w : mod 000 r/m : immediate data	2	
VERR – Verify a Segment for Reading			
register	0000 1111 : 0000 0000 : 11 100 reg	7	
memory	0000 1111 : 0000 0000 : mod 100 r/m	7	
VERW – Verify a Segment for Writing			
register	0000 1111 : 0000 0000 : 11 101 reg	7	
memory	0000 1111 : 0000 0000 : mod 101 r/m	7	
WAIT – Wait	1001 1011	1/1	
WBINVD – Write-Back and Invalidate Data Cache	0000 1111 : 0000 1001	2000+	
WRMSR – Write to Model-Specific Register	0000 1111 : 0011 0000	30–45	MN/MX
XADD – Exchange and Add			
reg1, reg2	0000 1111 : 1100 000w : 11 reg2 reg1	3	
memory, reg	0000 1111 : 1100 000w : mod reg r/m	4	U/L
XCHG – Exchange Register/Memory with Register			
reg1 with reg2	1000 011w : 11 reg1 reg2	3	2
accumulator with reg	1001 0 reg	2	2
memory with reg	1000 011w : mod reg r/m	3	2
XLAT/XLATB – Table Look-up Translation	1101 0111	4	
XOR – Logical Exclusive OR			
reg1 to reg2	0011 000w : 11 reg1 reg2	1	
reg2 to reg1	0011 001w : 11 reg1 reg2	1	
memory to register	0011 001w : mod reg r/m	2	
register to memory	0011 000w : mod reg r/m	3	U/L
immediate to register	1000 00sw : 11 110 reg : immediate data	1	
immediate to accumulator	0011 010w : immediate data	1	
immediate to memory	1000 00sw : mod 110 r/m : immediate data	3	U/L

x86 Architecture Instruction Set *(Cont.)*

Instruction	Format	Clocks	Notes
Prefix Bytes			
address size	0110 0111	1	
LOCK	1111 0000	1	
operand size	0110 0110	1	
CS segment override	0010 1110	1	
DS segment override	0011 1110	1	
ES segment override	0010 0110	1	
FS segment override	0110 0100	1	
GS segment override	0110 0101	1	
SS segment override	0011 0110	1	
External Interrupt		INT + 14	21
NMI – Non-Maskable Interrupt		INT + 6	21
Page Fault		INT + 40	21
Virtual 8086 Mode Exceptions			
CLI		INT + 9	21
STI		INT + 9	21
INT n		INT + 9	
PUSHF		INT + 9	21
POPF		INT + 9	21
IRET		INT + 9	
IN			
fixed port		INT + 34	21
variable port		INT + 34	21
OUT			
fixed port		INT + 34	21
variable port		INT + 34	21
INS		INT + 34	21
OUTS		INT + 34	21
REP INS		INT + 34	21
REP OUTS		INT + 34	21

I/O Instructions

Instruction	Format	Real Mode	Protected Mode CPL≤ IOPL	Protected Mode CPL>I OPL	Virtual 8086 Mode	Notes
IN – Input from:						
fixed port number	1110 010w : port	7	4	21	19	
variable port	1110 110w	7	4	21	19	
OUT – Output to:						
fixed port number	1110 011w : port	12	9	26	24	
variable port	1110 111w	12	9	26	24	
INS – Input from DX Port	0110 110w	9	6	24	22	
OUTS – Output to DX Port	0110 111w	13	10	27	25	1
REP INS – Input String 110w	1111 0011 : 0110	11 + 3c	8 + 3c	25 + 3c	23 + 3c	2
REP OUTS – Output String 111w	1111 0011 : 0110	13 + 4c	10 + 4c	27 + 4c	25 + 4c	3

NOTES:

1. Two clock cache miss penalty in all cases.
2. c = count in CX or ECX
3. Cache miss penalty in all modes: Add 2 clocks for every 16 bytes. Entire penalty on second operation.

Floating-Point Instructions

Instruction	Format	Clocks	Notes
F2XM1 – Compute 2$^{ST(0)}$ – 1	11011 001 : 1111 0000	13–57	
FABS – Absolute Value	11011 001 : 1110 0001	1/1	
FADD – Add			
ST(0) ← ST(0) + 32-bit memory	11011 000 : mod 000 r/m	3/1	
ST(0) ← ST(0) + 64-bit memory	11011 100 : mod 000 r/m	3/1	
ST(d) ← ST(0) + ST(i)	11011 d00 : 11 000 ST(i)	3/1	
FADDP – Add and Pop			
ST(0) ← ST(0) + ST(i)	11011 110 : 11 000 ST(i)	3/1	
FBLD – Load Binary Coded Decimal	11011 111 : mod 100 r/m	48–58	
FBSTP – Store Binary Coded Decimal and Pop	11011 111 : mod 110 r/m	148–154	
FCHS – Change Sign	11011 001 : 1110 0000	1/1	
FCLEX – Clear Exceptions	11011 011 : 1110 0010	9/9	2
FCOM – Compare Real			
32-bit memory	11011 000 : mod 010 r/m	4/1	
64-bit memory	11011 100 : mod 010 r/m	4/1	
ST(i)	11011 000 : 11 010 ST(i)	4/1	
FCOMP – Compare Real and Pop			
32-bit memory	11011 000 : mod 011 r/m	4/1	
64-bit memory	11011 100 : mod 011 r/m	4/1	
ST(i)	11011 000 : 11 011 ST(i)	4/1	
FCOMPP – Compare Real and Pop Twice	11011 110 : 11 011 001	4/1	
FCOS – Cosine of ST(0)	11011 001 : 1111 1111	18–124	
FDECSTP – Decrement Stack-Top Pointer	11011 001 : 1111 0110	1/1	
FDIV – Divide			
ST(0) ← ST(0) ÷ 32-bit memory	11011 000 : mod 110 r/m	39	1
ST(0) ← ST(0) ÷ 64-bit memory	11011 100 : mod 110 r/m	39	1
ST(d) ← ST(0) ÷ ST(i)	11011 d00 : 1111 R ST(i)	39	1
FDIVP – Divide and Pop			
ST(0) ← ST(0) ÷ ST(i)	11011 110 : 1111 1 ST(i)	39	1
FDIVR – Reverse Divide			
ST(0) ← 32-bit memory ÷ ST(0)	11011 000 : mod 111 r/m	39	1
ST(0) ← 64-bit memory ÷ ST(0)	11011 100 : mod 111 r/m	39	1
ST(d) ← ST(i) ÷ ST(0)	11011 d00 : 1111 R ST(i)	39	1
FDIVRP – Reverse Divide and Pop			
ST(0) ← ST(i) ÷ ST(0)	11011 110 : 1111 0 ST(i)	39	1
FFREE – Free ST(i) Register	11011 101 : 1100 0 ST(i)	1/1	
FIADD – Add Integer			
ST(0) ← ST(0) + 16-bit memory	11011 110 : mod 000 r/m	7/4	
ST(0) ← ST(0) + 32-bit memory	11011 010 : mod 000 r/m	7/4	

Floating-Point Instructions (*Cont.*)

Instruction	Format	Clocks	Notes
FICOM – Compare Integer			
16-bit memory	11011 110 : mod 010 r/m	8/4	
32-bit memory	11011 010 : mod 010 r/m	8/4	
FICOMP – Compare Integer and Pop			
16-bit memory	11011 110 : mod 011 r/m	8/4	
32-bit memory	11011 010 : mod 011 r/m	8/4	
FIDIV			
ST(0) ← ST(0) + 16-bit memory	11011 110 : mod 110 r/m	42	1
ST(0) ← ST(0) + 32-bit memory	11011 010 : mod 110 r/m	42	1
FIDIVR			
ST(0) ← ST(0) + 16-bit memory	11011 110 : mod 111 r/m	42	1
ST(0) ← ST(0) + 32-bit memory	11011 010 : mod 111 r/m	42	1
FILD – Load Integer			
16-bit memory	11011 111 : mod 000 r/m	3/1	
32-bit memory	11011 011 : mod 000 r/m	3/1	
64-bit memory	11011 111 : mod 101 r/m	3/1	
FIMUL			
ST(0) ← ST(0) + 16-bit memory	11011 110 : mod 001 r/m	7/4	
ST(0) ← ST(0) + 32-bit memory	11011 010 : mod 001 r/m	7/4	
FINCSTP – Increment Stack Pointer	11011 001 : 1111 0111	1/1	
FINIT – Initialize Floating-Point Unit	11011 011 : 1110 0011	16/12	2
FIST – Store Integer			
16-bit memory	11011 111 : mod 010 r/m	6/6	
32-bit memory	11011 011 : mod 010 r/m	6/6	
FISTP – Store Integer and Pop			
16-bit memory	11011 111 : mod 011 r/m	6/6	
32-bit memory	11011 011 : mod 011 r/m	6/6	
64-bit memory	11011 111 : mod 111 r/m	6/6	
FISUB			
ST(0) ← ST(0) + 16-bit memory	11011 110 : mod 100 r/m	7/4	
ST(0) ← ST(0) + 32 bit memory	11011 010 : mod 100 r/m	7/4	
FISUBR			
ST(0) ← ST(0) + 16-bit memory	11011 110 : mod 101 r/m	7/4	
ST(0) ← ST(0) + 32-bit memory	11011 010 : mod 101 r/m	7/4	
FLD – Load Real			
32-bit memory	11011 001 : mod 000 r/m	1/1	
64-bit memory	11011 101 : mod 000 r/m	1/1	
80-bit memory	11011 011 : mod 101 r/m	3/3	
ST(i)	11011 001 : 11 000 ST(i)	1/1	
FLD1 – Load +1.0 into ST(0)	11011 001 : 1110 1000	2/2	

Floating-Point Instructions (*Cont.*)

Instruction	Format	Clocks	Notes
FLDCW – Load Control Word	11011 001 : mod 101 r/m	7/7	
FLDENV – Load FPU Environment	11011 001 : mod 100 r/m		
real and v86 modes, 16-bit address		37	
real and v86 modes, 32-bit address		37	
protected mode, 16-bit address		32	
protected mode, 32-bit address		33	
FLDL2E – Load $\log_2(\varepsilon)$ into ST(0)	11011 001 : 1110 1010	5/3	
FLDL2T – Load $\log_2(10)$ into ST(0)	11011 001 : 1110 1001	5/3	
FLDLG2 – Load $\log_{10}(2)$ into ST(0)	11011 001 : 1110 1100	5/3	
FLDLN2 – Load $\log_\varepsilon(2)$ into ST(0)	11011 001 : 1110 1101	5/3	
FLDPI – Load π into ST(0)	11011 001 : 1110 1011	5/3	
FLDZ – Load +0.0 into ST(0)	11011 001 : 1110 1110	2/2	
FMUL – Multiply			
ST(0) ← ST(0) × 32-bit memory	11011 000 : mod 001 r/m	3/1	4
ST(0) ← ST(0) × 64-bit memory	11011 100 : mod 001 r/m	3/1	4
ST(d) ← ST(0) × ST(i)	11011 d00 : 1100 1 ST(i)	3/1	4
FMULP – Multiply			
ST(0) ← ST(0) × ST(i)	11011 110 : 1100 1 ST(i)	3/1	
FNOP – No Operation	11011 001 : 1101 0000	1/1	
FPATAN – Partial Arctangent	11011 001 : 1111 0011	19–134	
FPREM – Partial Remainder	11011 001 : 1111 1000	16–64	
FPREM1 – Partial Remainder (IEEE)	11011 001 : 1111 0101	20–70	
FPTAN – Partial Tangent	11011 001 : 1111 0010	17–173	
FRNDINT – Round to Integer	11011 001 : 1111 1100	9–20	
FRSTOR – Restore FPU State	11011 101 : mod 100 r/m		
real and v86 modes, 16-bit address		75/75	
real and v86 modes, 32-bit address		95/95	
protected mode, 16-bit address		70/70	
protected mode, 32-bit address		70/70	
FSAVE – Store FPU State	1101 101 : mod 110 r/m		
real and v86 modes, 16-bit address		127/127	2
real and v86 modes, 32-bit address		151/151	2
protected mode, 16-bit address		124/124	2
protected mode, 32-bit address		124/124	2
FSCALE – Scale	11011 001 : 1111 1101	20–31	
FSIN – Sine	11011 001 : 1111 1110	16–126	
FSINCOS – Sine and Cosine	11011 001 : 1111 1011	17–137	
FSQRT – Square Root	11011 001 : 1111 1010	70/70	
FST – Store Real			
32-bit memory	11011 001 : mod 010 r/m	2/2	
64-bit memory	11011 101 : mod 010 r/m	2/2	
ST(i)	11011 101 : 11 010 ST(i)	1/1	

Floating-Point Instructions (*Cont.*)

Instruction	Format	Clocks	Notes
FSTCW – Store Control Word	11011 001 : mod 111 r/m	2/2	2
FSTENV – Store FPU Environment	11011 001 : mod 110 r/m		
real and v86 modes, 16-bit address		50/50	2
real and v86 modes, 32-bit address		48/50	2
protected mode, 16-bit address		49/50	2
protected mode, 32-bit address		50/50	2
FSTP – Store Real and Pop			
32-bit memory	11011 001 : mod 011 r/m	2/2	
64-bit memory	11011 101 : mod 011 r/m	2/2	
80-bit memory	11011 011 : mod 111 r/m	3/3	
ST(i)	11011 101 : 11 011 ST(i)	1/1	
FSTSW – Store Status Word into AX	11011 111 : 1110 0000	6/2	2
FSTSW – Store Status Word into Memory	11011 101 : mod 111 r/m	5/2	2
FSUB – Subtract			
ST(0) ← ST(0) – 32-bit memory	11011 000 : mod 100 r/m	3/1	
ST(0) ← ST(0) – 64-bit memory	11011 100 : mod 100 r/m	3/1	
ST(d) ← ST(0) – ST(i)	11011 d00 : 1110 R ST(i)	3/1	
FSUBP – Subtract and Pop			
ST(0) ← ST(0) – ST(i)	11011 110 : 1110 1 ST(i)	3/1	
FSUBR – Reverse Subtract			
ST(0) ← 32 bit memory – ST(0)	11011 000 : mod 101 r/m	3/1	
ST(0) ← 64-bit memory – ST(0)	11011 100 : mod 101 r/m	3/1	
ST(d) ← ST(i) – ST(0)	11011 d00 : 1110 R ST(i)	3/1	
FSUBRP – Reverse Subtract and Pop			
ST(i) ← ST(i) – ST(0)	11011 110 : 1110 0 ST(i)	3/1	
FTST – Test	11011 001 : 1110 0100	4/1	
FUCOM – Unordered Compare Real	11011 101 : 1110 0 ST(i)	4/1	
FUCOMP – Unordered Compare and Pop	11011 101 : 1110 1 ST(i)	4/1	
FUCOMPP – Unordered Compare and Pop Twice	11011 010 : 1110 1001	4/1	
FXAM – Examine	11011 001 : 1110 0101	21/21	
FXCH – Exchange ST(0) and ST(i)	11011 001 : 1100 1 ST(i)	1	3
FXTRACT – Extract Exponent and Significand	11011 001 : 1111 0100	13/13	
FYL2X – ST(1) × $\log_2$(ST(0))	11011 001 : 1111 0001	22–111	
FYL2XP1 – ST(1) × $\log_2$(ST(0) + 1.0)	11011 001 : 1111 1001	22–103	
FWAIT – Wait until FPU Ready	1001 1011	1/1	

Floating-Point Instructions (*Cont.*)

L/NL	Clocks apply to loop and no loop cases respectively
MN/MX	Clocks shown define a range from minimum to maximum
P	Clocks apply to protected mode
R	Clocks apply to real-address mode
RV/P	First clock applies to real and V86 mode; second applies to protected mode
T/NT	Clocks apply to taken and not taken cases respectively
U/L	Clocks apply to unlocked and locked cases respectively
1.	Assuming that the operand address and stack address fall in different cache interleaves.
2.	Always locked. Always forced to miss cache.
4.	Clocks = {quotient(count/operand length)}*7 + 9 = 8 if count ≤ operand length (8/16/32).
5.	Clocks = {quotient(count/operand length)}*7 + 9 = 9 if count ≤ operand length (8/16/32).
8.	Penalty for cache miss: add 2 clocks for every stack value copied to the new stack frame.
9.	Add 8 clocks for each load of an unaccessed descriptor.
10.	Refer to Task Switch Clock Counts Table for value of TS.

For notes 12 − 13:$b = 0 − 3$, nonzero byte number;
$i = 0 − 1$, nonzero nibble number;
$n = 0 − 3$, nonzero bit number in nibble.

12.	Clocks= $8 + 4(b + 1) + 3(i + 1) + 3(n + 1)$ = 6 if second operand = 0.
13.	Clocks= $9 + 4(b + 1) + 3(i + 1) + 3(n + 1)$ = 7 if second operand = 0.

For notes 14 − 15: n = bit position $(0 − 31)$.

14.	Clocks= $7 + 2(32 − n)$ = 6 if second operand = 0.
15.	Clocks= $8 + 2(32 − n)$ = 7 if second operand = 0.
16.	Assuming that the two string addresses fall in different cache interleaves.
21.	Refer to the Interrupt Clock Counts Table for value of INT.
23.	Add $r + 3b$ for instruction cache miss. Add 3 for branch misprediction.
24.	Clocks shown define a range from minimum to maximum.
25.	Add $r + 3b$ for instruction cache miss.

Instructions of the Intel x86 architecture consist of one or two primary opcode bytes, possibly an address specifier consisting of the "mod r/m" byte and "scaled index" byte, a displacement if required, and an immediate data field if required. Within the primary opcode or opcodes, smaller encoding fields may be defined. These fields vary according to the class of operation. The fields define such information as direction of the operation, size of the displacements, register encoding, or sign extension. Almost all instructions referring to an operand in memory have an addressing mode byte following the primary opcode byte(s). This byte, called the mod r/m byte, specifies the addressing mode to be used. Certain encodings of the mod r/m byte indicate a second addressing byte, the scale-index-base byte, which follows the mod r/m byte to fully specify the addressing mode.

Addressing modes can include a displacement immediately following the mod r/m byte, or scaled index byte. If a displacement is present, the possible sizes are 8, 16, or 32 bits. If the instruction specifies an immediate operand, the immediate operand follows the displacement bytes (if any). The immediate operand, if specified, is always the last field of the instruction.

A general x86 architecture instruction format is shown in Fig. 2.A.1 [Pnt3 93]. Not all the indicated fields appear in every instruction. Table 2.A.1 lists the different fields that may appear in x86 instructions. Further ahead, detailed tables for each field are presented.

The encodings of the register fields and of the w bit are shown in Table 2.A.2. The sreg field in certain instructions is a 2-bit field allowing one of the four 80286 segment registers to be specified. The sreg field in other instructions is a 3-bit field, allowing the FS and GS segment registers to be specified, as shown in Table 2.A.3.

The s-i-b byte (scale-index-base byte) is specified when using 32-bit addressing modes and the mod r/m byte has r/m = 100 and mod = 00, 01, or 10. When the s-i-b byte is present, the 32-bit addressing mode is a function of the mod, ss, index, and base fields. When calculating an EA, either 16- or 32-bit addressing is used, implementing 16- or 32-bit address components to calculate the EA, respectively. When 16-bit addressing is used, the mod r/m byte is interpreted as a 16-bit addressing mode specifier. When 32-bit addressing is used, the mod r/m byte is interpreted as a 32-bit addressing mode specifier. Table 2.A.4 defines all encodings of all 16- and 32-bit addressing modes.

The d field indicates which operand is the source and which is the destination:

d = 0: reg field is the source, mod r/m is the destination

d = 1: reg is the destination, mod r/m is the source

The s field has an effect only if the size of the immediate data is 8 bits and it is being placed in a 16- or 32-bit destination. If s = 1, the 8-bit data item will be sign-extended to fill a 16- or 32-bit destination. The s = 0 value has no effect.

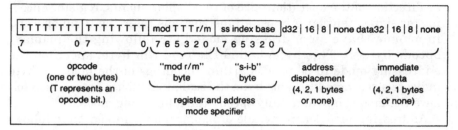

Figure 2.A.1 General instruction format. (*Courtesy of Intel Corp.*)

TABLE 2.A.1   Fields Within x86 Architecture Instructions

Field Name	Description	Number of Bits
w	Specifies if Data is Byte or Full Size (Full Size is either 16 or 32 Bits	1
d	Specifies Direction of Data Operation	1
s	Specifies if an Immediate Data Field Must be Sign-Extended	1
reg	General Register Specifier	3
mod r/m	Address Mode Specifier (Effective Address can be a General Register)	2 for mod; 3 for r/m
ss	Scale Factor for Scaled Index Address Mode	2
index	General Register to be used as Index Register	3
base	General Register to be used as Base Register	3
sreg2	Segment Register Specifier for CS, SS, DS, ES	2
sreg3	Segment Register Specifier for CS, SS, DS, ES, FS, GS	3
tttn	For Conditional Instructions, Specifies a Condition Asserted or a Condition Negated	4

TABLE 2.A.2 Register and w Bit Encoding

w Field	Operand Size During 16-Bit Data Operations	Operand Size During 32-Bit Data Operations
0	8 Bits	8 Bits
1	16 Bits	32 Bits

Encoding of reg Field When w Field is not Present in Instruction

reg Field	Register Selected During 16-Bit Data Operations	Register Selected During 32-Bit Data Operations
000	AX	EAX
001	CX	ECX
010	DX	EDX
011	BX	EBX
100	SP	ESP
101	BP	EBP
101	SI	ESI
101	DI	EDI

Encoding of reg Field When w Field is Present in Instruction

reg	Register Specified by reg Field During 16-Bit Data Operations: Function of w Field	
	(when w = 0)	(when w = 1)
000	AL	AX
001	CL	CX
010	DL	DX
011	BL	BX
100	AH	SP
101	CH	BP
110	DH	SI
111	BH	DI

reg	Register Specified by reg Field During 32-Bit Data Operations: Function of w Field	
	(when w = 0)	(when w = 1)
000	AL	EAX
001	CL	ECX
010	DL	EDX
011	BL	EBX
100	AH	ESP
101	CH	EBP
110	DH	ESI
111	BH	EDI

TABLE 2.A.3   Segment Register Encoding

2-Bit sreg2 Field	
2-Bit sreg2 Field	Segment Register Selected
00	ES
01	CS
10	SS
11	DS

3-Bit sreg3 Field	
3-Bit sreg3 Field	Segment Register Selected
000	ES
001	CS
010	SS
011	DS
100	FS
101	GS
110	do not use
111	do not use

TABLE 2.A.4   Addressing Mode Encoding

Encoding of 16-bit Address Mode with "mod r/m" Byte

mod r/m	Effective Address	mod r/m	Effective Address
00 000	DS:[BX + SI]	10 000	DS:[BX + SI + d16]
00 001	DS:[BX + DI]	10 001	DS:[BX + DI + d16]
00 010	SS:[BP + SI]	10 010	SS:[BP + SI + d16]
00 011	SS:[BP + DI]	10 011	SS:[BP + DI + d16]
00 100	DS:[SI]	10 100	DS:[SI + d16]
00 101	DS:[DI]	10 101	DS:[DI + d16]
00 110	DS:d16	10 110	SS:[BP + d16]
00 111	DS:[BX]	10 111	DS:[BX + d16]
01 000	DS:[BX + SI + d8]	11 000	register—see below
01 001	DS:[BX + DI + d8]	11 001	register—see below
01 010	SS:[BP + SI + d8]	11 010	register—see below
01 011	SS:[BP + DI + d8]	11 011	register—see below
01 100	DS:[SI + d8]	11 100	register—see below
01 101	DS:[DI + d8]	11 101	register—see below
01 110	SS:[BP + d8]	11 110	register—see below
01 111	DS:[BX + d8]	11 111	register—see below

mod r/m	Register Specified by r/m During 16-Bit Data Operations	
	Function of w Field	
	(when w = 0)	(when w = 1)
11 000	AL	AX
11 001	CL	CX
11 010	DL	DX
11 011	BL	BX
11 100	AH	SP
11 101	CH	BP
11 110	DH	SI
11 111	BH	DI

mod r/m	Register Specified by r/m During 32-Bit Data Operations	
	Function of w Field	
	(when w — 0)	(when w = 1)
11 000	AL	EAX
11 001	CL	ECX
11 010	DL	EDX
11 011	BL	EBX
11 100	AH	ESP
11 101	CH	EBP
11 110	DH	ESI
11 111	BH	EDI

Encoding of 32-bit Address Mode with "mod r/m" byte (no "s-i-b" byte present):

mod r/m	Effective Address	mod r/m	Effective Address
00 000	DS:[EAX]	10 000	DS:[EAX + d32]
00 001	DS:[ECX]	10 001	DS:[ECX + d32]
00 010	DS:[EDX]	10 010	DS:[EDX + d32]
00 011	DS:[EBX]	10 011	DS:[EBX + d32]
00 100	s-i-b is present	10 100	s-i-b is present
00 101	DS:d32	10 101	SS:[EBP + d32]
00 110	DS:[ESI]	10 110	DS:[ESI + d32]
00 111	DS:[EDI]	10 111	DS:[EDI + d32]
01 000	DS:[EAX + d8]	11 000	register—see below
01 001	DS:[ECX + d8]	11 001	register—see below
01 010	DS:[EDX + d8]	11 010	register—see below
01 011	DS:[EBX + d8]	11 011	register—see below
01 100	s-i-b is present	11 100	register—see below
01 101	SS:[EBP + d8]	11 101	register—see below
01 110	DS:[ESI + d8]	11 110	register—see below
01 111	DS:[EDI + d8]	11 111	register—see below

TABLE 2.A.4   Addressing Mode Encoding *(Cont.)*

Register Specified by reg or r/m during 16-Bit Data Operations:				Register Specified by reg or r/m during 32-Bit Data Operations:		
mod r/m	Function of w field			mod r/m	Function of w field	
	(when w = 0)	(when w = 1)			(when w = 0)	(when w = 1)
11 000	AL	AX		11 000	AL	EAX
11 001	CL	CX		11 001	CL	ECX
11 010	DL	DX		11 010	DL	EDX
11 011	BL	BX		11 011	BL	EBX
11 100	AH	SP		11 100	AH	ESP
11 101	CH	BP		11 101	CH	EBP
11 110	DH	SI		11 110	DH	ESI
11 111	BH	DI		11 111	BH	EDI

Encoding of 32-bit Address Mode ("mod r/m" byte and "s-i-b" byte present):

mod base	Effective Address
00 000	DS:[EAX + (scaled index)]
00 001	DS:[ECX + (scaled index)]
00 010	DS:[EDX + (scaled index)]
00 011	DS:[EBX + (scaled index)]
00 100	SS:[ESP + (scaled index)]
00 101	DS:[d32 + (scaled index)]
00 110	DS:[ESI + (scaled index)]
00 111	DS:[EDI + (scaled index)]
01 000	DS:[EAX + (scaled index) + d8]
01 001	DS:[ECX + (scaled index) + d8]
01 010	DS:[EDX + (scaled index) + d8]
01 011	DS:[EBX + (scaled index) + d8]
01 100	SS:[ESP + (scaled index) + d8]
01 101	SS:[EBP + (scaled index) + d8]
01 110	DS:[ESI + (scaled index) + d8]
01 111	DS:[EDI + (scaled index) + d8]
10 000	DS:[EAX + (scaled index) + d32]
10 001	DS:[ECX + (scaled index) + d32]
10 010	DS:[EDX + (scaled index) + d32]
10 011	DS:[EBX + (scaled index) + d32]
10 100	SS:[ESP + (scaled index) + d32]
10 101	SS:[EBP + (scaled index) + d32]
10 110	DS:[ESI + (scaled index) + d32]
10 111	DS:[EDI + (scaled index) + d32]

ss	Scale Factor
00	x1
01	x2
10	x4
11	x8

index	Index Register
000	EAX
001	ECX
010	EDX
011	EBX
100	no index reg**
101	EBP
110	ESI
111	EDI

**IMPORTANT NOTE:
When index field is 100, indicating "no index register," then ss field MUST equal 00. If index is 100 and ss does not equal 00, the effective address is undefined.

NOTE:
Mod field in "mod r/m" byte; ss, index, base fields in "s-i-b" byte.

Problems to Part 2

1. Write a simple program to load the CR0 register as follows: paging enabled, cache enabled, write-through enabled, alignment check enabled, write protect enabled, numerics exception enabled, emulate coprocessor disabled, monitor coprocessor set, protection mechanism enabled. Subsequently, transfer the contents of CR0 to a memory location starting at address AA8800H.

2. A page fault was detected at the execution of an instruction whose linear address is 78E00H. Write a short program that transfers this address, using the appropriate control register, into register ECX, and subsequently storing this information in the memory location FF00H.

3. The page directory table starts at the address 1A86000H. Write a short program loading the CR3 appropriately, while enabling the on-chip cache and allowing a write-through policy for the external cache. Subsequently, transfer the contents of CR3 into a memory location at the address 58AB00H.

4. Write a short program storing all 32-bit general-purpose registers on stack. If the contents of ESP was AC00H before, what will it be after execution?

5. Transfer the top of the stack into all 16-bit general-purpose registers. If (ESP) = 8400H before, what will it be after execution?

6. Write a short program to transfer a word from an input device whose address is 10A0H, into register DX.

7. Write a short program to transfer all bytes stored in EBX, ECX, and EDX into an 8-bit output device, whose address is 8.

8. Write a short program that would set the carry flag, clear the direction flag, set the interrupt flag, transfer the low byte of EFLAGS into CL, and push EFLAGS on stack.

9. Given two eight-element 16-bit arrays in memory: A1 and A2. Treating A1 and A2 as $8 \times 16 = 128$-bit numbers, write a short program that would add them. Store the 128-bit sum in a 16-bit array A3. The first 16-bit element in any array is interpreted as the least significant bits of the 128-bit number, and so on.

10. Write a short program using the SBB instruction to subtract (EBX, ECX) from (EAX, EDX), placing the 64-bit difference in (EAX, EDX).

11. Transfer a string of 20 word elements from a starting source address 0400H to as destination address starting at 2000H.

12. Write a short program that locates all doubleword 10-element string elements equal to the value stored in EAX. The string being compared starts at the address 0200H. The address of each value found equal should be transmitted out through the output port at the address FF00H.

13. Write a short program that transfers a string of 50 words from an input port address FF00H into a memory location starting at address 8000H. While transferring each element, it should be checked for being strictly positive. All positive elements should be also extracted and stored in a separate string starting at address A000H.

14. Write a short program with two levels of nested procedures, using 10-byte stack frames for passing parameters for both. Use ENTER and LEAVE instructions.

15. Interface the 80486 with eight I/O devices having interrupt privileges. Use the Intel 8259A Programmable Interrupt Controller chip.*

16. Design an 80486-based microcomputer with 8-kbyte EPROM and 64-Mbyte dynamic RAM (DRAM) memory, and an interface to eight I/O devices.

17. Extend the system in Problem 16 by adding to it a second level of external 64-kbyte cache. Select the external cache parameters and design any necessary control logic.

18. Initiate a stack in memory location 7700H of the i486 and push all general-purpose registers on it, using a minimal number of instructions. Assume a no-segmentation mode of operation.

19. Signed divide on the i486 the 32-bit number 640000H by 8000H placing the result in memory location A000H, with minimum instructions. Assume operation without segmentation.

20. Assume in the i486 FPU that R0 is the current top of the stack.

a. Describe in detail what will be accomplished by

```
FLD ST(3)
```

Note: Problems 15 to 17 require the use of Intel specification sheets and manuals.

b. How would you copy the top of stack into ST(6), without affecting the stack structure?

21.   Define by an assembly directive a CALL GATE, called CALL00, with the following properties: the target offset is 0FA000H, the target segment is pointed to by a descriptor, identified as number 7 in the GDT. The target PL = 0, and it is nonconforming. The CALL GATE is valid, its DPL = 3, and it is a 486 CALL GATE type. No parameter passing to the target stack is involved.

22.   List all possible procedures with different PLs that can access the CALL GATE in Problem 21, and/or its target, in any possible way. Illustrate by a circular PL diagram.

23.   A task is allowed to use ten 16-bit I/O devices. The devices' addresses start at address 0F0H and continue consecutively. Draw and explain the I/O permission map of the task. Fill in all appropriate bits in it.

24.   In the i486 PROC5 is the fifth nested procedure. Provide a complete memory map of its stack frame. Assume that 20 bytes are allocated for this procedure's parameters. What is the first and last two instructions of PROC5? Present all numbers in hexadecimal.

25.   In an i486 program a data segment using ES is referenced by a descriptor in LDT. Provide a complete memory map diagram with all CPU registers involved in locating operands in both segments.

26.   Initiate with the appropriate assembly directives an i486 valid CALL GATE named CGATE, whose privilege level is 2, and the number of parameters to copy from caller's stack is 16. The target code segment is pointed to by a descriptor number 128 in the GDT. The target code segment is at privilege level 0. The target procedure in the target code segment starts at an offset FF00H. Present your results in hexadecimal.

27.   The i486 is running in the full 32-bit mode. A procedure PRC is called. It will be using 20 bytes of procedure parameters.

a. Provide the ENTER instruction for PRC.

b. Assuming the contents of ESP and EBP before the CALL was ESP: 0008 D040H, EBP: 0000 0000H, provide a table showing their contents after CALL, after ENTER, after LEAVE, and after RET.

c. Provide a memory map sketch corresponding to the above procedure.

28.   Provide a complete interpretation of the content of CS: 0050H, including an appropriate diagram.

29.   The content of the descriptor corresponding to the code segment is: FC08 FF0A 0000 0000H. What is the base address and the size of the segment?

30. Explain in detail the following instruction:

```
ADD EAX, [EBX + EDI*2 + 64H].
```

If EBX: 0000 2A00H, EDI: 0000 000AH, what is the effective address?

31. In an i486 system we have to set up a CALL GATE permitting a user running in the applications region with PL = 3, to access a procedure in region PL = 0.

a. Draw an appropriate diagram illustrating the protection regions and the gate placement.

b. Draw and fill in completely the CALL GATE's access rights byte.

c. Assuming the called procedure at PL = 0 has a descriptor in GDT slot 88H, starts at offset FF00H within its segment, and there are eight parameters to copy from caller's stack to the called procedure's stack, fill in the rest of the CALL GATE in hexadecimal.

32. A task can use eight I/O ports of 8 bits, starting at address 16, four 16-bit ports, starting at address 32, and two 32-bit ports, starting at address 64. Present the appropriate I/O permission bitmap.

33. An i486 gate descriptor is defined by the assembly statement:

```
GATE1 DD 0040 FFA0H, 0000 EC04H
```

Provide a complete analysis of what is involved. Which procedures can use the gate?

The Motorola M68000 Family

The MC680x0 Architecture

11.1 Introduction

The Motorola M68000 family is one of the most widely used microprocessor families. Its applications include the well-known Apple computers (MacIntosh series), Sun workstations, and many other PCs, multiprocessors, and other computing systems. The development stages of the M68000 family are listed in Table 11.1.

An introductory historical overview of the evolution of the M68000 family microprocessors was given in Chap. 1. Some technical details involved in this evolution can be seen in Table 11.1. It should be noted that Motorola was one of the first microprocessor manufacturers to feature a dual cache on its CISC-type family, even though it was a modest 256-byte code and 256-byte data cache on the MC68030, announced in 1987, 2 years before the appearance of a dual cache on a

TABLE 11.1   The Motorola M68000 Family

M68000 CPU	68000	68020	68030	68040	68060
MIPS	2.4	6.5	12	39	100+
MFLOPS	—	0.25	0.5	3.5	12
Address range	16 Mbytes	4 Gbytes	4 Gbytes	4 Gbytes	4 Gbytes
Data bus	16-bit	32-bit	32-bit	32-bit	32-bit
Clock MHz	8–16	16–33	16–50	25,33,40	50–66
I-cache	—	256 bytes Direct	256 bytes Direct	4 kbytes Four-way	8 kbytes Four-way
D-cache	—	—	256 bytes Direct	4 kbytes Four-way	8 kbytes Four-way
Burst fill caches	—	—	16 B R	16 B R/W	16 B R/W
On-chip MMU	—	—	Y	Y	Y
FPU	68881	68881	68882	On-chip	On-chip

Note: B = byte; D-cache = data cache; FPU = Floating-point unit; I-cache = instruction cache; MMU = memory management unit; R = read; W = write; Y = yes.

RISC-type CPU chip (Intel i860 in 1989). The modest dual cache of the MC68030 was soon replaced by 4-kbyte (each cache) on the MC68040 (announced in 1989), and 8-kbyte (each cache) on the MC68060 (announced in 1993, available in 1994; see Chap. 12). Another significant organizational step-up occurred with the inclusion of the FPU on the CPU chip starting with the MC68040. The latest member of the M68000 family, the MC68060, is a two-issue superscalar.

All of the members of the M68000 family share essentially the same architecture, to be discussed in detail in this chapter. A considerable jump in architectural features occurred with the announcement of the first 32-bit member of the family, the MC68020, in 1984. The architecture of all 32-bit microprocessors of the M68000 family is practically identical with very few minor differences.

Subsequent to the discussion of the M68000 family architecture in this chapter, the organization of the various family members will be discussed in the two next chapters of this part: MC68060 and MC68040 in Chap. 12, and the previous ones, MC68030, MC68020, MC68010, and MC68000, briefly in Chap. 13, along with feature comparison tables.

11.2 CPU Registers

The M68000 architecture register file is subdivided into two main parts:

1. The user programming model, shown in Fig. 11.1
2. The supervisor programming model, shown in Fig. 11.2

The M68000 architecture microprocessors operate in one of two privilege modes: user or supervisor. Supervisor mode is at a higher privilege level. It is used exclusively by system programmers to implement OS functions. User mode is intended for regular applications programs. The user programming model of CPU registers can be freely handled by the regular programmer. A program running in user mode cannot access any of the registers in the supervisor programming model. Programs running in supervisor mode have access to all registers.

The user programming model (Fig. 11.1) consists of the following:

I. Integer part: identical in all M68000 family microprocessors.
    A. 16 general-purpose 32-bit registers
        1. Eight data registers, D0 to D7
        2. Eight address registers, A0 to A7
    B. 32-bit program counter (PC)
    C. 8-bit condition code register (CCR)
II. Floating-point part: identical to that of the MC68881 or MC68882

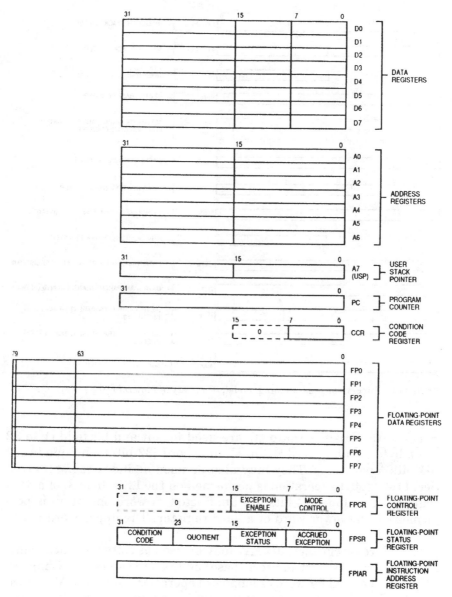

Figure 11.1   User programming model. (*Courtesy of Motorola, Inc.*)

floating-point coprocessors. On the same chip with the CPU start-ing with the MC68040.

A. Eight 80-bit floating-point data registers, FP0 to FP7

B. 16-bit floating-point control register (FPCR)

C. 32-bit floating-point status register (FPSR)

D. 32-bit floating-point instruction address register (FPIAR)

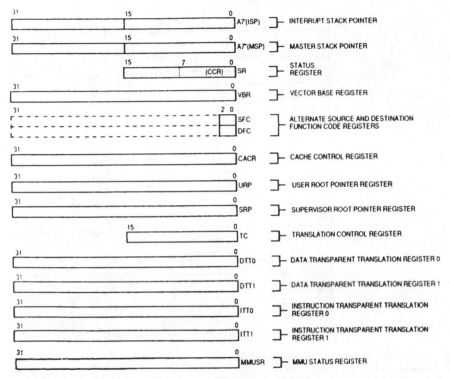

Figure 11.2 Supervisor programming model. (*Courtesy of Motorola, Inc.*)

The data registers D0 to D7 are used for bit and bit field (1 to 32 bits), byte (8 bits), word (16 bits), longword (32 bits), and quadword (64 bits) operations. They may also be used as index registers (see Sec. 11.4). A byte operation is performed on the LSB (byte 3) of a data register. The upper 24 bits are unaffected. A word operation is performed on the lower word of a data register. The upper word is not affected.

The address registers A0 to A7 may be used as software stack pointers, index registers, or indirect (base) address registers (see Chap. 3). They may be used for word or longword operations. Register A7 is used as a hardware stack pointer (SP) during stacking for subroutine calls and exception handling. The register designation A7 refers to three different registers in hardware:

User stack pointer (USP) in the user programming model or A7

Interrupt stack pointer (ISP), or A7′

Master stack pointer (MSP), or A7″, both in the supervisor programming model

The active stack pointer, at any instant of execution in the supervisor mode (ISP or MSP), is called the supervisor stack pointer.

The PC contains the address of the instruction to be fetched next. It was automatically incremented to point to the next instruction after the current instruction was fetched. It is also used as a pointer in PC-relative addressing (see Sec. 11.4).

The condition code register (CCR) is the lower byte of the status register (SR) (Fig. 11.2). It will be discussed later in this section.

The floating-point data registers FP0 to FP7 always contain 80-bit extended-precision numbers (see Chap. 3). All external operands, regardless of the data format, are converted to external-precision 80-bit values before being used in any calculation or stored in a floating-point data register.

The floating-point control register (FPCR) contains an exception enable byte, shown in Fig. 11.3, that enables or disables traps for each class of floating-point exceptions, and a mode control byte, shown in Fig. 11.4, that sets the user-selectable modes of rounding and rounding precision. The FPCR can be read or written to by the user. It is cleared by the reset function.

The floating-point status register (FPSR) is subdivided into 4 bytes as shown in Fig. 11.1. The floating-point condition code (FPCC) byte,

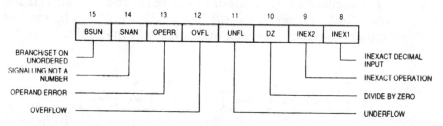

Figure 11.3 FPCR exception enable byte. (*Courtesy of Motorola, Inc.*)

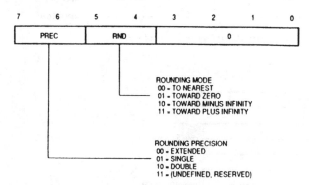

Figure 11.4 FPCR mode control byte. (*Courtesy of Motorola, Inc.*)

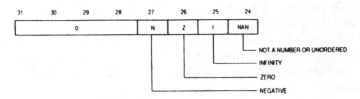

Figure 11.5 FPSR condition code byte. (*Courtesy of Motorola, Inc.*)

shown in Fig. 11.5, contains four condition code bits that are set at the end of all arithmetic instructions involving the FPU registers. The quotient byte contains the seven LSBs of the quotient and the sign (in its most significant bit, bit 23 of FPSR) of the entire quotient. The quotient bits can be used in argument reduction for transcendentals and other functions. For example, they can be used to determine the quadrant of a circle in which an operand resides.

The exception status (EXC) byte, shown in Fig. 11.6, contains a bit for each floating-point exception that may have occurred during the most recent arithmetic instruction or move operation. It can be used by an exception handler to determine which floating-point exception(s) caused a trap. The accrued exception (AEXC) byte, shown in Fig. 11.7, contains five exception bits required by the IEEE 754 standard [IEEE 85] for trap-disabled operations. All bits in the FPSR can be read or written into by the user. The reset function clears the FPSR.

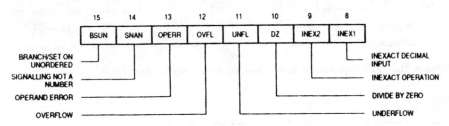

Figure 11.6 FPSR exception status byte. (*Courtesy of Motorola, Inc.*)

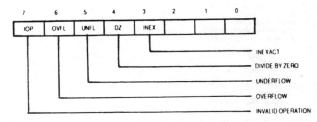

Figure 11.7 FPSR accrued exception byte. (*Courtesy of Motorola, Inc.*)

The floating-point instruction address register (FPIAR) is loaded with the logical address of the FPU instruction before the instruction is executed. This address can then be used by a floating-point exception handler to locate a floating-point instruction that has caused an exception. The FPIAR is cleared by a reset operation.

The supervisor programming model consists of the registers available to the user, discussed above, as well as the following control registers, shown in Fig. 11.2:

- Two 32-bit supervisor stack pointers ISP and MSP
- A 16-bit SR
- A 32-bit vector base register (VBR)
- Two 32-bit alternate function code registers:

    Source function code (SFC)

    Destination function code (DFC)

- A 32-bit cache control register (CACR)
- A 32-bit user root pointer (URP)
- A 32-bit supervisor root pointer (SRP)
- A 16-bit translation control register (TC)
- Two 32-bit data transparent translation registers (DTT0, DTT1)
- Two 32-bit instruction transparent translation registers (ITT0, ITT1)
- A 32-bit MMU status register (MMUSR)

Registers URP, SRP, TC, DTT0, DTT1, ITT0, ITT1, and MMUSR will be discussed in Sec. 11.6 on memory management. Register CACR will be discussed in Sec. 11.7 on caches. Registers ISP, MSP, and VBR will be discussed in Sec. 11.8 on exception processing.

The SR, which stores the processor status, is shown in Fig. 11.8. The SR bits are the following:

Carry (C), bit 0, set on a carry out

Overflow (V), bit 1, set on an overflow

Zero (Z), bit 2, set on a zero result

Negative (N), bit 3, set on a negative result (sign bit = 1)

Extend (X), bit 4, set on a carry out, used for multiword operations

Interrupt priority mask (I2, I1, I0), bits 10, 9, 8, indicates the level of the interrupt (000 to 111; 0 to 7) currently being handled

Master/interrupt state (M), bit 12, indicates the current use of the supervisor stack pointer:

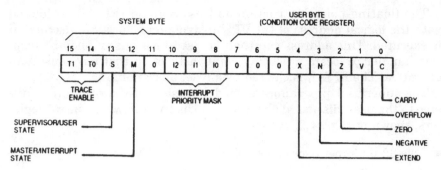

Figure 11.8 Status register. (*Courtesy of Motorola, Inc.*)

M	SP
0	ISP
1	MSP

Supervisor/user state (S), bit 13, indicates the currently running mode of operation:

S	Mode
0	User
1	Supervisor

Trace enable (T1, T0), bits 15, 14, indicates the tracing options:

T1	T2	Tracing option
0	0	No tracing
0	1	Trace on change of flow
1	0	Trace on instruction execution
1	1	Reserved

Only the CCR byte is available in the user mode; it is referenced as CCR in user programs. The full SR can be accessed in the supervisor mode only.

The alternate function code registers SFC and DFC contain 3-bit function codes. Function codes are automatically generated by the processor to select address spaces for data and programs at the user and supervisor modes. The SFC and DFC registers are used by certain instructions to explicitly specify the function codes for operations.

11.3   Data Formats

The data types recognized by the M68000 architecture are summarized in Table 11.2. The following data formats are supported uniformly by all arithmetic operations:

B—byte integer, 8 bits

W—word integer, 16 bits

L—longword integer, 32 bits

S—single-precision real, 32 bits

D—double-precision real, 64 bits

X—extended-precision real, 80 bits

The three integer data formats that are common to both the integer unit and the FPU (byte, word, and longword) are the standard 2's-complement data formats defined in the M68000 family architecture. Whenever an integer is used in a floating-point operation, the integer is automatically converted by the FPU to an extended-precision floating-point number before being used. The M68000 architecture features the big-endian addressing of bytes within a longword, that is, the bytes are numbered from 0 (the most significant byte) to 3 (the least significant byte).

TABLE 11.2   Data Types

Operand data type	Size, bits	Supported byte	Notes
Bit	1	IU	—
Bit field	1–32	IU	Field of consecutive bit
BCD	8	IU	Packed: 2 digits byte Unpacked: 1 digit byte
Byte integer	8	IU, FPU	—
Word integer	16	IU, FPU	—
Longword integer	32	IU, FPU	—
Quadword integer	64	IU	Any two data registers
16-Byte	128	IU	Memory-only, aligned to 16-byte boundary
Single-precision real	32	FPU	1-bit sign, 8-bit exponent, 23-bit mantissa
Double-precision real	64	FPU	1-bit sign, 11-bit exponent, 52-bit mantissa
Extended-precision real	80	FPU	1-bit sign, 15-bit exponent, 64-bit mantissa

Note: IU = integer unit.
SOURCE: Courtesy of Motorola, Inc.

Single- and double-precision floating-point data formats are implemented in the FPU as defined by the IEEE 754-1985 standard [IEEE 85]. These data formats are used for most calculations with real numbers. The extended-precision data format is also in conformance with the IEEE standard, but the standard does not specify this format to the bit level as it does for single and double precision. The memory format for the FPU consists of 96 bits (three longwords). Only 80 bits are actually used; the other 16 bits are for future expansibility and for longword alignment of the floating-point data structures in memory. The extended-precision format has a 15-bit exponent, a 64-bit mantissa, and a single-bit mantissa sign (see Chap. 3). Extended-precision numbers are intended for use as temporary variables, intermediate values, or where extra precision is needed.

11.4   Addressing Modes

The M68000 architecture addressing modes are summarized in Table 11.3. The addressing modes categories, along with the appropriate binary encodings, are listed in Table 11.4. The addressing mode categories are:

Data: refers to data operands

Memory: refers to memory operands

Alterable: refers to alterable (writable) operands

Control: refers to memory operands without an associated size

These categories are sometimes combined, forming new categories that are more restrictive. Two combined classifications are alterable memory or data alterable. The former refers to those addressing modes that are both alterable and memory addresses, and the latter refers to addressing modes that are both data and alterable.

The 18 addressing modes of the M68000 family will now be discussed in detail [Clem 94, M680 89, MC40 89, Skin 88, Wake 89].

Register direct modes

1. Data register direct, Dn: the operand is in the specified data register D0 to D7.

2. Address register direct, An: the operand is in the specified address register A0 to A7.

Register indirect modes

3. Address register indirect (ARI), (An): the specified address register An contains the address of the operand in memory.

TABLE 11.3   Addressing Modes

Addressing modes	Syntax
Register direct	
Data register direct	Dn
Address register direct	An
Register indirect	
Address register indirect	(An)
Address register indirect with postincrement	(An)+
Address register indirect with predecrement	− (An)
Address register indirect with displacement	(d_{16},An)
Register indirect with index	
Address register indirect with index (8-bit displacement)	$d_8,An,Xn)$
Address register indirect with index (base displacement)	(bd,An,Xn)
Memory indirect	
Memory indirect postindexed	([bd,An],Xn,od)
Memory indirect preindexed	([bd,An,Xn],od)
Program counter indirect with displacement	(d_{16},PC)
Program counter indirect with index	
PC indirect with index (8-bit displacement)	(d_8,PC,Xn)
PC indirect with index (base displacement)	(bd,PC,Xn)
Program counter memory indirect	
PC memory indirect postindexed	([bd,PC],Xn,od)
PC memory indirect preindexed	([bd,PC,Xn],oc)
Absolute	
Absolute short	xxx.W
Absolute long	xxx.L
Immediate	# data

Notes:

An	Address register, A7–A0.
bd	A 2's-complement base displacement; when present, size can be 16 or 32 bits.
data	Immediate value of 8, 16, or 32 bits.
Dn	Data register, D7–D0.
d_8, d_{16}	A 2's-complement or sign-extended displacement; added as part of the effective address calculation; size is 8 (d_8) or 16 (d_{16}) bits; when omitted, assemblers use a value of zero.
od	Outer displacement, added as part of effective address calculation after any memory indirection; use is optional with size of 16 or 32 bits.
PC	Program counter.
Xn	Address or data register used as an index register; form is Xn.SIZE SCALE, where SIZE is .W or .L (indicates index register size) and SCALE is 1, 2, 4, or 8 (index register is multiplied by SCALE); use of SIZE and or SCALE is optional.
( )	Effective address.
[ ]	Used as indirect access to longword address.

SOURCE: Courtesy of Motorola, Inc.

4. ARI with postincrement, (An)+: As in the ARI mode (3), the specified An register contains the address of the operand in memory. After the address in An is used to fetch the operand, the content of An is incremented by the size of the operand in bytes: by 1 for a byte, by 2 for a word, and by 4 for a longword operation.

TABLE 11.4 Effective Addressing Mode Categories

Address Modes	Mode Field	Register	Data	Memory	Control	Alterable	Assembler syntax
Data register direct	000	Reg. no.	X	—	—	X	Dn
Address register direct	001	Reg. no.	—	—	—	X	An
Address register indirect	010	Reg. no.	X	X	X	X	(An)
Address register indirect with postincrement	011	Reg. no.	X	X	—	X	(An) +
Address register indirect with predecrement	100	Reg. no.	X	X	—	X	– (An)
Address register indirect with displacement	101	Reg. no.	X	X	X	X	(d_{16},An)
Address register indirect with index (8-bit displacement)	110	Reg. no.	X	X	X	X	(d_8,An,Xn)
Address register indirect with index (base displacement)	110	Reg. no.	X	X	X	X	(bd,An,Xn)
Memory indirect postindexed	110	Reg. no.	X	X	X	X	([bd,An],Xn,od)
Memory indirect preindexed	110	Reg. no.	X	X	X	X	([bd,An,Xn],od)
Absolute short	111	000	X	X	X	X	(xxx).W
Absolute long	111	001	X	X	X	X	(xxx).L
Program counter indirect with displacement	111	010	X	X	X	—	(d_{16},PC)
Program counter indirect with index (8-bit displacement)	111	011	X	X	X	—	(d_8,PC,Xn)
Program counter indirect with index (base displacement)	111	011	X	X	X	—	(bd,PC,Xn)
PC memory indirect postindexed	111	011	X	X	X	—	([bd,PC],Xn,od)
PC memory indirect preindexed	111	011	X	X	X	—	([bd,PC,Xn],od)
Immediate	111	100	X	X	—	—	#<data>

SOURCE: Courtesy of Motorola, Inc.

Example In a word operation the operand address in memory is specified by (A1)+. The content of A1 before the operation is $2008 ($ denotes hexadecimal for the M68000 family). The word operand is fetched from the memory location at the address $2008. After that, the content of A1 is incremented by 2, and the content of A1 becomes $200A.

5. ARI with predecrement, − (An): As in ARI mode (3), the specified An register will contain the address of the operand in memory. However, before the content of An is used as an address, it is decremented by the size of the operand in bytes.

Example In a longword operation the operand address in memory is specified by − (A5). The content of A5 before the operation is $30016. At the beginning of the execution the content of A5 is decremented by 4, becoming $30012. The operand is fetched from memory at address $30012.

6. ARI with displacement, (d16,An): The EA of the operand in memory is the sum of the content of An and the sign-extended (to 32 bits) 16-bit displacement d16:

```
EA = (An) + d16
```

Example An operand is specified by ($100, A2). The content of A2 is $32004. The operand EA is $32004+$100 = $32104.

Register indirect with index modes

7. ARI with index and 8-bit displacement, (d8, An, Xn): The EA of the operand in memory is the sum of the content of An, the content of the index register Xn, and the sign-extended (to 32 bits) 8-bit displacement d8:

```
EA = (An) + (Xn) + d8
```

Any data or address register can be used as an index register. The index register indicator includes the size and scale information:

```
Xn.SIZE*SCALE
```

SIZE can be W for word or L for longword. SCALE can be 1, 2, 4, or 8. By default, Xn will be taken to be the whole 32-bit Xn register.

Example An operand is specified by ($10, A0, D0*2). (A0) = $10000, (D0) = $8. The EA is

```
EA = (A0) + (D0)*2 + $10 = $10000 + $8*2 +$10 = $10020
```

8. ARI with index and base displacement, (bd, An, Xn): Similarly to mode (7),

```
EA = (An) + (Xn) + bd
```

where bd is a 16-bit sign-extended or 32-bit base displacement.

Example   An operand is specified by ($200A, A2, A4*4). (A2) = $2000, (A4) = $4. The operand EA is

```
EA = (A2) + (A4)*4 + $200A = $2000 + $4*4 + $200A = $401A
```

Memory indirect modes

9. Memory indirect postindexed, ([bd, An], Xn, od): The operand and its address are stored in memory. The processor calculates an intermediate indirect memory address equal to the sum of the contents of An and bd. The processor then accesses a longword at this address and adds to it the index operand Xn.SIZE*SCALE and the outer displacement (od) to yield the EA. Both bd and od and the Xn contents are sign-extended to 32 bits. The EA can be expressed as follows:

```
EA = (bd + (An)) + (Xn)*SCALE + od
```

Example   An operand is specified by ([$50, A0], D1, $6). (A0) = $10000, (D1) = $22000. The intermediate address calculated by the processor is

```
(A0) + $50 = $10000 + $50 = $10050
```

The 32-bit value at location $10050 is $120000. The EA is thus:

```
EA = $120000 + (D1) + $6 = $120000 + $22000 + $6 = $142006
```

10. Memory indirect preindexed, ([bd, An, Xn], od): In this case,

```
EA = (bd + (An) + (Xn)*SCALE) + od
```

The index register is used in the initial calculation of the intermediate address, where a 32-bit value is stored, to be added to od to obtain the operand EA. This is why this mode is called preindexed. In the postindexed mode, mode 9, the index register value is used after the intermediate address was calculated.

Example   An operand is specified by ([A6, D6], $100). (A6) = $5000, (D6) = $2000. The intermediate memory address is $5000+$2000 = $7000. The 32-bit value stored at address $7000 is $3000. Thus,

```
EA = $3000 + $100 = $3100
```

PC indirect modes

11. PC indirect with displacement, (d16, PC): In analogy to mode 6, with PC replacing An:

```
EA = (PC) + d16
```

The value in the PC is the address of the extension word, the second word of the instruction.

12. PC indirect with index and 8-bit displacement, (d8, PC, Xn): In analogy to mode 7, with PC replacing An:

```
EA = (PC) + (Xn) + d8
```

13. PC indirect with index and base displacement, (bd, PC, Xn): In analogy to mode 8, with PC replacing An:

```
EA = (PC) + (Xn) + bd
```

PC memory indirect modes

14. PC memory indirect postindexed, ([bd, PC], Xn, od): In analogy to mode 9, with PC replacing An:

```
EA = (bd + (PC)) + (Xn) + od
```

15. PC memory indirect preindexed, ([bd, PC, Xn], od): In analogy to mode 10, with PC replacing An:

```
EA = (bd + (PC) + (Xn)) + od
```

In all the PC modes 11 to 15, the rules governing bd, od, and Xn are the same as in modes 7 to 10.

Absolute modes

16. Absolute short, (xxx).W: The address of the operand is in the extension word (second word) of the instruction. This mode saves a word in the program memory space if the address is short enough to be contained in 16 bits.

17. Absolute long, (xxx).L: The address of the operand is in two extension words (32-bit address) of the instruction.

Immediate mode

18. Immediate, #xxx: The operand is a part of the instruction, located in one or two extension words, depending on its size (16 or 32 bits).

The M68000 architecture EA specification formats are shown in Table 11.5. All abbreviations and notations are explained in the table. The first word of the instruction is called the operation word, or opword. Its most significant bits contain the instruction opcode. The subsequent words of the instruction are called extension words. The M68000 architecture instructions may have up to 10 extension words.

It should be noted that the 16-bit members of the M68000 family (see Chap. 13) have a reduced set of addressing modes. The scaled index and memory indirect features are not available. The displacement can be only 16-bit [Clem 94, Wake 89].

11.5 Instruction Set and Assembly Directives

One of the main decisions of the design team, when the MC68040 was added to the M68000 family [Eden 90], was to optimize the instruc-

TABLE 11.5 Effective Address Specificaton Format

Single Effective Address Instruction Format

15	14	13	12	11	10	9	8	7	6	5	4	3	2	1	0
X	X	X	X	X	X	X	X	X	X	EFFECTIVE ADDRESS					
										MODE			REGISTER		

Brief Format Extension Word

15	14	13	12	11	10	9	8	7	6	5	4	3	2	1	0
D/A	REGISTER			W/L	SCALE		0	DISPLACEMENT							

Full Format Extension Word(s)

15	14	13	12	11	10	9	8	7	6	5	4	3	2	1	0
D/A	REGISTER			W/L	SCALE		1	BS	IS	BD SIZE		0	I/IS		
BASE DISPLACEMENT (0, 1, OR 2 WORDS)															
OUTER DISPLACEMENT (0, 1, OR 2 WORDS)															

Field	Definition	Field	Definition
Instruction		BS	Base Register Suppress:
Register	General Register Number		0 Base Register Added
Extensions			1 Base Register Suppressed
Register	Index Register Number	IS	Index Suppress.
D/A	Index Register Type		0 Evaluate and Add Index Operand
	0 Dn		1 Suppress Index Operand
	1 An	BD SIZE	Base Displacement Size:
W/L	Word Long Word Index Size		00 Reserved
	0 Sign Extended Word		01 Null Displacement
	1 Long Word		10 Word Displacement
Scale	Scale Factor		11 Long Displacement
	00 1	I/IS	Index Indirect Selection.
	01 2		Indirect and Indexing Operand Deter-
	10 4		mined in Conjunction with Bit 6, Index
	11 8		Suppress

SOURCE: Courtesy of Motorola, Inc.

tions that executed most frequently and to allow other instructions to be implemented conveniently. A reasonable subset of the M68000 instruction set was selected, based on trace data and user input. This instruction subset was examined, and it was discovered that most of the instructions would execute in one clock cycle if the IU was pipelined. This is indeed the case on the MC68040 with its six-stage pipeline (see Chap. 12). Furthermore, the IU's ALU cycle was matched to the cache access time. This allowed the control unit to access each cache once per clock cycle and fixed the peak instruction execution rate to the ALU cycle rate. Thus, effective, optimized execution of most instructions was attained.

The M68000 architecture instruction set is briefly summarized in Table 11.6. A more detailed tabulation of the M68000 instructions is given in App. 3.A. The instruction set includes the following types of operation:

Data movement

Integer arithmetic

Floating-point arithmetic

Logical

Shift and rotate

Bit manipulation

Bit-field manipulation

BCD arithmetic

Program control

System control

Memory management

Cache maintenance

Multiprocessor communications

The above groups of instructions will be discussed in more detail in this section, presenting each group by a separate table. The notation used in the instruction tables is summarized in Table 11.7.

The data movement instructions are summarized in Table 11.8. MOVE instructions transfer byte, word, and longword operands, memory-to-memory, memory-to-register, register-to-memory, and register-to-register.

Examples

MOVE.B D0,D1   Move byte from D0 to D1, upper 24 bits of the registers are not affected.

TABLE 11.6   Instruction Set Summary

Mnemonic	Description
ABCD	Add Decimal with Extend
ADD	Add
ADDA	Add Address
ADDI	Add Immediate
ADDQ	Add Quick
ADDX	Add with Extend
AND	Logical AND
ANDI	Logical AND Immediate
ASL, ASR	Arithmetic Shift Left and Right
Bcc	Branch Conditionally
BCHG	Test Bit and Change
BCLR	Test Bit and Clear
BFCHG	Test Bit Field and Change
BFCLR	Test Bit Field and Clear
BFEXTS	Signed Bit Field Extract
BFEXTU	Unsigned Bit Field Extract
BFFFO	Bit Field Find First One
BFINS	Bit Field Insert
BFSET	Test Bit Field and Set
BFTST	Test Bit Field
BRA	Branch
BSET	Test Bit and Set
BSR	Branch to Subroutine
BTST	Test Bit
CAS	Compare and Swap Operands
CAS2	Compare and Swap Dual Operands
CHK	Check Register Against Bounds
CHK2	Check Register Against Upper and Lower Bounds
*CINV	Invalidate Cache Entries
CLR	Clear
CMP	Compare
CMPA	Compare Address
CMPI	Compare Immediate
CMPM	Compare Memory to Memory
CMP2	Compare Register Against Upper and Lower Bounds
*CPUSH	Push then Invalidate Cache Entries
DBcc	Test Condition, Decrement and Branch
DIVS, DIVSL	Signed Divide
DIVU, DIVUL	Unsigned Divide
EOR	Logical Exclusive OR
EORI	Logical Exclusive OR Immediate
EXG	Exchange Registers
EXT, EXTB	Sign Extend
*FABS	Floating-Point Absolute Value
*FADD	Floating-Point Add
FBcc	Floating-Point Branch
FCMP	Floating-Point Compare
FDBcc	Floating-Point Decrement and Branch
*FDIV	Floating-Point Divide
*FMOVE	Move Floating-Point Register
FMOVEM	Move Multiple Floating-Point Registers
*FMUL	Floating-Point Multiply
*FNEG	Floating-Point Negate
FRESTORE	Restore Floating-Point Internal State
FSAVE	Save Floating-Point Internal State
FScc	Floating-Point Set According to Condition
*FSQRT	Floating-Point Square Root

Mnemonic	Description
*FSUB	Floating-Point Subtract
FTRAPcc	Floating-Point Trap-On Condition
FTST	Floating-Point Test
ILLEGAL	Take Illegal Instruction Trap
JMP	Jump
JSR	Jump to Subroutine
LEA	Load Effective Address
LINK	Link and Allocate
LSL, LSR	Logical Shift Left and Right
MOVE	Move
*MOVE16	16-Byte Block Move
MOVEA	Move Address
MOVE CCR	Move Condition Code Register
MOVE SR	Move Status Register
MOVE USP	Move User Stack Pointer
*MOVEC	Move Control Register
MOVEM	Move Multiple Registers
MOVEP	Move Peripheral
MOVEQ	Move Quick
*MOVES	Move Alternate Address Space
MULS	Signed Multiply
MULU	Unsigned Multiply
NBCD	Negate Decimal with Extend
NEG	Negate
NEGX	Negate with Extend
NOP	No Operation
NOT	Logical Complement
OR	Logical Inclusive OR
ORI	Logical Inclusive OR Immediate
PACK	Pack BCD
PEA	Push Effective Address
*PFLUSH	Flush Entry(ies) in the ATCs
*PTEST	Test a Logical Address
RESET	Reset External Devices
ROL, ROR	Rotate Left and Right
ROXL, RORX	Rotate with Extend Left and Right
RTD	Return and Deallocate
RTE	Return from Exception
RTR	Return and Restore Codes
RTS	Return from Subroutine
SBCD	Subtract Decimal with Extend
Scc	Set Conditionally
STOP	Stop
SUB	Subtract
SUBA	Subtract Address
SUBI	Subtract Immediate
SUBQ	Subtract Quick
SUBX	Subtract with Extend
SWAP	Swap Register Words
TAS	Test Operand and Set
TRAP	Trap
TRAPcc	Trap Conditionally
TRAPV	Tap on Overflow
TST	Test Operand
UNLK	Unlink
UNPK	Unpack BCD

*MC68040 additions or alterations to the MC68030 and MC68881 M68882 instruction set.

SOURCE: Courtesy of Motorola, Inc.

TABLE 11.7   Instruction Table Notation

The following notations are used in this section. In the operand syntax statements of the instruction definitions, the operand on the right is the destination operand.

An = any address register, A7–A0
Dn = any data register, D7–D0
Rn = any address or data register
CCR = condition code register (lower byte of status register)
cc = condition codes from CCR
SR = status register
SP = active stack pointer
USP = user stack pointer
ISP = supervisor/interrupt stack pointer
MSP = supervisor/master stack pointer
SSP = supervisor (master or interrupt) stack pointer
DFC = destination function code register
SFC = source function code register
Rc = control register (VBR, SFC, DFC, CACR)
MRc = MMU control register (SRP, URP, TC, DTT0, DTT1, ITT0, ITT1, MMUSR)
MMUSR = MMU status register
B, W, L = specifies a signed integer data type (twos complement) of byte, word, or long word
S = single precision real data format (32 bits)
D = double precision real data format (64 bits)
X = extended precision real data format (96 bits, 16 bits unused)
P = packed BCD real data format (96 bits, 12 bytes)
FPm, FPn = any floating-point data register FP7–FP0
FPcr = floating-point system control register (FPCR, FPSR, or FPIAR)
k = a twos complement signed integer (– 64 to + 17) that specifies the format of a number to be stored in the packed decimal format
d = displacement; d_{16} is a 16-bit displacement
< ea > = effective address
list = list of registers, for example D3–D0
# < data > = immediate data; a literal integer
{offset:width} = bit field selection
label = assemble program label
[m] = bit m of an operand
[m:n] = bits m through n of operand
X = extend (X) bit in CCR
N = negative (N) bit in CCR
Z = Zero (Z) bit in CCR
V = overflow (V) bit in CCR
C = carry (C) bit in CCR
+ = arithmetic addition or postincrement indicator
– = arithmetic subtraction or predecrement indicator
× = arithmetic multiplication
÷ = arithmetic division or conjunction symbol
~ = invert; operand is logically complemented
\ = logical AND
V = logical OR
⊕ = logical exclusive OR
Dc = data register, D7–D0 used during compare
Du = data register, D7–D0 used during update
Dr, Dq = data registers, remainder or quotient of divide
Dh, Dl = data registers, high or low order 32 bits of product
MSW = most significant word
LSW = least significant word
MSB = most significant bit
FC = function code
{R W} = read or write indicator
[An] = address extensions

SOURCE: Courtesy of Motorola, Inc.

TABLE 11.8   Data Movement Operations

Instruction	Operand syntax	Operand size	Operation
EXG	Rn,Rn	32	Rn ↔ Rn
FMOVE	FPm,FPn	X	Source → destination
	ea,FPn	B,W,L,S,D,X,P	
	FPm,ea	B,W,L,S,D,X,P	
	ea,FPcr	32	
	FPcr,ea	32	
FSMOVE,	FPm,FPn	X	Source → destination, round destination
FDMOVE	ea,FPn	B,W,L,S,D,X	to single- or double-precision
FMOVE M	ea,list[1]	32,X	Listed registers → destination
	ea,Dn	X	
	list[1],ea	32,X	Source → listed registers
	Dn,ea	X	
LEA	ea,An	32	ea → An
LINK	An,# d	16, 32	Sp-4 → SP, An → (SP), SP → An, SP + D → SP
MOVE	ea,ea	8, 16, 32	Source → destination
MOVE 16	ea,ea	16 bytes	aligned 16-byte block → destination
MOVE A	ea,An	$16,32 \rightarrow 32$	
MOVE M	list, ea	16,32	Listed registers → destination
	ea,list	$16,32 \rightarrow 32$	source → listed registers
MOVE P	Dn, (d_{16},An)	16,32	Dn[31:24] → (An + d), Dn[23:16] → An + d + 2);
	$(d_{16},An),Dn$		Dn[15:8] → (An + d + 4), Dn[7:0] → (An + d + 6)
			(An + d) → Dn[31:24], (An + d + 2) → Dn[23:16];
			(An + d + 4) → Dn[15:8], (An + d + 6) → Dn[7:0]
MOVE Q	# data,Dn	$8 \rightarrow 32$	Immediate data → destination
PEA	ea	32	SP-4 → SP; <ea> → (SP)
UNLK	An	32	An → SP, (SP) → An, SP + 4 → SP

Note: The register list may include any combination of the eight floating-point data registers, or it may contain any combination of the three control registers (FPCR, FPSR, and FPIAR). If the register list mask resides in a data register, only floating-point data registers may be specified.

SOURCE: Courtesy of Motorola, Inc.

MOVE.W A1,A5	Move word, upper 16 bits are not affected.
MOVE A1,A5	Move word (by default).
MOVE.L (A0),D1	Move a longword from a memory location, pointed to by A0, to D1.
MOVE (A1),(A2)	Move word memory-to-memory.

The FMOVE instructions move operands into, between, and from the floating-point data registers. Data format conversion functions for the FPU instructions are implicitly supported. For operands moved into a floating-point data register, FSMOVE and FDMOVE explicitly select single and double-precision rounding of the result, respectively. FMOVEM moves any combination of either floating-point data registers or floating-point control registers.

The integer arithmetic instructions are summarized in Table 11.9. Instructions ADDX, SUBX, EXT, and NEGX, which use the X flag (see Sec. 11.2), provide for multiprecision and mixed-size arithmetic [M680 89, MC40 89].

TABLE 11.9   Integer Arithmetic Operations

Instruction	Operand syntax	Operand size	Operation
ADD	Dn,<ea>	8, 16, 32	Source + destination → destination
	<ea>,Dn	8, 16, 32	
ADDA	<ea>,An	16, 32	
ADDI	#<data>,<ea>	8, 16, 32	Immediate data + destination → destination
ADDQ	#<data>,<ea>	8, 16, 32	
ADDX	Dn,Dn	8, 16, 32	Source + destination + X → destination
	(An),(An)	8, 16, 32	
CLR	<ea>	8, 16, 32	0 → destination
CMP	<ea>,Dn	8, 16, 32	Destination − source
CMPA	<ea>,An	16, 32	
CMPI	#(data),<ea>	8, 16, 32	Destination − immediate data
CMPM	(An)+,(An)+	8, 16, 32	Destination − source
CMP2	<ea>,Rn	8, 16, 32	Lower bound ≤ Rn ≤ upper bound
DIVS DIVU	<ea>,Dn	32:16 → 16:16	Destination:source → destination (signed or unsigned)
	<ea>,Dr:Dq	64:32 → 32:32	
	<ea>,Dq	32:32 → 32	
DIVSL DIVUL	<ea>,Dr:Dq	32:32 → 32:32	
EXT	Dn	8 → 16	Sign extended destination → destination
	Dn	16 → 32	
EXTB	Dn	8 → 32	
MULS MULU	<ea>,Dn	16 × 16 → 32	Source × destination → destination (signed or unsigned)
	<ea>,Dl	32 × 32 → 32	
	<ea>,Dh Dl	32 × 32 → 64	
NEG	<ea>	8, 16, 32	0 − destination → destination
NEGX	<ea>	8, 16, 32	0 − destination − X → destination
SUB	<ea>,Dn	8, 16, 32	Destination − source → destination
	Dn,<ea>	8, 16, 32	
SUBA	<ea>,An	16, 32	
SUBI	#<data>,<ea>	8, 16, 32	Destination − immediate data → destination
SUBQ	#<data>,<ea>	8, 16, 32	
SUBX	Dn,Dn	8, 16, 32	Destination − source − X → destination
	−(An),−(An)	8, 16, 32	

SOURCE: Courtesy of Motorola, Inc.

Examples

```
MULS.L D0,D4:D5; signed multiply, (D0)*(D5) → (D4,D5)
DIVS.L D0,D1:D2; signed divide, (D1,D2)/(D0), quotient → D2, remain-
der → D1
```

The floating-point arithmetic instructions are an enhanced subset of the MC68881/MC68882 coprocessor instructions, used in conjunction with M68000 family members up to MC68030 (starting with the MC68040 the FPU is on the same chip with the CPU). There are two basic types of floating-point instructions:

Dyadic—two operands

Monadic—single operand

For dyadic operations the source can be in memory, in an integer data register Dn, or in a floating-point data register FPm. The destination is always a floating-point data register FPm (also the second operand). All operations support any data format (see Sec. 11.3). The general form of a dyadic operation is:

```
Fdop ea,FPm; (source) function (FPm) → FPm
```

where Fdop is any one of the dyadic operation specifiers (such as, for the FADD instruction, dop = ADD).

The dyadic operations are (see also App. 3.A):

```
FADD, FSADD, FDADD        Add
FCMP                      Compare
FDIV, FSDIV, FDDIV        Divide
FMUL, FSMUL, FDMUL        Multiply
FSUB, FSSUB, FDSUB        Subtract
```

The "S" and "D" after "F" (wherever applicable) specify the rounding precision of the result, S for single- and D for double-precision. Otherwise, the result is rounded to the extended precision.

The monadic operations perform a function on a single operand, located in a source location, and store the result in a destination location. The source location may be anywhere as in the dyadic operations; however, the destination location must be in an FPm register. A single FPm register may be specified as an operand. The monadic operation formats are:

```
Fmop ea,FPm; function(ea) → FPm
Fmop FPm; function(FPm) → FPm
```

where Fmop is any one of the monadic operation specifiers (such as in the FABS instruction, mop = ABS). When the source is not an FPm register, all data formats are supported. The data format is always extended precision for register-to-register operations.

The monadic operations are:

```
FABS, FSABS, FDABS          Absolute value
FNEG, FSNEG, FDNEG          Negate
FSQRT, FSSQRT, FDSQRT       Square root
```

The "S" and "D" after "F" have the same meaning as in the dyadic operations.

The logical instructions are summarized in Table 11.10, and the shift and rotate instructions are listed in Table 11.11. The arithmetic shift instructions (ASR and ASL) and logical shift instructions (LSR and LSL) provide shift operations in both directions. The ROR, ROL, ROXR, and ROXL instructions perform rotate operations, without or with the extend bit, respectively. All shift and rotate operations can be performed on either registers or memory. Register shift and rotate operations handle all operand sizes. The shift count may be specified in the instruction opword (to shift from 1 to 8 bits) or in a register (modulo 64 shift count).

Memory shift and rotate operations handle word-length operands by one bit only. The SWAP instruction exchanges the 16-bit halves of a register.

Bit manipulation operations are accomplished using the following instructions: bit test (BTST), bit test and set (BSET), bit test and clear (BCLR), and bit test and change (BCHG). All bit manipulation operations can be performed on either registers or memory. The bit number is specified as immediate data or in a register. Register operands are

TABLE 11.10   Logical Operations

Instruction	Operand syntax	Operand size	Operation
AND	<ea>,Dn	8, 16, 32	Source $\wedge$ destination → destination
	Dn,<ea>	8, 16, 32	
ANDI	# data , ea	8, 16, 32	Immediate data $\wedge$ destination → destination
EOR	Dn, data , ea	8, 16, 32	Source $\oplus$ destination → destination
EORI	#<data>,<ea>	8, 16, 32	Immediate data $\oplus$ destination → destination
NOT	<ea>	8, 16, 32	Destination → destination
OR	<ea>,Dn	8, 16, 32	Source $\vee$ destination → destination
	Dn,<ea>	8, 16, 32	
ORI	#<data>,<ea>	8, 16, 32	Immediate data $\vee$ destination → destination

SOURCE: Courtesy of Motorola, Inc.

TABLE 11.11   Shift and Rotate Operations

Instruction	Operand Syntax	Operand Size	Operation
ASL	Dn,Dn #⟨data⟩,Dn ⟨ea⟩	8, 16, 32 8, 16, 32 16	
ASR	Dn,Dn #⟨data⟩,Dn ⟨ea⟩	8, 16, 32 8, 16, 32 16	
LSL	Dn,Dn #⟨data⟩,Dn ⟨ea⟩	8, 16, 32 8, 16, 32 16	
LSR	Dn,Dn #⟨data⟩,Dn ⟨ea⟩	8, 16, 32 8, 16, 32 16	
ROL	Dn,Dn #⟨data⟩,Dn ⟨ea⟩	8, 16, 32 8, 16, 32 16	
ROR	Dn,Dn #⟨data⟩,Dn ⟨ea⟩	8, 16, 32 8, 16, 32 16	
ROXL	Dn,Dn #⟨data⟩,Dn ⟨ea⟩	8, 16, 32 8, 16, 32 16	
ROXR	Dn,Dn #⟨data⟩,Dn ⟨ea⟩	8, 16, 32 8, 16, 32 16	
SWAP	Dn	32	

SOURCE: Courtesy of Motorola, Inc.

32 bits long, and memory operands are 8 bits long. In Table 11.12, the summary of the bit manipulation operations, Z refers to the zero bit of the status register (see Sec. 11.2).

The M68000 architecture supports variable-length bit-field operations on fields of up to 32 bits. The bit-field insert (BFINS) instruction inserts a value into a bit field. Bit-field extract unsigned (BFEXTU) and bit-field extract signed (BFEXTS) extract a value from the field. Bit-field find first one (BFFFO) finds the first bit that is set in a bit field. Also included are instructions that are analogous to the bit manipulation operations: bit-field test (BFTST), bit-field test and set (BFSET), bit-field test and clear (BFCLR), and bit-field test and change (BFCHG). Table 11.13 is a summary of the bit-field operations.

TABLE 11.12   Bit Manipulation Operations

Instruction	Operand syntax	Operand size	Operation
BCHG	Dn,<ea>	8, 32	~(<bit number> of destination) → Z → bit of destination
	#<data>,<ea>	8, 32	
BCLR	Dn,<ea>	8, 32	~(<bit number> of destination) → Z;
	#<data>,<ea>	8, 32	0 → bit of destination
BSET	Dn,<ea>	8, 32	~(<bit number> of destination) → Z;
	#<data>,<ea>	8, 32	1 → bit of destination
BTST	Dn,<ea>	8, 32	~(<bit number> of destination) → Z
	#<data>,<ea>	8, 32	

SOURCE: Courtesy of Motorola, Inc.

TABLE 11.13   Bit-Field Operations

Instruction	Operand syntax	Operand size	Operation
BFCHG	<ea> I offset:width I	1–32	~Field → field
BFCLR	<ea> I offset:width I	1–32	0's → field
BFEXTS	<ea> I offset:width I ,Dn	1–32	Field → Dn; sign extended
BFEXTU	<ea> I offset:width I ,Dn	1–32	Field → Dn; zero extended
BFFFO	<ea> I offset:width I ,Dn	1–32	Scan for first bit set in field; offset → Dn
BFINS	Dn,<ea> I offset:width I	1–32	Dn → field
BFSET	<ea> I offset:width I	1–32	1's → field
BFTST	<ea> I offset:width I	1–32	Field MSB → N; ~(OR of all bits in field) → Z

Note: All bit-field instructions set the N and Z bits as shown for BFTST before performing the specified operation.

SOURCE: Courtesy of Motorola, Inc.

Five instructions support operations on BCD numbers. The arithmetic operations on packed BCD numbers are add decimal with extend (ABCD), subtract decimal with extend (SBCD), and negate decimal with extend (NBCD). PACK and UNPACK instructions aid in the conversion of byte-encoded numerical data, such as ASCII or EBCDIC strings, to BCD data and vice versa. Table 11.14 is a summary of the BCD operations.

A set of subroutine call and return instructions and conditional and unconditional branch instructions perform program control operations. Also included are test operand instructions (TST and FTST) that set the integer or floating-point condition codes for use by the other program and system control instructions, and a no-operation instruction (NOP) that may be used to force synchronization of the internal pipelines. Table 11.15 summarizes these instructions.

TABLE 11.14   Binary Coded Decimal Operations

Instruction	Operand syntax	Operand size	Operation
ABCD	Dn,Dn	8	$Source_{10} + destination_{10} + X \rightarrow destination$
	–(An), –(An)	8	
NBCD	<ea>	8	$0 - Destination_{10} - X \rightarrow destination$
PACK	–(An), –(An),#<data>	$16 \rightarrow 8$	Unpacked source + immediate data $\rightarrow$ packed destination
	Dn,Dn,#<data>	$16 \rightarrow 8$	
SBCD	Dn,Dn	8	$Destination_{10} - source_{10} - X \rightarrow destination$
	–(An), –(An)	8	
UNPK	–(An), –(An),#<data>	$8 \rightarrow 16$	Packed source $\rightarrow$ unpacked source
	Dn,Dn#<data>	$8 \rightarrow 16$	Unpacked source + immediate data $\rightarrow$ unpacked destination

SOURCE: Courtesy of Motorola, Inc.

The conditional mnemonics for the floating-point conditional instructions are shown in Table 11.16, along with the conditional test function. The FPU supports 32 conditional tests that are separated into two groups; 16 that cause an exception if an unordered condition is present when the conditional test is attempted, and 16 that do not cause an exception in such a case (an unordered condition occurs when an input to an arithmetic operation is a NaN, i.e., not a number).

Privileged instructions, trapping instructions, and instructions accessing the CCR provide system control operations, summarized in Table 11.17. FSAVE and FRESTORE instructions save and restore the non user visible portion of the FPU during context switches in a virtual memory (see Chap. 4) or multitasking system. The conditional trap instructions (TRAPcc, FTRAPcc) use the same conditional tests as their corresponding program control instructions (see Table 11.15) and allow an optional 16- or 32-bit immediate operand to be included as part of the instruction for passing parameters to the OS. All of these trap instructions cause the processor to flush the instruction pipeline.

The PFLUSH instructions flush the address translation caches (ATCs) and can optionally select only nonglobal entries for flushing. PTEST performs a search of the address translation tables, storing results in the MMUSR (see Sec. 11.6) and loading the entry into the ATC. Table 11.18 summarizes these MMU instructions.

The cache instructions, summarized in Table 11.19, provide maintenance function for managing the instruction and data caches. CINV invalidates cache entries in both caches, and CPUSH pushes dirty (modified) data from the data cache to update memory. Both instructions can operate on either or both caches, and can select a single cache line, all lines in a page, or the entire cache.

TABLE 11.15   Program Control Operations

Instruction	Operand syntax	Operand size	Operation
		Integer and Floating-Point Conditional	
Bcc, FBcc	Label	8, 16, 32	If condition true, then PC + d → PC
DBcc, FDBcc	Dn, label	16	If condition false, then Dn − 1 → Dn if Dn ≠ − 1, then PC + d → PC
Scc, FScc	EA	8	If condition true, then 1's → destination; else 0's → destination
		Unconditional	
BRA	Label	8, 16, 32	PC + d → PC
BSR	Label	8, 16, 32	SP − 4 → SP; PC → (SP); PC + d → PC
JMP	EA	None	Destination → PC
JSR	EA	None	SP − 4 → SP; PC → (SP); destination → PC
NOP	None	None	PC + 2 → PC
FNOP	None	None	PC + 4 → PC
		Returns	
RTD	# d	16	(SP) → PC; SP + 4+d → SP
RTR	None	None	(SP) → CCR; SP + 2 → SP; (SP) → PC; SP + 4 → SP
RTS	None	None	(SP) → PC; SP + 4 → SP
		Test Operand	
TST	EA	8, 16, 32	Set integer condition codes
FTST	EA FPn	B,W,L,S,D,X,P X	Set floating-point condition codes

Letters cc in the integer instruction mnemonics Bcc, DBcc, and Scc specify testing one of the following conditions:

CC—Carry clear	LS—Lower or same
CS—Carry set	LT—Less than
EQ—Equal	MI—minus
F—Never true*	NE—Not equal
GE—Greater or equal	PL—Plus
GT—Greater than	T—Always true*
HI—Higher	VC—Overflow clear
LE—Less or equal	VS—overflow set

*Not applicable to the Bcc instructions.

SOURCE: Courtesy of Motorola, Inc.

The TAS, CAS, and CAS2 instructions coordinate the operations of processors in multiprocessing systems (see also Sec. 3.2). These instructions use read-modify-write (RMW) bus cycles to ensure uninterrupted updating of memory. Table 11.20 lists these instructions.

Table 11.21 lists the condition names, encodings, and tests for the conditional branch and test instructions. The test associated with each condition is a logical formula using the current states of the condition

TABLE 11.16   FPU Conditional Test Mnemonics

	Exception on unordered		No exception on unordered
GE	Greater than or equal to	OGE	Ordered greater than or equal to
GL	Greater than or less than	OGL	Ordered greater than or less than
GLE	Greater than or less	OR	Ordered
GT	Greater than	OGT	Ordered greater than
LE	Less than or equal to	OLE	Ordered less than or equal to
LT	Less than	OLT	Ordered less than
NGE	Not (greater than or equal to)	UGE	Unordered or greater than equal to
NGL	Not (greater than or less than)	UEQ	Unordered or equal to
NGLE	Not (greater than or less than or equal to)	UN	Unordered
NGT	Not greater than	UGT	Unordered or greater than
NLE	Not (less than or equal to)	ULE	Unordered or less than or equal to
NLT	Not less than	ULT	Unordered or less than
SEQ	Signaling equal to	EQ	Equal to
SNE	Signaling not equal to	NE	Not equal to
SF	Signaling always false	F	Always false
ST	Signaling always true	T	Always true

SOURCE: Courtesy of Motorola, Inc.

codes. If this formula evaluates to 1, the condition is true. If the formula evaluates to 0, the condition is false. For example, the T condition is always true, and the EQ condition is true only if the Z flag is set ($Z = 1$).

The immediate data in the M68000 architecture are identified by the symbol # in front of the number. The base is decimal by default. Otherwise, the base indicators in front of the number are [M680 89]:

```
%  Binary
@  Octal
$  Hexadecimal
```

Assembly directives

Definition of constants is done using the define constant (DC) directive:

```
LABEL: DC.SIZE LIST
```

The SIZE can be B (byte), W (word), or L (longword). It is W by default if omitted.

Examples

```
NUMBER: DC.L 100
ARRAY: DC.W $A, $B, $C, $D, $E, $F
```

TABLE 11.17   System Control Operations

Instruction	Operand syntax	Operand size	Operation
		Privileged	
ANDI	# data ,SR	16	Immediate data $\wedge$ SR $\rightarrow$ SR
EORI	# data ,SR	16	Immediate data $\oplus$ SR $\rightarrow$ SR
FRESTORE	EA	None	State frame $\rightarrow$ internal floating-point registers
FSAVE	EA	None	Internal floating-point registers $\rightarrow$ state frame
MOVE	EA ,SR	16	source $\rightarrow$ SR
	SR, EA	16	SR $\rightarrow$ destination
MOVE	USP,An	32	USP $\rightarrow$ An
	An,USP	32	An $\rightarrow$ USP
MOVEC	Rc,Rn	32	Rc $\rightarrow$ Rn
	Rn,Rc	32	Rn $\rightarrow$ Rc
MOVES	Rn, EA	8,16,32	Rn $\rightarrow$ destination using DFC
	EA, Rn		source using SFC $\rightarrow$ Rn
ORI	# data, SR	16	Immediate data $\vee$ SR $\rightarrow$ SR
RESET	None	None	Assert RSTO line
RTE	None	None	(SP) $\rightarrow$ SR; SP + 2 $\rightarrow$ SP, (SP) $\rightarrow$ PC, SP + 4 $\rightarrow$ SP. Restore stack according to format
STOP	# data	16	Immediate data $\rightarrow$ SR; STOP
		Trap Generating	
BKPT	# data	None	Run breakpoint cycle, then trap as illegal instruction
CHK	EA, Dn	16,32	If Dn <0 or Dn> <ea>, then CHK exception
CHK2	EA, Rn	8.16,32	If Rn <lower bound or Rn> upper bound, then CHK exception
ILLEGAL	None	None	SSP − 2 $\rightarrow$ SSP, Vector Offset $\rightarrow$ (SSP). SSP − 4 $\rightarrow$ SSP, PC $\rightarrow$ (SSP); SSP − 2 $\rightarrow$ SSP, SR $\rightarrow$ (SSP); Illegal Instruction Vector Address $\rightarrow$ PC
TRAP	# data	None	SSP − 2 $\rightarrow$ SSP; Format and Vector Offset $\rightarrow$ (SSP) SSP − 4 $\rightarrow$ SSP; PC $\rightarrow$ (SSP); SSP − 2 $\rightarrow$ SSP. SR $\rightarrow$ (SSP), Vector Address $\rightarrow$ PC
TRAPcc	None	None	If cc true, then TRAP exception
	# data	16,32	
FTRAPcc	None	None	If floating-point cc true, then TRAP exception
	# data	16,32	
TRAPV	None	None	If V then take overflow TRAP exception
		Condition Code Register	
ANDI	# data ,CCR	8	Immediate data $\wedge$ CCR $\rightarrow$ CCR
EORI	# data ,CCR	8	Immediate data $\oplus$ CCR $\rightarrow$ CCR
MOVE	EA ,CCR	16	source $\rightarrow$ CCR
	CCR ,EA	16	CCR $\rightarrow$ destination
ORI	# data ,CCR	8	Immediate data $\vee$ CCR $\rightarrow$ CCR

SOURCE: Courtesy of Motorola, Inc.

TABLE 11.18   MMU Instructions

Instruction	Operand syntax	Operand size	Operation
PFLUSHA	None	None	Invalidate all ATC entries
PFLUSHA.N	None	None	Invalidate all nonglobal ATC entries
PFLUSH	(An)	None	Invalidate ATC entries at effective address
FFLUSH.N	(An)	None	Invalidate nonglobal ATC entries at effective address
PTEST	(An)	None	Information about logical address → MMU status register

SOURCE: Courtesy of Motorola, Inc.

TABLE 11.19   Cache Instructions

Instruction	Operand syntax	Operand size	Operation
CINVL	Caches,(An)	None	Invalidate cache line
CINVP	Caches,(An)	None	Invalidate cache page
CINVA	Caches	None	Invalidate entire cache
CPUSHL	Caches,(An)	None	Push selected dirty data cache lines,
CPUSHP	Caches,(An)	None	then invalidate selected cache lines
CPUSHA	Caches	None	

SOURCE: Courtesy of Motorola, Inc.

TABLE 11.20   Multiprocessor Operations (Read-Modify-Write)

Instruction	Operand syntax	Operand size	Operation
CAS	Dc,Du, ea	8,16,32	Destination—Dc → CC; if Z then Du → destination else destination → Dc
CAS2	Dc1:Dc2, Du1:Du2, (Rn):(Rn)	8,16,32	Dual operand CAS
TAS	ea	8	Destination—0; set condition codes; 1 → destination [7]

SOURCE: Courtesy of Motorola, Inc.

If the value of some data is unknown in advance, but we would like to save space for them in memory, the define storage (DS) directive can be used:

```
LABEL: DS.SIZE ITEMS
```

Examples

```
ARRAY: DS.L 100; an array of 100 longwords or 400 bytes is saved
NUMBER: DS.W 1; a word (2 bytes) is saved
```

TABLE 11.21 Conditional Tests

Mnemonic	Condition	Encoding	Test
T*	True	0000	1
F*	False	0001	0
HI	High	0010	$\overline{C} \cdot \overline{Z}$
LS	Low or same	0011	$C + Z$
CC(HS)	Carry clear	0100	$\overline{C}$
CS(LO)	Carry set	0101	C
NE	Not equal to	0110	$\overline{Z}$
EQ	Equal to	0111	Z
VC	Overflow clear	1000	$\overline{V}$
VS	Overflow set	1001	V
PL	Plus	1010	$\overline{N}$
MI	Minus	1011	N
GE	Greater or equal to	1100	$N \cdot V + \overline{N} \cdot \overline{V}$
LT	Less than	1101	$N \cdot \overline{V} + \overline{N} \cdot V$
GT	Greater than	1110	$N \cdot V \cdot \overline{Z} + \overline{N} \cdot \overline{V} \cdot \overline{Z}$
LE	Less or equal to	1111	$Z + N \cdot \overline{V} + \overline{N} \cdot V$

• Boolean AND.
+ Boolean OR.
$\overline{N}$ Boolean NOT N.
*Not available for the Bcc instruction.

SOURCE: Courtesy of Motorola, Inc.

We can assign values to symbols using the equate (EQU) directive:

```
LABEL: EQU EXPRESSION
```

Examples

```
NUMBER: EQU 10
A: EQU 10
B: EQU A + 10; B assigned the value of 20
```

An assembly program must be terminated by the END directive.

11.6 Memory Management

One of the almost unique features of the M68000 architecture is that starting with the MC68040 it has an on-chip dual MMU, one for code and the other for data. As argued in Chap. 5, this permits parallel access of instructions and data by the CPU, enhancing the overall pipeline performance. This is even more important for the MC68060, which is a two-issue superscalar (see Chaps. 5 and 12). Each MMU contains an address translation cache (ATC), called TLB in other systems (see Chap. 4), in which recently used virtual to physical address translations are stored. If the translation pair is not in the ATC, the processor searches the translation tables in memory for the translation information. Each ATC has 64 entries and it is four-way set-associative.

The page size can be either 4 or 8 kbytes, selectable by software. In the earlier M68000 microprocessors one could select eight page sizes on the MMU coprocessor: 256 bytes, 512 bytes, or 1, 2, 4, 8, 16, or 32 kbytes.

All M68000 memories are byte-addressable, stored by a big-endian byte ordering. The address N of a longword data item corresponds to the address of its most significant byte. Its least significant byte address is $N+3$. Data alignment along word boundaries is not required, however, the most efficient data transfers occur when data is aligned on the same boundary as its operand size. Thus, words should be aligned on even address boundaries, and longwords on addresses divisible by 4, for better efficiency. Instruction words must be aligned on word boundaries at even addresses.

As mentioned before, when there is an ATC miss, the translation table mechanism must be invoked. The general translation table structure is a three-level tree structure, illustrated by an example in Fig. 11.9. The pointer tables in the upper levels (levels A and B) contain the base addresses of the tables at the next level. The page tables, at the lowest level (level C), contain either the physical address for the translation or a pointer to the memory location containing the address. Only a portion of the translation table for the entire logical (virtual) address space is required to be resident in memory at any time; specifically, only the portion of the table that translates the logical addresses of the currently executing process must be resident. Portions of translation tables can be dynamically allocated as the process requires additional memory.

The current privilege mode selects the SRP or URP for translation of the address (see Figs. 11.2 and 11.9). Each root pointer contains the base address of the first level table (level A) for a translation level tree. The base address for each table is indexed by a field extracted from the logical address. An example of a logical address, subdivided into table index fields, is shown at the top of Fig. 11.9. The table index A (TIA) field, which is 7 bits wide, is used to index into the first level (level A) pointer table and select 1 of 128 (27) pointer descriptors. At this level, each descriptor corresponds to a 32-mbyte block of memory and points to the base of a second-level table (level B).

Table index B (TIB) selects 1 of 128 pointer descriptors in the selected second-level table; each of these descriptors points to a page table in the third-level (level C) and corresponds to a 256-kbyte block of memory. Table index C (TIC) selects 1 of either 32 (for 8-kbyte pages) or 64 (for 4-kbyte pages) descriptors in the third-level page table. Descriptors in the page tables contain either a page descriptor

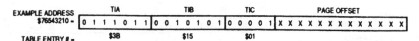

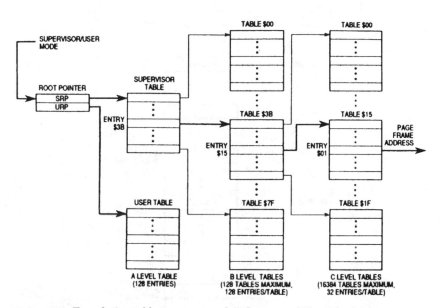

Figure 11.9 Translation table tree—example. (*Courtesy of Motorola, Inc.*)

for the translation or an indirect descriptor that points to a memory location containing the page descriptor. The page size, either 4 or 8 kbytes, is selected by a bit in the TC register (Fig. 11.2).

Figure 11.9 shows an example of an access to address $7654 3210 in supervisor mode with a page size of 8 kbytes. In this example the page offset field of the logical address has 13 bits (2^{13} = 8K) and the TIC field has 5 bits. For a 4-kbyte page, the page offset would have 12 bits and the TIC field would have 6 bits (bit 12 of the page offset would belong to the TIC field in this case). Figure 11.10 shows a possible layout of the example in Fig. 11.9 translation tree in memory.

After becoming familiar with the details of the translation tables, we are in a position to appreciate the considerations of the Motorola designers that led to the establishment of the translation table mechanism [Eden 90]. The main design goal in this case was to minimize the physical memory space overhead taken up by the tables at different levels (see Fig. 11.9). The designers have opted for a relatively small size table (compared to Intel; see Chap. 7) for the following reasons:

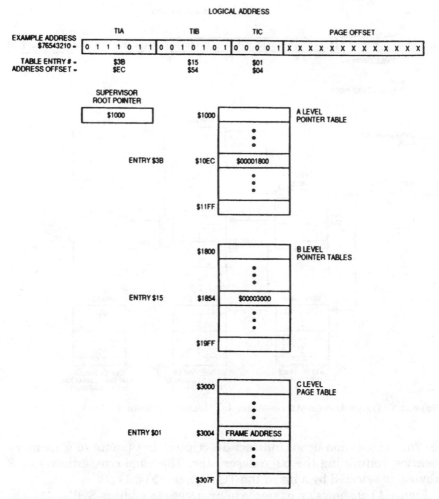

Figure 11.10 Translation tree layout in memory—example. (*Courtesy of Motorola, Inc.*)

1. A smaller table takes up less memory space and allows faster access and initialization.

2. A smaller table allows a greater number of tables to be allocated to a process. A recent trend in OS design is to use "lightweight" processes that are small and of short duration. Such a policy reduces process creation and deletion overhead, improving the overall performance. Smaller tables facilitate the implementation of lightweight processes.

For a 4-kbyte page size, a 7, 7, 6, 12 (bits) logical address partitioning is used, as explained earlier. All table entries are 4 bytes long.

Thus, the size of the first and second-level tables is

```
2⁷ * 4 = 512 bytes
```

The size of a third-level table is

```
2⁶ * 4 = 256 bytes
```

Compared to the above, the Intel page table size (see Chap. 7) is uniformly 4 kbytes, the size of any Intel system page.

A typical UNIX implementation allocates a minimum of four segments per process [Eden 90]. Thus, the Motorola table structure would require an overhead of

```
512 + 512 + 4 * 256 = 2 kbytes
```

while the Intel partitioning would require

```
4 kbytes + 4 * 4 kbytes = 20 kbytes of overhead
```

The Motorola 7, 7, 6, 12 partitioning provides a maximum of

```
2¹⁴ = 16K segments per process
```

while the Intel 10, 10, 12 partitioning provides a maximum of

```
2¹⁰ = 1024 = 1K segments
```

The above considerations led the designers to adopt the 7, 7, 6, 12 partitioning for 4-kbyte pages and the 7, 7, 5, 13 for 8-kbyte pages. This three-level structure results in less table overhead and permits more lightweight processes to be attached to a parent process. Of course, an extra level tends to increase the delay penalty in case of an ATC miss. Also, a nonuniform size of page tables complicates the page mechanism hardware circuitry. On the other hand, the small size of the tables (and hence their faster access) tends to compensate for the timing deficiencies.

There has been for some time an overall trend in computer design to use a 4-kbyte page size (see also Chap. 4). This was one of the factors that prompted Motorola designers to feature it. The lower page sizes featured in early M68000 systems (256 to 2048 bytes) were dropped because the ATC and physical cache lookup must occur simultaneously in the code and data caches. The ATC uses the upper address bits, while the cache uses the lower ones, and the same bits cannot be used for both lookups, since the ATC bits get translated and therefore changed. This imposes a constraint on the minimum page size. There are other general considerations against too small page sizes, discussed in Chap. 4. The Motorola designers also believed that there is a general trend toward the use of larger page size (this is supported by some recent research results discussed in Chap. 4), and they have retained the 8-kbyte page option for this reason. On the other hand, if a page is too large, its transmission overhead is increased and there may be more memory space lost if portions of pages are unused. For this reason the former 16- and 32-kbyte page options were dropped. More research and experimental results are needed to resolve the question of an optimal page size. Recent research points to a feature of several optional page sizes, as implemented on M68000 (see Chap. 4).

Four independent transparent translation registers (DTT0 and DTT1 in the data MMU, ITT0 and ITT1 in the instruction MMU; see also Fig. 11.2) optionally define four blocks of the logical address space that are directly translated to the physical address spaces. The blocks of addresses defined by these registers include at least 16 Mbytes of logical address space. The 4 blocks can overlap or they can be separate.

If either of the mTTx (m = D or I, x = 0 or 1) registers match during an access to a memory unit (either code or data), the access is transparently translated. If both registers match, the mTT0 status bits are used for the access. Transparent translation can also be implemented by the translation tables of the translation trees if the physical addresses of pages are set equal to their logical addresses.

Each ATC is 64-entry, four-way set-associative. Figure 11.11 shows its organization. The 4 bits of the logical address, located just above the page offset (bits 16 to 13 for 8-kbyte pages, bits 15 to 12 for 4-kbyte pages), index into the ATC's 16 sets of entries. The tags are compared against the remaining upper bits of the logical address and bit FC2 (when FC2 = 1, the processor operates in the supervisor mode, when FC2 = 0, in user mode). If one of the tags matches and is valid, then the corresponding entry is chosen by the multiplexer to produce the physical address and status information. If no tag matches, we have a miss and a table search is required. The ATC address translation occurs in parallel with indexing into the on-chip caches.

The 16-bit translation control (TC) register, shown in Fig. 11.2, has only 2 bits in current use. Bit 15, enable (E), enables (when E = 1) and disables (when E = 0) paged address translation. When translation is disabled, logical addresses are used as physical addresses. This occurs in simple configurations where virtual memory is not implemented. A reset operation clears this bit. Bit 14, page size (P), selects the memory page size:

P = 0, 4 kbytes

P = 1, 8 kbytes

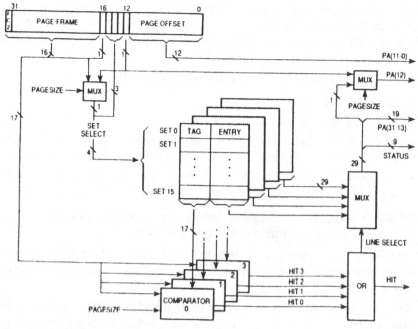

Figure 11.11 ATC organization. (*Courtesy of Motorola, Inc.*)

The 32-bit MMU status register (MMUSR) contains the status information returned by execution of the PTEST instruction. The PTEST instruction searches the translation tables to determine status information about the translation of a specified logical address. The MMUSR is shown in Fig. 11.12.

The fields of the MMUSR are as follows:

Bits 31 to 12, physical address: Contains the upper 20 bits of the translated physical address.

Bit 11, bus error (B): Bit B is set if a transfer error is encountered during the table search. If B is set, all other bits are zero.

Bit 10, global (G): When G is set, it indicates that the entry is global. Global entries are not invalidated by the PFLUSH instruction.

Bits 9,8, user page attributes (U1,U0): These bits are user-defined. Applications include extended addressing and snoop protocol selection.

31	12	11	10	9	8	7	6	5	4	3	2	1	0
PHYSICAL ADDRESS		B	G	U1	U0	S	CM		M	0	W	T	R

Figure 11.12 MMU status register. (*Courtesy of Motorola, Inc.*)

Bit 7, supervisor protection (S): When S is set, only programs operating in the supervisor mode are allowed to access the portion of the logical address space mapped by this descriptor. If S is clear, both supervisor and user accesses are allowed.

Bits 6,5, cache mode (CM): The encoding of the CM field is

- 00 Cacheable, write-through.
- 01 Cacheable, write-back.
- 10 Cache inhibited, serialized; the sequence of read and write accesses to the page matches the sequence of instruction ordering.
- 11 Cache inhibited, nonserialized; read accesses are allowed to occur before completion of a write-back for a prior instruction.

Bit 4, modified (M): Bit M is set when a valid write access to the logical address corresponding to the entry occurs.

Bit 2, write protected (W): When W is set, a write or a RMW access to the logical address corresponding to this entry causes a bus error exception.

Bit 1, transparent translation register hit (T): If T is set, it means that the PTEST instruction address matched an instruction or data TTR. In this case the R bit is set, and all other bits are zero.

Bit 0, resident (R): Bit R is set if the PTEST address matches a TTR or if the table search successfully completes by obtaining a valid page descriptor.

The page descriptors have a structure almost identical to that of the MMUSR. The upper bits (20 for 4-kbyte pages, 19 for 8-kbyte pages) contain the upper bits of the physical page base address. Bits 10 to 4 and bit 2 are identical to the bits with the same numbers in the MMUSR. If these bits are set in the descriptor, they are also set in the MMUSR. They have the same interpretation. Bit 3 in the descriptor is called used (U). Bit U is set when the descriptor is accessed.

11.7   Instruction and Data Caches

The instruction and data caches are a part of their respective MMUs. The caches are organized in an identical form, illustrated in Fig. 11.13. Both caches are four-way set-associative with 16 bytes/line. The size of each cache on the MC68060 is 8 kbytes (total 16 kbytes), and on the MC68040: 4 kbytes (total 8 kbytes). The cache shown in Fig. 11.13 is that of the MC68040. Except for the size, the cache parameters on both microprocessors are the same.

Each cache line contains an address tag (TAG), status information, and four longwords of data (D0 to D3), as shown in Fig. 11.13. The TAG contains the upper 22 bits of the physical address. The status

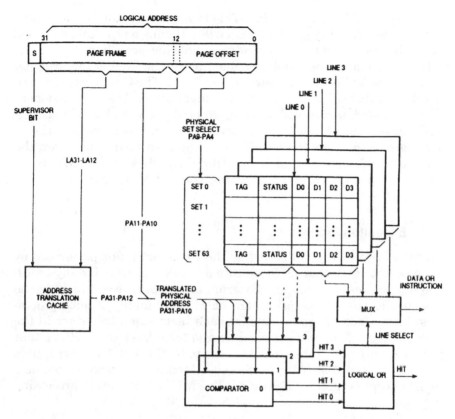

Figure 11.13 Internal caches. (*Courtesy of Motorola, Inc.*)

information for the instruction cache consists of a single valid bit for the entire line. The status information for the Dcache contains a valid bit, as well as 4 additional bits to indicate dirty status (modified status) for each longword in the line. Since entry validity is provided on a line basis, an entire line must be loaded from system memory in order for the cache to store an entry. Only burst mode accesses that successfully read four longwords can be cached.

While the upper bits of the logical address go to the ATC for translation to a physical address, the lower bits are used for direct access to the cache. The lowest 4 bits (bits 3 to 0) select a byte out of 16 bytes in the line. Bits 9 to 4 (6 bits) select one of the 64 sets (on the MC68040). In an 8-kbyte cache there are 128 sets for the same parameters, values, and the set field is 7 bits wide (bits 10 to 4).

Motorola designers decided to use physical addressing on both caches. Using this approach makes the caches reflect the physical memory. Therefore, when a process switch occurs, the caches do not have to be flushed and valuable time can be saved. Physically addressed caches allow external logic access to them without requiring a reverse physi-

cal-to-logical address translation. This is very important for bus snooping in cache coherency protocols [Eden 90]. Having a physical addressed
cache also solves the problem of address aliasing (see Chap. 4).

The caches are individually enabled using the MOVEC instruction
to access the 32-bit CACR shown in Fig. 11.14. The CACR contains two
enable bits that allow the instruction and data caches to be independently enabled or disabled. Setting an enable bit enables the associated cache without affecting the state of any lines within the cache. A
hardware reset clears the CACR, disabling both caches; however, the
tags, state information, and data within the caches are not affected by
reset and must be cleared by using the CINV instruction before
enabling the caches.

11.8   Exception Processing

Exception processing is defined as the set of activities performed by
the processor in preparing to execute a handler routine for any condition that causes an exception. External interrupts are considered to be
a particular case of exceptions on the M68000 family of microprocessors. Exceptions are recognized at each instruction boundary in the
execute stage of the integer pipeline and force later instructions that
have not yet reached the execute stage to be flushed. Uninterruptible
instructions (such as TAS, CAS, and CAS2) are allowed to complete
before exception processing begins. The M68000 exception processing
steps are:

1. Copy SR internally, set the S bit in SR, changing to supervisor
mode. Inhibit tracing by clearing the T1 and T0 bits in SR. For reset
and interrupt exceptions, the interrupt priority mask in SR should be
updated.

2. Determine the vector number of the exception, provided by internal logic. For interrupts, an interrupt acknowledge cycle to obtain the
vector number should be performed.

3. Save the current processor context by creating an exception stack
frame on the active supervisor stack. If it is an interrupt and the M bit
of SR is set, the M bit should be cleared and a second stack frame on
the interrupt stack should be built.

4. Initiate execution of the exception handler. Multiply the vector
number by 4 to determine the exception vector offset. Add the offset to

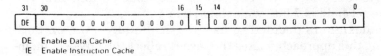

DE   Enable Data Cache
IE   Enable Instruction Cache

Figure 11.14 Cache control register. (*Courtesy of Motorola, Inc.*)

the value stored in the VBR to obtain the memory address of the exception vector. Load the PC (and the ISP for the reset exception) from the exception vector table entry.

All exception vectors are located in supervisor address space. Since the VBR provides the base address of the vector table, it can be located anywhere in memory. The M68000 architecture supports a 1024-byte (1-kbyte) vector table containing 256 exception vectors, shown in Table 11.22. The first 64 vectors are defined by Motorola, and 192 vectors are

TABLE 11.22 Exception Vector Assignments

Vector Number(s)	Vector Offset (Hex)	Assignment
0	000	Reset Initial Interrupt Stack Pointer
1	004	Reset Initial Program Counter
2	008	Access Fault
3	00C	Address Error
4	010	Illegal Instruction
5	014	Integer Divide by Zero
6	018	CHK, CHK2 Instruction
7	01C	FTRAPcc, TRAPcc, TRAPV Instructions
8	020	Privilege Violation
9	024	Trace
10	028	Line 1010 Emulator (Unimplemented A-Line Opcode)
11	02C	Line 1111 Emulator (Unimplemented F Line Opcode)
12	030	(Unassigned, Reserved)
13	034	Defined for MC68020 and MC68030, not used by MC68040
14	038	Format Error
15	03C	Uninitialized Interrupt
16-23	040-05C	(Unassigned, Reserved)
24	060	Spurious Interrpt
25	064	Level 1 Interrupt Autovector
26	068	Level 2 Interrupt Autovector
27	06C	Level 3 Interrupt Autovector
28	070	Level 4 Interrupt Autovector
29	074	Level 5 Interrupt Autovector
30	078	Level 6 Interrupt Autovector
31	07C	Level 7 Interrupt Autovector
32-47	080-0BC	TRAP #0-15 Instruction Vectors
48	0C0	FP Branch or Set on Unordered Condition
49	0C4	FP Inexact Result
50	0C8	FP Divide by Zero
51	0CC	FP Underflow
52	0D0	FP Operand Error
53	0D4	FP Overflow
54	0D8	FP Signaling NAN
55	0DC	FP Unimplemented Data Type
56	0E0	Defined for MC68030 and MC68851, not used by MC68040
57	0E4	Defined for MC68851, not used by MC68040
58	0E8	Defined for MC68851, not used by MC68040
59-63	0EC-0FC	(Unassigned, Reserved)
64-255	100-3FC	User Defined Vectors (192)

SOURCE: Courtesy of Motorola, Inc.

reserved for user-defined interrupt vectors. External devices may use vectors reserved for internal purposes at the discretion of the system designer.

For regular user interrupts (vector number region 64 and up), the processor reads the vector number off the low byte of the data bus (D0 to D7). There are, however, seven levels of autovector external interrupts (vector numbers 25 to 31, Table 11.22) for which the level (1 to 7) is read (encoded) from the 3 interrupt priority level (IPL) external signals. The external control signal AVEC# is activated by external logic for an autovector interrupt.

When several exceptions occur simultaneously, they are processed according to a fixed priority. Table 11.23 lists the exceptions, grouped by characteristics. Each group has a priority, with 0 as the highest priority.

11.9   Concluding Comment

The architecture of the Motorola M68000 microprocessor family was presented in detail in this chapter. The M68000 family microproces-

TABLE 11.23   Exception Priority Groups

Group/ Priority	Exception and Relative Priority	Characteristics
0	Reset	Aborts all processing (instruction or exception) and does not save old context.
1	Data Access Error (ATC Fault or Bus Error)	Aborts current instructions — can have pending trace, FP post instruction, or unimplemented FP instruction exceptions.
2	Floating-Point Pre Instruction	Exception processing begins before current floating point instruction is executed. Instruction is restarted on return from exception.
3	BKPT #n, CHK, CHK2, Divide by Zero, FTRAPcc, RTE, TRAP #n, TRAPV	Exception processing is part of instruction execution.
	Illegal Instruction, Unimplemented Line A and Line F, Privilege Violation	Exception processing begins before instruction is executed.
	Unimplemented Floating-Point Instruction	Exception processing begins after memory operands are fetched and before instruction is executed.
4	Floating-Point Post-Instruction	Only reported for FMOVE to memory. Exception processing begins when FMOVE instruction and previous exception processing is completed.
5	Address Error	Reported after all previous instructions and associated exceptions complete.
6	Trace	Exception processing begins when current instruction or previous exception processing is completed.
7	Instruction Access Error (ATC Fault or Bus Error)	Reported after all previous instructions and associated exceptions complete.
8	Interrupt	Exception processing begins when current instruction or previous exception processing is completed.

SOURCE: Courtesy of Motorola, Inc.

sors are widely used in numerous workstations and other computing systems. This architecture will continue to be implemented in years to come. The existing implementations of this architecture will be discussed in the next chapters of this part. The most recent systems, the MC68060 and MC68040 will be presented in Chap. 12, and the earlier M68000 implementations will be briefly reviewed in Chap. 13. The reason for the brief coverage of earlier M68000 systems is the existence of a vast amount of books amply covering this topic.

12

The MC68060 and MC68040

The organization of the current top M68000 family systems, the MC68060 and the MC68040, will be presented in this chapter.

The MC68060

The MC68060 is a 2.5-million-transistor, 0.5-micron CMOS, three-layer metal, 223-pin PGA (pin grid array) microprocessor. Its initial implementation is planned to run at 50 MHz, with 66 MHz operation targeted for the near future. It is a two-issue superscalar system. It is fully compatible with other microprocessors of the M68000 family. A block diagram of the MC68060 is shown in Fig. 12.1. The left upper part in Fig. 12.1 shows the four-stage instruction fetch pipeline (IFP) for prefetching instructions. Subsequent to the IFP there are two four-stage pipelines: the primary operand execution pipeline (pOEP) and the secondary OEP (sOEP), for decoding the instructions, fetching the required operands, and actual execution of the instructions. Since the IFP and OEP pipelines are decoupled by a 96-byte FIFO instruction buffer, the IFP is able to prefetch instructions in advance of their actual use by the OEPs. Instructions are not executed out of order.

The MC68060 is composed of the following subsystems:

1. Instruction fetch unit
2. Dual operand pipeline units
3. Execution units
   a. Dual integer execution unit
   b. Floating-point execution unit
4. Memory units
   a. Instruction memory unit, including
      (1) Instruction address translation cache (ATC)
      (2) Icache, 8 kbytes
      (3) Instruction memory controller

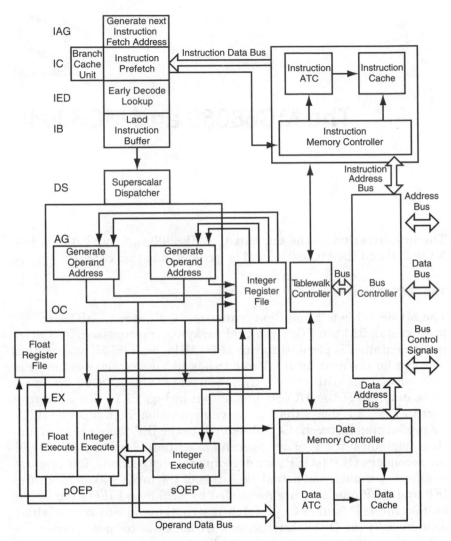

Figure 12.1   68060 internal block diagram. (*Courtesy of Motorola, Inc.*)

   b. Data memory unit
      (1) Data ATC
      (2) Dcache, 8 kbytes
      (3) Data memory controller
5. Bus interface unit (BIU)

The dual memory unit permits the CPU to simultaneously deal with instruction and data address translations and cache access, thus enhancing the pipeline operation. This is particularly important for superscalar operation (see Chap. 5).

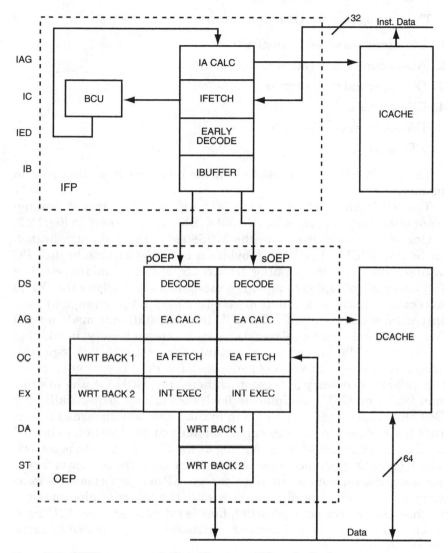

Figure 12.2 68060 processor pipeline. (*Courtesy of Motorola, Inc.*)

The MC68060 pipelines, defined above, are illustrated in more detail in Fig. 12.2.

The four stages of the IFP pipeline are

1. IAG—instruction address generation
2. IC—instruction cache access
3. IED—instruction early decode
4. IB—instruction buffer

The six stages of the OEP pipeline are

1. DS—decode and select instructions
2. AG—address generation of the operand
3. OC—operand cache access
4. EX—execute
5. DA—data available
6. ST—store

The last two stages are used when the processor is writing data to memory.

The pOEP pipeline branches out for a separate execution of floating-point instructions in stages 3 (OC) and 4 (EX), as can be seen in Fig. 12.2.

One of the new features on the MC68060 is the 256-entry branch cache unit (BCU). The BCU provides a table associating branch PC values with the corresponding branch target virtual addresses. The fundamental concept is to provide a mechanism that allows the IFP to detect and change instruction streams before the change of flow instructions enter an OEP. The BCU implementation is made up of a five-state prediction model based on past execution history, in addition to the current PC/branch target virtual address association logic. For each instruction fetch address generated, the BCU is examined to see if a valid branch entry is present. If there is no BCU hit, the instruction fetch unit (IFU) continues to fetch instructions sequentially. If a BCU hit occurs, indicating a taken branch, the IFU discards the current instruction stream and begins fetching at the location indicated by the branch target address. As long as the BCU prediction is correct, which happens a significant percentage of the time, the change of flow of the instruction stream is "invisible" to the OEP and performance is maximized. If the BCU prediction is wrong, the internal pipelines are flushed and the correct instruction flow is established. The BCU must be cleared by the OS on all context switches. It is also cleared by cache invalidate instructions.

The MC68060 architecture differs from the general M68000 architecture (fully implemented on the MC68040), described in Chap. 11, in the following points:

1. The MC68060 has only one supervisor stack pointer (as it was originally on the MC68000).
2. In the status register there is only one trace bit T (as on the MC68000).
3. There is no MMUSR.
4. There are two new system registers on the MC68060:
   a. A processor configuration register (PCR), an 8-bit register, with only 3 bits implemented:

(1) Bit 7—enable debug features (EDEBUG), when set.
(2) Bit 1—disable FPU (DFP), when set.
(3) Bit 0—enable superscalar dispatch (ESS), when set.
   b. A bus control register (BUSCR), a 32-bit register, with only 2
   bits used:
   (1) Bit 31—lock bit (L). When L = 1, assert external LOCK signal. Negate for L = 0.
   (2) Bit 30—shadow copy lock bit (SL). When SL = 1, there is a
   locked sequence at time of exception. No locked sequence
   when SL = 0.

The MC68060 instruction set is listed in Table 12.1. The superscalar classification of MC68060 floating-point instructions is listed
in Table 12.2, and the superscalar classification of the privileged
instructions is listed in Table 12.3.

TABLE 12.1   68060 Superscalar Classification of M680X0 Integer Instructions

Mnemonic	Instruction	Superscalar classification
ABCD	Add decimal with extend	pOEP-only
ADD	Add	pOEP | sOEP
ADDA	Add address	pOEP | sOEP
ADDI,Dx	Add immediate	pOEP | sOEP
ADDI,-(Ax)+	Add immediate	pOEP | sOEP
Remaining ADDI	Add immediate	pOEP-until-last
ADDQ	Add quick	pOEP | sOEP
ADDX	Add extended	pOEP-only
AND	AND logical	pOEP | sOEP
ANDI,Dx	AND immediate	pOEP | sOEP
ANDI,-(Ax)+	AND immediate	pOEP | sOEP
Remaining ANDI	AND immediate	pOEP-until-last
ANDI to CCR	AND immediate to condition codes	pOEP-only
ASL	Arithmetic shift left	pOEP | sOEP
ASR	Arithmetic shift right	pOEP | sOEP
Bcc	Branch conditionally	pOEP-only[1]
BCHG Dy,	Test a bit and change	pOEP-only
BCHG #<imm>,	Test a bit and change	pOEP-until-last
BCLR Dy,	Test a bit and clear	pOEP-only
BCLR #<imm>,	Test a bit and clear	pOEP-until-last
BFCHG	Test bit field and change	pOEP-only
BFCLR	Test bit field and clear	pOEP-only
BFEXTS	Extract bit field signed	pOEP-only
BFEXTU	Extract bit field unsigned	pOEP-only
BFFFO	Find first one in bit field	pOEP-only
BFINS	Insert bit field	pOEP-only
BFSET	Set bit field	pOEP-only
BFTST	Test bit field	pOEP-only
BRA	Branch always	pOEP-only
BSET Dy,	Test a bit and set	pOEP-only

TABLE 12.1   68060 Superscalar Classification of M680X0 Integer Instructions *(Cont.)*

Mnemonic	Instruction	Superscalar classification
BSET #<imm>	Test a bit and set	pOEP-until-last
BSR	Branch to subroutine	pOEP-only
BTST Dy,	Test a bit	pOEP-only
BTST #<imm>.	Test a bit	pOEP-until-last
CAS	Compare and swap with operand	pOEP-only
CHK	Check register against bounds	pOEP-only
CLR	Clear an operand	pOEP | sOEP
CMP	Compare	pOEP | sOEP
CMPA	Compare address	pOEP | sOEP
CMPI,Dx	Compare immediate	pOEP | sOEP
CMPI,-(Ax)+	Compare immediate	pOEP | sOEP
Remaining CMPI	Compare immediate	pOEP-until-last
CMPM	Compare memory	pOEP-until-last
DBcc	Test condition, decrement and branch	pOEP-only
DIVS.L	Signed divide long	pOEP-only
DIVS.W	Signed divide word	pOEP-only
DIVU.L	Unsigned long divide	pOEP-only
DIVU.W	Unsigned divide word	pOEP-only
EOR	Exclusive OR logical	pOEP | sOEP
EORI,Dx	Exclusive OR immediate	pOEP | sOEP
EORI,-(Ax)+	Exclusive OR immediate	pOEP | sOEP
Remaining EORI	Exclusive OR immediate	pOEP-until-last
EORI to CCR	Exclusive OR immediate to condition codes	pOEP-only
EXG	Exchange registers	pOEP-only
EXT	Sign extend	pOEP | sOEP
EXTB.L	Sign extend byte to long	pOEP | sOEP
ILLEGAL	Take illegal instruction trap	pOEP | sOEP
JMP	Jump	pOEP-only
JSR	Jump to subroutine	pOEP-only
LEA	Load effective address	pOEP | sOEP
LINK	Link and allocate	pOEP-until-last
LSL	Logical shift left	pOEP | sOEP
LSR	Logical shift right	pOEP | sOEP
MOVE,Rx	Move data from source to destination	pOEP | sOEP
MOVE Ry,	Move data from source to destination	pOEP | sOEP
MOVE <mem>y,<mem>x	Move data from source to destination	pOEP-until-last
MOVE #<imm>,<mem>x	Move data from source to destination	pOEP-until-last
MOVEA	Move address	pOEP | sOEP
MOVE from CCR	Move from condition codes	pOEP-only
MOVE to CCR	Move to condition codes	pOEP | sOEP
MOVE16	Move 16-byte block	pOEP-only
MOVEM	Move multiple registers	pOEP-only

TABLE 12.1   68060 Superscalar Classification of M680X0 Integer Instructions *(Cont.)*

Mnemonic	Instruction	Superscalar classification
MOVEQ	Move quick	pOEP	sOEP
MULS.L	Signed multiply long	pOEP-only
MULS.W	Signed multiply word	pOEP-only
MULU.L	Unsigned multiply long	pOEP-only
MULU.W	Unsigned multiply word	pOEP-only
NBCD	Negate decimal with extend	pOEP-only
NEG	Negate	pOEP	sOEP
NEGX	Negate with extend	pOEP-only
NOP	No operation	pOEP-only
NOT	Logical complement	pOEP	sOEP
OR	Inclusive OR logical	pOEP	sOEP
ORI,Dx	Inclusive OR immediate	pOEP	sOEP
ORI,-(Ax)+	Inclusive OR immediate	pOEP	sOEP
Remaining ORI	Inclusive OR immediate	pOEP-until-last
ORI to CCR	Inclusive OR immediate to condition codes	pOEP-only
PACK	Pack BCD digit	pOEP-only
PEA	Push effective address	pOEP-only
ROL	Rotate without extend left	pOEP	sOEP
ROR	Rotate without extend right	pOEP	sOEP
ROXL	Rotate with extend left	pOEP-only
ROXR	Rotate with extend right	pOEP-only
RTD	Return and deallocate parameters	pOEP-only
RTR	Return and restore condition codes	pOEP-only
RTS	Return from subroutine	pOEP-only
SBCD	Subtract decimal with extend	pOEP-only
Scc	Set according to condition	pOEP-but allows-sOEP
SUB	Subtract	pOEP	sOEP
SUBA	Subtract address	pOEP	sOEP
SUBI,Dx	Subtract immediate	pOEP	sOEP
SUBI,-(Ax)+	Subtract immediate	pOEP	sOEP
Remaining SUBI	Subtract immediate	pOEP-until-last
SUBQ	Subtract quick	pOEP	sOEP
SUBX	Subtract with extend	pOEP-only
SWAP	Swap register halves	pOEP-only
TAS	Test and set an operand	pOEP-only
TRAP	Trap	pOEP	sOEP
TRAPF	Trap on false	pOEP	sOEP
Remaining TRAPcc	Trap on condition	pOEP-only
TRAPV	Trap on overflow	pOEP-only
TST	Test an operand	pOEP	sOEP
UNLK	Unlink	pOEP-only
UNPK	Unpack BCD digit	pOEP-only

[1]A Bcc instruction is pOEP-but-allows-sOEP if it is not predicted from the branch cache and the direction of the branch is forward or if the Bcc is predicted as a "not-taken" branch.

SOURCE: Courtesy of Motorola, Inc.

TABLE 12.2   68060 Superscalar Classification of M680X0 Floating-Point Instructions

Mnemonic	Instruction	Superscalar Classification
FABS, FDABS, FSABS	Absolute value	pOEP-but-allows-sOEP[1]
FADD, FDADD, FSADD	Add	pOEP-but-allows-sOEP[1]
FBcc	Branch conditionally	pOEP-only
FCMP	Compare	pOEP-but-allows-sOEP[1]
FDIV, FDDIV, FSDIV, FSGLDIV	Divide	pOEP-but-allows-sOEP[1]
FINT, FINTRZ	Integer part, round-to-zero	pOEP-but-allows-sOEP[1]
FMOVE, FDMOVE, FSMOVE	Move floating-point data register	pOEP-but-allows-sOEP[1]
FMOVE	Move system control register	pOEP-only
FMOVEM	Move multiple data registers	pOEP-only
FMUL, FDMUL, FSMUL, FSGLMUL	Multiply	pOEP-but-allows-sOEP[1]
FNEG, FDNEG, FSNEG	Negate	pOEP-but-allows-sOEP[1]
FNOP	No operation	pOEP-only
FSQRT	Square root	pOEP-but-allows-sOEP[1]
FSUB, FDSUB, FSSUB	Subtract	pOEP-but-allows-sOEP[1]
FTST	Test operand	pOEP-but-allows-sOEP[1]

[1]These floating-point instructions are pOEP-but-allows-sOEP except for the following:

F<op>   Dm,FPn
F<op>   &imm,FPn
F<op>.x <mem>,FPn
which are classified as pOEP-only.

SOURCE: Courtesy of Motorola, Inc.

TABLE 12.3   68060 Superscalar Classification of M680X0 Privileged Instructions

Mnemonic	Instruction	Superscalar Classification
ANDI to SR	AND immediate to status register	pOEP-only
CINV	Invalidate cache lines	pOEP-only
CPUSH	Push and invalidate cache lines	pOEP-only
EORI to SR	Exclusive OR immediate to status register	pOEP-only
MOVE from SR	Move from status register	pOEP-only
MOVE to SR	Move to status register	pOEP-only
MOVE USP	Move user stack pointer	pOEP-only
MOVEC	Move control register	pOEP-only
MOVES	Move address space	pOEP-only
ORI to SR	Inclusive OR immediate to status register	pOEP-only
PFLUSH	Flush ATC entries	pOEP-only
PLPA	Load physical address	pOEP-only
RESET	Reset external devices	pOEP-only
RTE	Return from exception	pOEP-only
STOP	Load status register and stop	pOEP-only

SOURCE: Courtesy of Motorola, Inc.

The differences between the MC68060 and the MC68040 hardware-supported instruction sets are summarized in Table 12.4. The main differences are expansion of the MOVEC instruction, revision of PFLUSH functionality, and introduction of new LPSTOP and PLPA instructions.

The MC68060 ATC is essentially identical to the ATC shown in Fig. 11.11. The MC68060 cache structure is illustrated in Fig. 12.3. The MC68060 cache parameters are the same as on the cache of MC68040: four-way set-associative with 16 bytes/line, except that the cache size

TABLE 12.4   68060/68040 Hardware-Supported Instruction Set Differences

Mnemonic	Description	Notes
CAS	Compare and swap with operand	Emulation support on 68060 for misaligned operand
CAS2	Compare and swap dual operands	Emulation support on 68060, not implemented in hardware
CHK2	Check register against upper and lower bound	Emulation support on 68060, not implemented in hardware
CMP2	Compare register against upper and lower bound	Emulation support on 68060, not implemented in hardware
DIVS.L	Signed divide	Emulation support on 68060 for 64/32
DIVU.L	Unsigned divide	Emulation support on 68060 for 64/32
FDBcc	FP decrement and branch	Emulation support on 68060, not implemented in hardware
FINT	FP integer part	Implemented in hardware on 68060, not on 68040
FINTRZ	FP integer part, round-to-zero	Implemented on hardware on 68060, not on 68040
FMOVEM	FP move multiple data registers	Emulation support on 68060 for dynamic register list
FScc	FP set according to condition	Emulation support on 68060, not implemented in hardware
FTRAPcc	FP trap on condition	Emulation support on 68060, not implemented in hardware
LPSTOP	Low-power stop	Implemented on 68060, not on 68040
MOVEC	Move control registers	Revised functionality
MOVEP	Move peripheral	Emulation support on 68060, not implemented in hardware
MULS.L	Signed long multiply	Emulation support on 68060 for $32 \times 32 \to 64$
MULU.L	Unsigned long multiply	Emulation support on 68060 for $32 \times 32 \to 64$
PTEST	Test a logical address	Not implemented on 68060, PLPA added on 68060 (equivalent functionality)
PLPA	Load physical address	Implemented on 68060, not on 68040

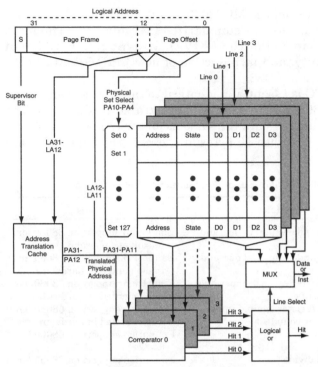

Figure 12.3   Cache structure.

of the MC68060 is double (8 kbytes as opposed to the 4 kbytes on
MC68040). As we can see, Fig. 12.3 is very similar to Fig. 11.13, except
for the double number of sets in Fig. 12.3.

The MC68060 external signals are shown in Fig. 12.4. Many of
these signals are identical to those on MC68040, described next in this
chapter.

The MC68040

The MC68040 is a 1.2-million transistor, 0.8-micron HCMOS, 179-pin
PGA microprocessor. Figure 12.5 shows the block diagram of the
MC68040. It runs at 25- to 40-MHz frequencies.

The MC68040 has a six-stage integer pipeline and a three-stage
FPU pipeline, coupled to the integer pipeline, as can be seen in Fig.
12.5. The MC68040 integer pipeline is illustrated in Fig. 12.6. The
MC68040 MMU structure is essentially identical to that of the
MC68060, except that the MC68040 caches are half the size (4 kbytes
each) of those on the MC68060 (8 kbytes each). The other cache parameters of both microprocessors are the same (four-way set-associative,
16 bytes/line).

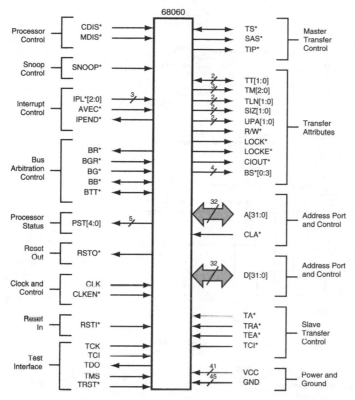

Figure 12.4   MC68060 functional signal groups. (*Courtesy of Motorola, Inc.*)

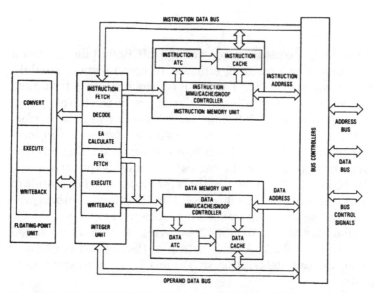

Figure 12.5   The MC68040 block diagram. (*Courtesy of Motorola, Inc.*)

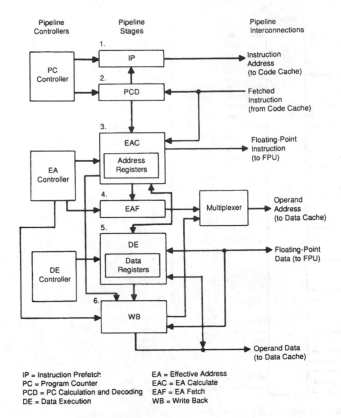

Figure 12.6   MC68040 IU pipeline.

The interface design considerations of the MC68040 design team [Eden 90] were based on providing efficient interconnection capabilities to low-cost memory systems. The main reason for this is that most potential users are interested in high performance at the lowest possible cost. Therefore, the MC68040 bus design maximized the performance available from low-cost memory systems (*dynamic random-access memory*—DRAMs [Uffe 91]) while minimizing the degradation to high-performance designs for the fastest and more expensive *static RAM* (SRAMs) [Uffe 91]. The bus was therefore interfaced to standard *transistor-transistor logic* (TTL) devices. The interface protocol was simplified to minimize the delay caused by cache misses.

The pinout diagram of the MC68040, showing its functional signal groups, is shown in Fig. 12.7. A brief listing of the signals is given in Table 12.5. The function description in Table 12.5 is self-explanatory for most signals.

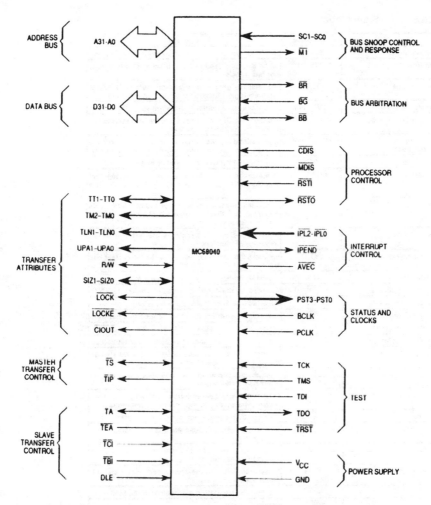

Figure 12.7 MC68040 functional signal groups. (*Courtesy of Motorola, Inc.*)

The MC68040 was successfully implemented in many notable workstations and other computing systems such as the NeXT systems, Apple's Macintosh Quadra 950, and Macintosh Centris 650, and many others. No doubt the top-level microprocessors of the Motorola M68000 family will continue to be used in future applications.

TABLE 12.5   Signal Index

Signal Name	Mnemonic	Function
Address Bus	A31–A0	32-bit address bus used to address any of 4 Gbytes.
Data Bus	D31–D0	32-bit data bus used to transfer up to 32 bits of data per bus transfer.
Transfer Type	TT1,TT0	Indicates the general transfer type: normal, MOVE16, alternate logical function code, and acknowledge.
Transfer Modifier	TM2,TM0	Indicates supplemental information about the access.
Transfer Line Number	TLN1,TLN0	Indicates which cache line in a set is being pushed or loaded by the current line transfer.
User Programmable Attributes	UPA1,UPA0	User-defined signals, controlled by the corresponding user attribute bits from the address translation entry.
Read/Write	R/W̄	Identifies the transfer as a read or write.
Transfer Size	SIZ1,SIZ0	Indicates the data transfer size. These signals, together with A0 and A1, define the active sections of the data bus.
Bus Lock	L̄O̅C̅K̄	Indicates a bus transfer is part of a read-modify-write operation, and that the sequence of transfers should not be interrupted.
Bus Lock End	L̄O̅C̅K̄E̅	Indicates the current transfer is the last in a locked sequence of transfers.
Cache Inhibit Out	C̄I̅O̅U̅T̄	Indicates the processor will not cache the current bus transfer.
Transfer Start	T̄S̄	Indicates the beginning of a bus transfer.
Transfer in Progress	T̄I̅P̄	Asserted for the duration of a bus transfer.
Transfer Acknowledge	T̄Ā	Asserted to acknowledge a bus transfer.
Transfer Error Acknowledge	T̄E̅Ā	Indicates an error condition exists for a bus transfer.
Transfer Cache Inhibit	T̄C̄I̅	Indicates the current bus transfer should not be cached.
Transfer Burst Inhibit	T̄B̄I̅	Indicates the slave cannot handle a line burst access.
Data Latch Enable	DLE	Alternate clock input used to latch input data when the processor is operating in DLE mode.
Snoop Control	SC1,SC0	Indicates the snooping operation required during an alternate master access.
Memory Inhibit	M̄I̅	Inhibits memory devices from responding to an alternate master access during snooping operations.
Bus Request	B̄R̄	Asserted by the processor to request bus mastership.
Bus Grant	B̄Ḡ	Asserted by an arbiter to grant bus mastership to the processor.
Bus Busy	B̄B̄	Asserted by the current bus master to indicate it has assumed ownership of the bus.
Cache Disable	C̄D̄I̅S̄	Dynamically disables the internal caches to assist emulator support.
MMU Disable	M̄D̄I̅S̄	Disables the translation mechanism of the MMUs.
Reset In	R̄S̄T̄I̅	Processor reset.
Reset Out	R̄S̄T̄O̅	Asserted during execution of a RESET instruction to reset external devices.
Interrupt Priority Level	I̅P̄L̄2–I̅P̄L̄0	Provides an encoded interrupt level to the processor.
Interrupt Pending	I̅P̄E̅N̄D̄	Indicates an interrupt is pending.
Autovector	A̅V̄E̅C̄	Used during an interrupt acknowledge transfer to request internal generation of the vector number.
Processor Status	PST3–PST0	Indicates internal processor status.
Bus Clock	BCLK	Clock input used to derive all bus signal timing.
Processor Clock	PCLK	Clock input used for internal logic timing. The PCLK frequency is exactly 2X the BCLK frequency.
Test Clock	TCK	Clock signal for the IEEE P1149.1 Test Access Port (TAP).
Test Mode Select	TMS	Selects the principle operations of the test-support circuitry.
Test Data Input	TDI	Serial data input for the TAP.
Test Data Output	TDO	Serial data output for the TAP.
Test Reset	T̄R̄S̄T̄	Provides an asynchronous reset of the TAP controller.
Power Supply	V$_{CC}$	Power supply.
Ground	GND	Ground connection.

Source: Courtesy of Motorola, Inc.

13

Earlier
M68000 Family
Microprocessors

The comparative details of the various properties of the M68000 family microprocessors are summarized in Table 13.1, listed by different topics. Although all chips have 32-bit CPU registers, the 68000, 68008, and 68010 are 16-bit systems, while the microprocessors starting with 68020 and on are 32-bit. The 68008 has the same architecture as the 68000, but an 8-bit external data bus. The 32-bit systems have essentially the same architecture. They have identical addressing modes and instruction formats and almost the same instruction set. Compared to the 68020, most of the new instructions on the 68030 are due to the inclusion of the MMU on the 68030 chip. It was previously on the MC68851 paged MMU (PMMU) chip, and the instructions beginning with a "P" are available on the 68851, used in conjunction with the 68020. The floating-point instructions (beginning with an "F"), new on the MC68040, were previously available on the MC68881/MC68882 coprocessors used with the 68030 or 68020. The whole M68000 family is upward object code compatible.

A block diagram of the MC68030 is shown in Fig. 13.1. As stated before, the 68030 and 68040 are almost identical architecturally. The main differences are essentially in their organization and implementation. These differences can be summarized as follows:

Feature	MC68030	MC68040
FPU	Off-chip,68881/68882	On-chip
MMU	On-chip, single	On-chip, dual
Cache, on-chip dual	256-byte code, data	4-kbyte code, data
Pipelines	1	3

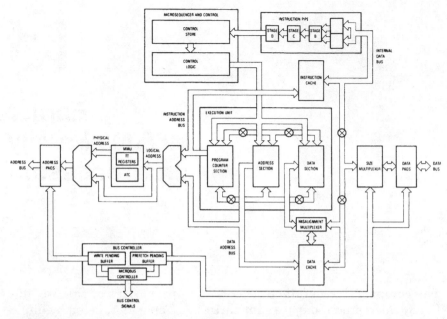

Figure 13.1   MC68030 block diagram. (*Courtesy of Motorola, Inc.*)

TABLE 13.1   M68000 Family Summary

Attribute	MC68000	MC68008	MC68010	MC68020	68030	MC68040
Data bus size (bits)	16	8	16	8, 16, 32	8, 16, 32	32
Address bus size (bits)	24	20	24	32	32	32
Instruction cache (in bytes)	—	—	3^1 (Words)	256	256	4096
Data cache (in bytes)	—	—	—	—	256	4096

Note 1: The MC68010 supports a three-word cache for the loop mode.

Virtual Interfaces

MC68010, MC68020, MC68030	Virtual memory machine
MC68040	Virtual memory
MC68010, MC68020, MC68030, MC68040	Provide bus error detection, fault recovery
MC68030, MC68040	On-chip MMU

Coprocessor Interface

MC68000, MC68008, MC68010	Emulated in software
MC68020, MC68030	In microcode
MC68040	Emulated in software (on-chip floating-point unit)

TABLE 13.1   M68000 Family Summary *(Cont.)*

Word/Long Word Data Alignment

MC68000, MC68008, MC68010	Word long data instructions, and stack must be word aligned
MC68020, MC68030, MC68040	Only instructions must be word aligned (data alignment improves performance)

Control Registers

MC68000, MC68008	None
MC68010	SFC, DFC, VBR
MC68020	SFC, DFC, VBR, CACR, CAAR
MC68030	SFC, DFC, VBR, CACR, CAAR,CRP, SRP, TC, TT0, TT1, MMUSR
MC68040	SFC, DFC, VBR, CACR, URP, SRP, TC, DTT0, DTT1, ITT0, ITT1, MMUSR

Stack Pointer

MC68000, MC68008, MC68010	USP, SSP
MC68020, MC68030, MC68040	USP, SSP, (MSP, ISP)

Status Register Bits

MC68000, MC68008, MC68010	T, S, I0 I1 I2, X N Z V C
MC68020, MC68030, MC68040	T0, T1, S, M, I0 I1 I2, X N Z V C

Function Code/Address Space

MC68000, MC68008	FC2-FC0 - 7 is Interrupt Acknowledge Only
MC68010, MC68020, MC68030, MC68040	FC2-FC0 - 7 is CPU space
MC68040	User, Supervisor, and Acknowledge

Indivisible Bus Cycles

MC68000, MC68008, MC68010	Use $\overline{AS}$ signal
MC68020, MC68030	Use $\overline{RMC}$ signal
MC68040	Use $\overline{LOCK}$ and $\overline{LOCKE}$ signal

Stack Frames

MC68000, MC68008	Supports original set
MC68010	Supports formats $0, $8
MC68020, MC68030	Supports formats $0, $1, $2, $9, $A, $B
MC68040	Supports formats $0, $1, $2, $3, $7

TABLE 13.1   M68000 Family Summary *(Cont.)*

Addressing Modes	
MC68020, MC68030, and MC68040 Extensions	Memory indirect addressing modes, scaled index, and larger displacements. Refer to specific data sheets for details.

MC68020, MC68030, and MC68040 Instruction Set Extensions		Applies to		
Instruction	Notes	MC68020	MC68030	MC68040
Bcc	Supports 32-bit Displacement	✓	✓	✓
BFxxxx	Bit field instructions (BCHG, BFCLR, BFEXTS, BFEXTU, BFFFO BFINS, BFSET, BFTST)	✓	✓	✓
BKPT	New instruction functionally	✓	✓	
BRA	Supports 32-bit displacement	✓	✓	✓
BSR	Supports 32-bit displacement	✓	✓	✓
CALLM	New instructions	✓		
CAS, CAS2	New instruction	✓	✓	✓
CHK	Supports 32-bit operands	✓	✓	✓
CHK2	New instruction	✓	✓	✓
CINV	Cache maintenance instruction			✓
CMPI	Supports program counter relative addressing modes	✓	✓	✓
CMP2	New instruction	✓	✓	✓
CPUSH	Cache maintenance instruction			✓
cp	Coprocessor instructions	✓	✓	
DIVS DIVU	Supports 32- and 64-bit operands	✓	✓	✓
EXTB	Supports 8-bit extend to 32 bits	✓	✓	✓
FABS	New instruction			✓
FADD	New instruction			✓
FBcc	New instruction			✓
FCMP	New instruction			✓
FDBcc	New instruction			✓
FDIV	New instruction			✓
FMOVE	New instruction			✓
FMOVEM	New instruction			✓
FMUL	New instruction			✓
FNEG	New instruction			✓
FRESTORE	New instruction			✓
FSAVE	New instruction			✓
FScc	New instruction			✓
FSQRT	New instruction			✓
FSUB	New instruction			✓
FTRAPcc	New instruction			✓
FTST	New instruction			✓
LINK	Supports 32-bit displacement	✓	✓	✓
MOVE16	New instruction			✓
MOVEC	Supports new control registers	✓	✓	✓
MULS/MULU	Supports 32-bit operands	✓	✓	✓

TABLE 13.1   M68000 Family Summary *(Cont.)*

MC68020, MC68030, and MC68040 instruction set extensions		Applies to		
Instruction	Notes	MC68020	MC68030	MC68040
PACK	New instruction	✓	✓	✓
PFLUSH	MMU instruction		✓	✓
PLOAD	MMU instruction		✓	
PMOVE	MMU instruction		✓	
PTEST	MMU instruction		✓	✓
RTM	New instruction	✓		
TST	Supports program counter relative addressing modes	✓	✓	✓
TRAPcc	New instruction	✓	✓	✓
UNPK	New instruction	✓	✓	✓

This table summarizes the characteristics of the microprocessors in the M68000 family. The M68000PM/AD, *M68000 Programmer's Reference Manual* includes more detailed information on the M68000 family differences. [M680 89]

SOURCE: Courtesy of Motorola, Inc.

The MC68020 does not have an MMU on-chip, nor an FPU. It uses the MC68851 PMMU coprocessor as an MMU. It has only a 256-byte code cache on-chip. The 68020 code cache is direct-mapped with 4 bytes/line [Clem 92, Clem 94]. It is always addressed on a word basis. Both the 68030 256-byte code and data caches are direct-mapped with 16 bytes/line. The data cache uses a write-through policy. The on-chip 68030 ATC address translation cache) has 22 entries.

The MC68030 is a 128-pin, 1.2-micron, 25-MHz HCMOS chip. The MC68020 is a 114-pin, 1.5-micron, 25-MHz HCMOS chip. The details of the 16-bit members of the M68000 family are well covered in detail in other books and therefore are not presented here [Clem 92, Clem 94, Wake 89].

3

M68000 Architecture
Instruction Set

Notation for operands:

        PC—Program counter
        SR—Status register
        V—Overflow condition code
  Immediate data—Immediate data from the instruction
      Source—Source contents
   Destination—Destination contents
     Vector—Location of exception vector
      + inf—Positive infinity
      − inf—Negative infinity
      <fmt>—Operand data format: byte (B), word (W), long (L), single (S),
             double (D), extended (X), or packed (P)
      FPm—One of eight floating-point data registers (always specifies
             the source register)
      FPn—One of eight floating-point data registers (always specifies
             the destination register)

Notation for subfields and qualifiers:

    <bit> of <operand>—Selects a single bit of the operand
     <ea>{offset:width}—Selects a bit field
        (<operand>)—The contents of the referenced location
        <operand>10—The operand is binary coded decimal, operations are per-
             formed in decimal
  (<address register>)—The register indirect operator
 − (<address register>)—Indicates that the operand register points to the memory
(<address register>) + —Location of the instruction operand—the optional mode
             qualifiers are −, +, (d), and (d,ix)
    #xxx or #<data>—Immediate data that follows the instruction word(s)

Notations for operations that have two operands, written <operand><op> <operand>, where <op> is one of the following:

→ —The source operand is moved to the destination operand

↔ —The two operands are exchanged

+ —the operands are added

− —The destination operand is subtracted from the source operand

× —The operands are multiplied

÷ —The source operand is divided by the destination operand

< —Relational test, true if source operand is less than destination operand

> —Relational test, true if source operand is greater than destination operand

V —Logical OR

⊕ —Logical exclusive OR

∧ —Logical AND

Shifted by, rotated by—The source operand is shifted or rotated by the number of positions specified by the second operand

Notation for single-operand operations:

~ <operand>—The operand is logically complemented

<operand>sign-extended—The operand is sign extended, all bits of the upper portion are made equal to the high order bit of the lower portion

<operand>tested—The operand is compared to zero and the condition codes are set appropriately

Notation for other operations:

TRAP—Equivalent to Format/Offset Word → (SSP); SSP − 2 → SSP; PC → (SSP); SSP − 4 → SSP; SR → (SSP); SSP − 2 → SSP; (Vector) → PC

STOP—Enter the stopped state, waiting for interrupts

If <condition> then—The condition is tested. If true, the operations
<operations> else   after "then" are performed. If the condition is
<operations>   false and the optional "else" clause is present, the operations after "else" are performed. If the condition is false and else is omitted, the instruction performs no operation. Refer to the Bcc instruction description as an example.

Instruction Set Summary (Sheet 1 of 8)

Opcode	Operation	Syntax
ABCD	$Source_{10} + destination_{10} + X \rightarrow destination$	ABCD Dy,Dx ABCD $-$ (Ay), $-$ (Ax)
ADD	Source + destination $\rightarrow$ destination	ADD <ea>,Dn ADD Dn,<ea>
ADDA	Source + destination $\rightarrow$ destination	ADDA <ea>,An
ADDI	Immediate data + destination $\rightarrow$ destination	ADDI #<data>,<ea>
ADDQ	Immediate data + destination $\rightarrow$ destination	ADDQ #<data>,<ea>
ADDX	Source + destination + X $\rightarrow$ destination	ADDX Dy,Dx ADDX $-$ (Ay), $-$ (Ax)
AND	Source $\wedge$ destination $\rightarrow$ destination	AND <ea>,Dn AND Dn,<ea>
ANDI	Immediate data $\wedge$ destination $\rightarrow$ destination	ANDI #<data>,<ea>
ANDI to CCR	Source $\wedge$ CCR $\rightarrow$ CCR	ANDI #<data>,CCR

ANDI to SR	If supervisor state
	  the source $\wedge$ SR $\rightarrow$ SR
	  else TRAP
		ANDI #<data>,SR

ASL,ASR	Destination shifted by <count> $\rightarrow$ destination	ASd Dx,Dy ASd #<data>,Dy ASd <ea>
Bcc	If (condition true) then PC + d $\rightarrow$ PC	Bcc <label>
BCHG	(<number> of destination) $\rightarrow$ Z; (<number> of destination) $\rightarrow$ <bit number> of destination	BCHG Dn,<ea> BCHG #<data>,<ea>
BCLR	(<bit number> of destination) $\rightarrow$ Z 0 $\rightarrow$ <bit number> of destination	BCLR Dn,<ea> BCLR #<data>,<ea>
BFCHG	(<bit field> of destination) $\rightarrow$ <bit field> of destination	BFCHG <ea>{offset:width}
BFCLR	0 $\rightarrow$ <bit field> of destination	BFCLR <ea>{offset width}
BFEXTS	<bit field> of source $\rightarrow$ Dn	BFEXTS <ea>{offset:width},Dn
BFEXTU	<bit offset> of source $\rightarrow$ Dn	BFEXTU <ea>{offset:width},Dn
BFFFO	<bit offset> of source bit scan $\rightarrow$ Dn	BFFFO <ea>{offset:width},Dn
BFINS	Dn $\rightarrow$ <bit field> of destination	BFINS Dn,<ea>{offset:width}
BFSET	1s $\rightarrow$ <bit field> of destination	BFSET <ea>{offset:width}
BFTST	<bit field> of destination	BFTST <ea>{offset:width}
BKPT	Run breakpoint acknowledge cycle: TRAP as illegal instruction	BKPT <data>

Instruction Set Summary (Sheet 2 of 8)

Opcode	Operation	Syntax
BRA	$PC + d \rightarrow PC$	BRA <label>
BSET	(<bit number> of destination) $\rightarrow$ Z	BSET Dn,<ea>
	$1 \rightarrow$ <bit number> of destination	BSET #<data>,<ea>
BSR	$SP - 4 \rightarrow SP$; $PC \rightarrow (SP)$; $PC + d \rightarrow PC$	BSR <label>
BTST	(<bit number> of destination) $\rightarrow$ Z;	BTST Dn,<ea> BTST #<data>,<ea>
CAS	CAS destination compare operand $\rightarrow$ cc;	CAS Dc,Du,<ea>
CAS2	if Z, update operand $\rightarrow$ destination else destination $\rightarrow$ compare operand CAS2 destination 1 $-$ compare 1 $\rightarrow$ cc; if Z, destination 2 $-$ compare $\rightarrow$ cc; if Z, update 1 $\rightarrow$ destination 1; update 2 $\rightarrow$ destination 2 else destination 1 $\rightarrow$ compare 1; destination 2 $\rightarrow$ compare 2	CAS2 Dc1:Dc2,Du1:Du2,(Rn1):(Rn2)
CHK	If Dn <0 or Dn> source then TRAP	CHK <ea>,Dn
CHK2	If Rn <lower bound or Rn> upper bound then TRAP	CHK2 <ea>,Rn
CINV	If supervisor state then invalidate selected cache lines else TRAP	CINVL caches [1],(An) CINVP caches [1],(An) CINVA caches [1]
CLR	$0 \rightarrow$ destination	CLR <ea>
CMP	Destination $-$ source $\rightarrow$ cc	CMP <ea>,Dn
CMPA	Destination $-$ source	CMPA <ea>,An
CMPI	Destination $-$ immediate data	CMPI #<data>,<ea>
CMPM	Destination $-$ source $\rightarrow$ cc	CMPM (Ay) + ,(Ax) +
CMP2	Compare Rn <lower bound or Rn> upper bound and set condition codes	CMP2 <ea>,Rn
CPUSH	If supervisor state then if data cache then push selected dirty data cache lines invalidate selected cache lines else TRAP	CPUSHL caches [1],(An) CPUSHP caches [1],(An) CPUSHA caches [1]
DBcc	If condition false then Dn $- 1 \rightarrow$ Dn; if Dn $\neq -1$ then PC $+ d \rightarrow$ PC	DBcc Dn,<label>
DIVS	Destination/source $\rightarrow$ destination	DIVS.W <ea>,Dn $32/16 \rightarrow 16r:16q$
DIVSL		DIVS.L <ea>,Dq $32/32 \rightarrow 32q$ DIVS.L <ea>,Dr:Dq $64/32 \rightarrow 32r:32q$ DIVSL.L <ea>,Dr:Dq $32:32 \rightarrow 32r:32q$

Instruction Set Summary (Sheet 3 of 8)

Opcode	Operation	Syntax	
DIVU DIVUL	Destination/source → destination	DIVU.W <ea>,Dn DIVU.L <ea>,Dq DIVU.L <ea>,Dr:Dq DIVUL.L <ea>,Dr:Dq	$32/16 \rightarrow 16r{:}16q$ $32/32 \rightarrow 32q$ $64/32 \rightarrow 32r{:}32q$ $32/32 \rightarrow 32r{:}32q$
EOR	Source $\oplus$ destination → destination	EOR Dn,<ea>	
EORI	Immediate data $\oplus$ destination → destination	EORI #<data>,<ea>	
EORI to CCR	Source $\oplus$ CCR → CCR	EORI <ea>,CCR	
EORI to SR	If supervisor state then Source $\oplus$ SR → SR else TRAP	EORI #<data>,SR	
EXG	Rx $\leftrightarrow$ Ry	EXG Dx,Dy EXG Ax,Ay EXG Dx,Ay EXG Ay,Dx	
EXT EXTB	Destination sign-extended → destination	EXT.W Dn extend byte to word EXT.L Dn extended word to longword EXTB.L Dn extend byte to longword	
FABS	Absolute value of source → FPn	FABS.<fmt> FABS.X FABS.X FrABS.<fmt> FrABS.X[2] FrABS.X[2]	<ea>,FPn FPm,FPn FPn <ea>,FPn FPm,FPn FPn
FADD	Source + FPn → FPn	FADD.<fmt> FADD.X FrADD.<fmt>[2] FrADD.X[2]	<ea>,FPn FPm,FPn <ea>,FPn FPm,FPn
FBcc	If condition true then PC + d → PC	FBcc.<size>	<label>
FCMP	FPn − source	FCMP.<fmt> FCMP.X	<ea>,FPn FPm,FPn

FDBcc	If condition true then no operation
	else Dn $-1 \rightarrow$ Dn
	  if Dn $\neq -1$
	    then PC + d → PC
	    else execute next instruction
		FDBcc Dn,<label>	

FDIV	FPn $\div$ source → FPn	FDIV.<fmt> FDIV.X FrDIV.<fmt>[2] FrDiv.X[2]	<ea>,FPn FPm,FPn <ea>,FPn FPm,FPn
FMOVE	Source → destination	FMOVE.<fmt> FMOVE.<fmt> FMOVE.P FMOVE.P FrMOVE.<fmt>[2]	<ea>,FPn FPm,<ea> FPm,<ea>{Dn} FPm,<ea>{#k} <ea>,FPn

Instruction Set Summary (Sheet 4 of 8)

Opcode	Operation	Syntax	
FMOVE	Source → destination	FMOVE.L	<ea>,FPcr
		FMOVE.L	FPcr,<ea>
FMOVEM	Register list → destination	FMOVEM.X	<list>[3],<ea>
	Source → register list	FMOVEM.X	Dn,<ea>
		FMOVEM.X	<ea>,<list>[3]
		FMOVEM.X	<ea>,Dn
FMOVEM	Register list → destination	FMOVEM.L	<list>[4],<ea>
	Source → register list	FMOVEM.L	<ea>,<list>[4]
FMUL	Source × FPn → FPn	FMUL.<fmt>	<ea>,FPn
		FMUL.X	FPm,FPn
		FrMUL.<fmt>[2]	<ea>,FPn
		FrMUL.X[2]	FPm,FPn
FNEG	~(Source) → FPn	FNEG.<fmt>	<ea>,FPn
		FNEG.X	FPm,FPn
		FNEG.X	FPn
		FrNEG.<fmt>[2]	<ea>,FPn
		FrNEG.X[2]	FPm,FPn
		FrNEG.X[2]	FPn
FNOP	None	FNOP	
FRESTORE	If in supervisor state then FPU state frame → internal state else TRAP	FRESTORE <ea>	
FSAVE	If in supervisor state then FPU internal state → state frame else TRAP	FSAVE <ea>	
FScc	If (condition true) then 1s → destination else 0s → destination	FScc.<size> <ea>	
FSQRT	Square root of source → FPn	FSQRT.<fmt>	<ea>,FPn
		FSQRT.X	FPm,FPn
		FSQRT.X	FPn
		FrSQRT.<fmt>[2]	<ea>,FPn
		FrSQRT[2]	FPm,FPn
		FrSQRT[2]	FPn
FSUB	FPn-source → FPn	FSUB.<fmt>	<ea>,FPn
		FSUB.X	FPm,FPn
		FrSUB.<fmt>	<ea>,FPn
		FrSUB.X[2]	FPm,FPn
FTRAPcc	If condition true, then TRAP	FTRAPcc	
		FTRAPcc.W #<data>	
		FTRAPcc.L #<data>	
FTST	Condition codes for operand → FPCC	FTST.<fmt>	<ea>
		FTST.X	FPm

Instruction Set Summary (Sheet 5 of 8)

Opcode	Operation	Syntax
ILLEGAL	SSP − 2 → SSP; vector offset → (SSP); SSP − 4 → SSP; PC → (SSP); SSP − 2 → SSP; SR → (SSP); illegal instruction vector address → PC	ILLEGAL
JMP	Destination address → PC	JMP <ea>
JSR	SP − 4 → SP; PC → (SP) Destination address → PC	JSR <ea>
LEA	<ea> → An	LEA <ea>,An
LINK	SP − 4 → SP; An → (SP) SP → An, SP + d → SP	LINK An,#<displacement>
LSL, LSR	Destination shifted by <count> → destination	LSd^{5}Dx,Dy LSd^5 <data>, Dy LSd^5 #<ea>
MOVE	Source → destination	MOVE <ea>,<ea>
MOVEA	Source → destination	MOVEA <ea>,An
MOVE to CCR	CCR → destination	MOVE CCR,<ea>
MOVE to CCR	Source → CCR	MOVE <ea>,CCR
MOVE from SR	If supervisor state then SR → destination else TRAP	MOVE SR,<ea>
MOVE to SR	If supervisor state then source → SR else TRAP	MOVE <ea>,SR
MOVE USP	If supervisor state then USP → An or An → USP else TRAP	MOVE USP,An MOVE An,USP
MOVE 16	Source block → destination block	MOVE 16 (Ax)+,(Ay)+ MOVE 16 xxx.L,(An) MOVE 16 xxx.L,(An)+ MOVE 16 (An),xxx.L MOVE 16 (An)+,xxx.L
MOVEC	If supervisor state then Rc → Rn or Rn → Rc else TRAP	MOVEC Rc,Rn MOVEC Rn,Rc
MOVEM	Registers → destination Source → registers	MOVEM register list,<ea> MOVEM <ea>,register list
MOVEP	Source → destination	MOVEP Dx,(d,Ay) MOVEP (d,Ay),Dx
MOVEQ	Immediate data → destination	MOVEQ #<data>,Dn

Instruction Set Summary (Sheet 6 of 8)

Opcode	Operation	Syntax	
MOVES	If supervisor state then Rn → destination [DFC] or source [SFC] → Rn else TRAP	MOVES Rn,<ea> MOVES <ea>,Rn	
MULS	Source × destination → destination	MULS.W <ea>,Dn MULS.L <ea>,DI MULS.L <ea>,Dh:DI	$16 \times 16 \to 32$ $32 \times 32 \to 32$ $32 \times 32 \to 64$
MULU	Source × destination → destination	MULU.W <ea>,Dn MULU.L <ea>,DI MULU.L <ea>,Dh:DI	$16 \times 16 \to 32$ $32 \times 32 \to 32$ $32 \times 32 \to 64$
NBCD	$0 - (Destination_{10}) - X \to$ destination	NBCD <ea>	
NEG	$0 - (Destination) \to$ destination	NEG <ea>	
NEGX	$0 - (Destination) - X \to$ destination	NEGX <ea>	
NOP	None	NOP	
NOT	~Destination → destination	NOT <ea>	
OR	Source $\vee$ destination → destination	OR <ea>,Dn OR Dn,<ea>	
ORI	Immediate data $\vee$ destination → destination	ORI #<data>,<ea>	
ORI to CCR	Source $\vee$ CCR → CCR	ORI #<data>,CCR	
ORI to SR	If supervisor state then source $\vee$ SR → SR else TRAP	ORI #<data>,SR	
PACK	Source (Unpacked BCD) + adjustment → destination (packed BCD)	PACK − (Ax), − (Ay),#<adjustment> PACK Dx,Dy,#<adjustment>	
PEA	$Sp - 4 \to SP$; <ea> → (SP)	PEA <ea>	
PFLUSH	If supervisor state then invalidate instruction and data ATC entries for destination address else TRAP	PFLUSH (An) PFLUSHN (An) PFLUSHA PFLUSHAN	
PTEST	If supervisor state then logical address status → MMUSR; entry → ATC else TRAP	PTESTR (An) PTESTW (An)	
RESET	If supervisor state then assert $\overline{\text{RSTO}}$ line else TRAP	RESET	
ROL,ROR	Destination rotated by <count> → destination	ROd^{5}Rx,Dy ROd^5 #<data>,Dy ROd^5 <ea>	

Instruction Set Summary (Sheet 7 of 8)

Opcode	Operation	Syntax
ROXL,ROXR	Destination rotated with X by <count> → destination	ROXd^{5}Dx,Dy ROXd^5 #<data>,Dy ROXd^5 <ea>
RTD	(SP) → PC; SP + 4 + d → SP	RTD #<displacement>
RTE	If supervisor state the (SP) → SR; SP + 2 → SP; (SP) → PC; SP + 4 → SP; restore state and deallocate stack according to (SP) else TRAP	RTE
RTR	(SP) → CCR; SP + 2 → SP; (SP) → PC; SP + 4 → SP	RTR
RTS	(SP) → PC; SP + 4 → SP	RTS
SBCD	Destination$_{10}$ − source$_{10}$ − X → destination	SBCD Dx,Dy SBCD − (Ax), − (Ay)
Scc	If condition true then 1s → destination else 0s → destination	Scc <ea>
STOP	If supervisor state then immediate data → SR; STOP else TRAP	STOP #<data>
SUB	Destination − source → destination	SUB <ea>,Dn SUB Dn,<ea>
SUBA	Destination − source → destination	SUBA <ea>,An
SUBI	Destination − immediate data → destination	SUBI #<data>,<ea>
SUBQ	Destination − immediate data → destination	SUBQ #<data>,<ea>
SUBX	Destination − source − X → destination	SUBX Dx,Dy SUBX − (Ax), − (Ay)
SWAP	Register [31:16] ↔ register [15:0]	SWAP Dn
TAS	Destination tested → condition codes; 1 → bit 7 of destination	TAS <ea>
TRAP	SSP − 2 → SSP; format/offset → (SSP); SSP − 4 → SSP; PC → (SSP); SSP − 2 → SSP SR → (SSP); vector address → PC	TRAP #<vector>
TRAPcc	If cc then TRAP	TRAPcc TRAPcc.W #<data> TRAPcc.L #<data>
TRAPV	If V then TRAP	TRAPV
TST	Destination tested → condition codes	TST <ea>

Instruction Set Summary (Sheet 8 of 8)

Opcode	Operation	Syntax
UNLK	An → SP; (SP) → An; SP + 4 → SP	UNLK An
UNPK	Source (packed BCD)+adjustment → destination (unpacked BCD)	UNPACK − (Ax), − (Ay),#\<adjustment> UNPACK Dx,Dy,#\<adjustment>

Notes:
1. Specifies either the instruction (IC), data (DC), or IC/DC caches.
2. Where r is rounding precision, S or D.
3. A list of any combination of the eight floating-point data registers, with individual register names separated by a slash (/), and/or contiguous blocks of registers specified by the first and last register names separated by a dash (—).
4. A list of any combination of the three floating-point system control registers (FPCR, FPSR, and FPIAR) with individual register names separated by a slash (/).
5. where d is direction, L or R.

source: Courtesy of Motorola, Inc.

Problems to Part 3

1. Write the instructions necessary to branch to label STOP if the contents of the D0 register is equal to 100.

2. Will the carry bit be set in the following cases?
   a. (D0) = 55;   ADD.B #27, D0
   b. (D0) = 150;   ADD.B #110, D0

3. Will the overflow bit be set in the following cases?
   a. (D0) = − 100;   ADD.B #50, D0
   b. (D0) = − 100;   ADD.B # − 50, D0

4. Write the instructions necessary to compare VAR1 and VAR2, branching to label EQUAL if they are equal and to label LESS if VAR1 < VAR2.

5. Use the address register indirect mode to add the first 20 longwords starting at address $2000. The result will be in D0.

6. Repeat Problem 5 using the address register indirect with postincrement mode.

7. Push value 100 onto the stack.

8. Push register D0 on stack.

9. Pop the stack into D1.

10. Write the instructions necessary to rotate to the left and right register D2 to 10 bits, not including the extend bit.

11. Perform the NOR operation of the D4 and D5 registers, placing the result in D5.

12. Test whether bit 15 of register D2 is 1.

13. Multiply the unsigned bytes in VAR1 and VAR2 locations, placing the result in VAR2.

14. Divide unsigned word VAR by 15, placing the result in VAR.

15. Check whether 25 < (D2) < 50. Branch to OUTOB if the inequality does not hold.

16. Assume S = 1, M = 1 in the SR. Write an instruction to initialize the MSP with the contents of A0.

17. Show the results of D0 and D1 after the following instruction executes:

```
BFEXTU D0{32:16}, D1
```

Before	After
D0 $ 55550000	
D1 $ 00000000	

18. The contents of D0 prior to the execution of the instructions EXT.W D0 is $FFFFFF5C. What will D0 contain after execution?

19. Write the instructions necessary to replace

```
LINK A3, # - $10
```

20. Write the instructions necessary to replace

```
UNLK A3
```

21. The contents of the MC68060 SR is $0519. Provide a complete interpretation of the processor status.

22. Write the necessary instructions to enable both caches.

23. Given the logical address on the MC68040: $FFFFA000. Assume a page size of 8 kbytes. Provide a drawing of all levels of the translation table tree and the appropriate logical address format partitioning. Assume supervisor mode operation.

24. Repeat Problem 23 for a 4-kbyte page size. List the differences.

25. The content of the MMUSR of the MC68040 is $FFA00424. Provide a complete interpretation.

26. Given an address on the MC68040: $ AAAA F040. Provide a complete analysis of the data cache access with this address. Which set in the cache is accessed? What is the tag? Which line in main memory is represented?

Advanced RISC
Microprocessors

14

The DEC Alpha AXP

14.1 Introduction

Digital Equipment Corporation (DEC), or simply Digital, is well known for its pioneering minicomputer products, such as the PDP-11 family, and its superminicomputer VAX family. Both the PDP and the VAX families have been widely used throughout the entire world for many years. The VAX is an obvious example of a CISC architecture (see Chap. 6), with its extensive instruction set, formats, and addressing modes. In parallel with extending the capabilities and the various models of the VAX, DEC researchers have been working on the development of their own RISC-type microprocessor. This development has culminated in the announcement of the Alpha AXP in 1991, the first Digital RISC. Prior to that, Digital has been using MIPS R2000 and R3000 RISC-type microprocessors (see Chap. 16) in its DECstation family of workstations [Tabk 90b]. The Alpha has all of the attributes of its other RISC-type contemporaries. It practices instruction-level parallelism; it is a two-issue superscalar (see Chap. 5). It has an on-chip FPU, MMU, and a dual cache (8 kbytes code, 8 kbytes data). In some sense it can be said that it is ahead of its time: starting at 150 MHz, and reaching 300 MHz, it surpasses all its contemporaries in its frequency of operation (at least during the period of 1992–1994). It could indeed be called the fastest microprocessor of the early nineties. The first product realizing the Alpha AXP architecture is labeled 21064. It starts a new Digital family of RISC-type microprocessors.

The architecture of the Alpha will be discussed in detail in the next section. Its physical realization and organization will be presented in the subsequent one.

14.2   The Alpha AXP Architecture

The Alpha is a 64-bit RISC-type microprocessor [McLe 93, Sits 92a,b, 93]. It essentially satisfies all the RISC properties (see Chap. 6). In particular, it features register-to-register operation and memory access by load and store instructions only. All the Alpha instructions have a fixed length of 32 bits, as do practically all other RISC-type processors. As most other RISC-type systems, Alpha features three-operand instructions. Since the Alpha is a 64-bit system, all its CPU general-purpose registers are 64-bit. The bits in all registers are numbered so that the LSB is 0, and the MSB is 63. Alpha has two sets of thirty-two 64-bit registers: one for the IU: R0, R1,..., R31, and one for the FPU: F0, F1,..., F31. Registers R31 and F31 are hardwired to contain a value of zero at all times, similar to most other RISCs (however, in other RISCs it is usually R0). Register R30 is designated as a SP. Otherwise, all other registers R0 to R29 and F0 to F30 do not have any prespecified tasks. The PC is a separate 64-bit register (unlike the preceding VAX architecture where the PC is R15). All Alpha operations are done between the CPU 64-bit registers. The architecture also supports 32-bit integer operations. Alpha is byte-addressable; its basic addressable unit is the 8-bit byte. Memory is accessed using 64-bit virtual little-endian (the lowest byte has the lowest address) byte addresses. The minimum virtual address is 43 bits. The Alpha architecture is also called Alpha AXP architecture.

Data types

Alpha architecture recognizes the following data types.

Integer data types

1. Byte, 8 bits. Basic addressable unit.
2. Word, 16 bits. Two contiguous bytes starting on an arbitrary byte boundary. A word is addressed by the address of its least significant byte (the byte that contains bit zero).
3. Longword, 32 bits. Four contiguous bytes starting on an arbitrary byte boundary. A longword is addressed by the address of its least significant byte. In a longword integer, bit 31 (MSB) is the sign bit.
4. Quadword, 64 bits. Eight contiguous bytes starting on an arbitrary byte boundary. A quadword is addressed by the address of its LSB. In a 64-bit integer, bit 63 is the sign bit.

The Alpha integer data types are shown in Fig. 14.1.

Although words, longwords, and quadwords may be stored at any byte address, better performance can be achieved if they are naturally aligned. That is, longwords are stored in addresses divisible by 4 (low-

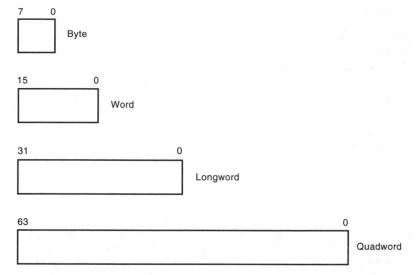

Figure 14.1 Alpha integer data types.

order 2 bits of the address are zero), and quadwords are stored in addresses divisible by 8 (low-order 3 bits are zero).

Floating-point data types

Alpha architecture features two groups of floating-point data types:

1. VAX floating-point formats, for backward compatibility with the VAX software

2. IEEE standard (ANSI/IEEE 754-1985) floating-point formats, as practiced in practically all other modern systems [IEEE 85]

VAX floating-point formats

Alpha architecture features three VAX floating-point formats:

1. F floating, 32 bits

2. G floating, 64 bits

3. D floating, 64 bits

The memory storage of the above formats is illustrated in Fig. 14.2, and their CPU floating register storage is illustrated in Fig. 14.3. Although A may be any address in memory, better performance will be attained if A is naturally aligned (divisible by 4 for F, and by 8 for G and D formats). The main difference between the G and D formats is

F_floating Datum

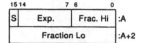

G_floating Datum

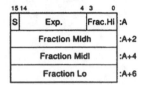

D_floating Datum

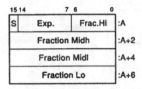

Figure 14.2 Memory storage of VAX floating-point formats. (*Courtesy of Digital Equipment Corp.*)

F_floating Register Format

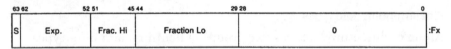

G_floating Format

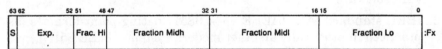

D_floating Register Format

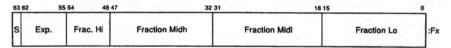

Figure 14.3 Register format of VAX floating-point formats. (*Courtesy of Digital Equipment Corp.*)

that G has an 11-bit exponent field, while D has an 8-bit one. Thus, the G format has a much higher range. The D format is not fully supported on the Alpha, and no D floating-point arithmetic operations are provided. For VAX compatibility, exact D floating-point arithmetic may be provided by software emulation.

In the F floating-point format the 8-bit exponent encodes the values of 0 to 255. An exponent value of 0, together with a sign bit (S) of 0, is interpreted as an F datum of zero value. Exponent values of 1 to 255 represent true binary exponents of -127 to 127, respectively. Thus, the bias is 128. An exponent value of 0, together with a sign bit of 1, is interpreted as a *reserved operand*. An attempt to use a reserved operand in a floating-point instruction causes an arithmetic exception. The range of an F datum is approximately $0.29(10)^{-38}$ to $1.7(10)^{38}$ decimal. The precision of an F datum is approximately 2^{-23}, typically seven decimal digits.

In the G floating-point format the 11-bit exponent encodes the values of 0 to 2047. An exponent value of 0, together with a sign 0, represents a zero value. Exponent values of 1 to 2047 represent true binary exponents of -1023 to 1023, respectively. An exponent value of 0, together with a sign bit of 1, represent's a reserved operand. The G datum range is approximately $0.56(10)^{-308}$ to $0.9(10)^{308}$. The precision of a G datum is approximately 2^{-52}, typically 15 decimal digits. The precision of a D datum is approximately 2^{-55}, typically 16 decimal digits.

IEEE floating-point formats

The IEEE standard features the single-precision 32-bit (S floating), and the double-precision 64-bit (T floating) formats (see Chap. 3 for details). Their memory and floating register storage are illustrated in Fig. 14.4. Location A may be anywhere in memory, but for better performance it should be naturally aligned, as in the case of the VAX formats.

Longword and quadword integers may be stored in FPU registers. Their storage in memory and in FPU registers is illustrated in Fig. 14.5.

Instruction formats

All Alpha instructions are 32 bits (longword) long. The Alpha architecture features five basic instruction formats illustrated in Fig. 14.6. The notation used in Fig. 14.6 is the following:

Ra, Rb, Rc are integer register operands.

Fa, Fb, Fc are floating-point register operands.

disp is a displacement, added to the value in Rb to form a virtual address.

SBZ should be zero.

LIT is an 8-bit literal value from 0 to 255.

S_floating Datum

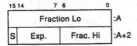

S_floating Register Format

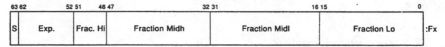

T_floating Datum

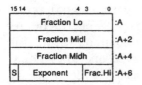

T_floating Register Format

63 62	52 51	48 47	32 31	16 15	0
S	Exp.	Frac. Hi	Fraction Midh	Fraction Midl	Fraction Lo	:Fx

Figure 14.4   Alpha IEEE floating-point formats. (*Courtesy of Digital Equipment Corp.*)

All instruction formats have a 6-bit (bits <31:26>) major opcode field. Any unused register field (5 bits) of an instruction (Ra, Rb, Fa, or Fb) must be set to a value of 31 (11111 binary).

The five instruction formats will now be discussed separately.

Memory instruction format.   This format is used to transfer information between registers and memory, to load an effective address, and for subroutine jumps. The Memory_disp field is a byte offset. It is sign extended and added to the contents of register Rb to form a virtual address. The virtual address is used as a memory load/store address or a result value, depending on the specific instruction. For some instructions, the Memory_disp field is replaced by the Function field. It serves as an extension of the opcode that designates a set of miscellaneous instructions.

Branch instruction format.   The branch format is used for conditional branch instructions (in which case the Ra field contains the condition encoding) and for PC-relative subroutine jumps. As each instruction is decoded, the PC value is advanced to point to the next sequential instruction. The new PC value is referred to as the *updated PC*.

Longword Integer Datum

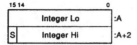

Longword Integer Floating-Register Format

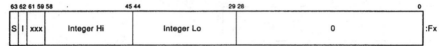

Quadword Integer Datum

Quadword Integer Floating-Register Format

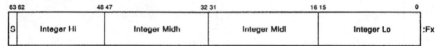

Figure 14.5 Integer data storage in memory and FPU registers. (*Courtesy of Digital Equipment Corp.*)

The Branch_disp field is treated as a longword offset. It is shifted left 2 bits (to address a longword boundary), sign-extended to 64 bits, and added to the updated PC value to form the target virtual address.

Operate instruction format. The operate instruction format is used for instructions that perform integer register-to-register operations. Fields Ra and Rb specify source operands. Field Rc specifies the destination. The Function field is an extension of the opcode. If bit 12 is 0, Rb specifies a source register operand. If bit 12 is 1, an 8-bit zero-extended literal constant is formed by bits <20:13> of the instruction. The literal is interpreted as a positive integer between 0 and 255 and is zero-extended to 64 bits.

Floating-point operate instruction format. This format is used for instructions that perform floating-point register-to-register operations. The Fa and Fb fields specify floating-point register source operands. The Fc field specifies the destination. Floating-point convert instructions use a subset of the floating-point operate format and perform register-to-register conversion operations. The Fb operand specifies the source, the Fa field must be F31 (i.e., zero), and Fc is naturally the destination.

Memory Instruction Format

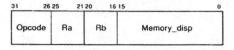

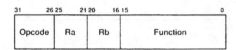

Branch Instruction Format

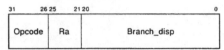

Operate Instruction Format

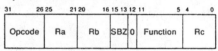

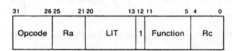

Floating-Point Operate Instruction Format

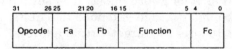

PALcode Instruction Format

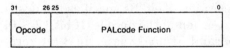

Figure 14.6 Alpha instruction formats. (*Courtesy of Digital Equipment Corp.*)

PALcode instruction format.   The privileged architecture library (PAL) code format is used to specify extended processor functions (more on PALcode later in this section). The 26-bit PALcode Function field specifies the particular PALcode operation. The source and destination operands for PALcode instructions are supplied in fixed registers that are specified in the individual instruction descriptions. An opcode of zero and a PALcode function of zero specify the HALT instruction.

Addressing modes

The Alpha architecture features four simple addressing modes (see Chap. 3), as practiced on RISC-type systems (see Chap. 6):

1. Register
2. Immediate
3. Register indirect with displacement
4. PC-relative

Instruction set

The Alpha architecture features the following types of instructions [Sits 92a,b]:

1. Integer load and store
2. Integer control
3. Integer arithmetic
4. Logical and shift
5. Byte manipulation
6. Floating-point load and store
7. Floating-point control
8. Floating-point operate
9. Miscellaneous

All the instruction types will be discussed in this section group by group. Prior to that, some notation must be introduced. An instruction operand is specified by the following attributes:

```
<name>.<access type><data type>
```

The <name> may be any of the registers: Ra, Rb, Rc, Fa, Fb, Fc, or

```
disp    The displacement field of the instruction
fnc     The PAL function field of the instruction
#b      An integer literal operand in the Rb field of the instruction
```

The <access type> is a letter denoting the operand access type. It may be one of the following:

Access type	Meaning
a	Used in address calculation "al" means scale by 4 (longwords) "aq" means scale by 8 (quadwords) "ab" means the operand is in byte units
i	The operand is an immediate literal
r	The operand is read only
m	The operand is both read and written
w	The operand is write only

The <data type> is a letter denoting the data type of the operand. It may be one of the following:

Data type	Meaning
b	Byte
f	F_floating
g	G_floating
l	Longword
q	Quadword
s	S_floating
t	T_floating
w	Word
x	Specified by the instruction

Integer load and store instructions

The memory access integer load and store instructions (a total of 12) are summarized in Table 14.1.

In all instructions in Table 14.1, except LDAH, the virtual memory address is computed by adding the content of register Rb to the sign-extended 16-bit displacement.

TABLE 14.1   Integer Load and Store Instructions

Mnemonic	Operands	Operation
LDA	Ra.wq, disp.ab(Rb.ab)	Load address
LDAH	Ra.wq, disp.ab(Rb.ab)	Load address high
LDL	Ra.wq, disp.ab(Rb.ab)	Load SE longword
LDQ	Ra.wq, disp.ab(Rb.ab)	Load quadword
LDQ_U	Ra.wq, disp.ab(Rb.ab)	Load quadword unaligned
LDL_L	Ra.wq, disp.ab(Rb.ab)	Load SE longword locked
LDQ_L	Ra.wq, disp.ab(Rb.ab)	Load quadword locked
STL	Ra.rq, disp.ab(Rb.ab)	Store longword
STQ	Ra.rq, disp.ab(Rb.ab)	Store quadword
STL_C	Ra.mq, disp.ab(Rb.ab)	Store longword conditional
STQ_C	Ra.mq, disp.ab(Rb.ab)	Store quadword conditional
STQ_U	Ra.rq, disp.ab(Rb.ab)	Store quadword unaligned

SE = sign-extended

```
Ra ← Rbv + SE(disp)
```

where Rbv is the value in Rb.

For LDAH the displacement is multiplied by $65536 = 2^{16}$ and added sign-extended to Rbv:

```
Ra ← Rbv + SE(disp*65536)
```

The load locked and store conditional instructions are intended for multiprocessor implementation and will not be discussed in this section.

Integer control instructions

Integer control instructions include conditional and unconditional branch, branch to subroutine, and jump instructions. The integer control instructions (a total of 14) are summarized in Table 14.2.

The displacement field of the instruction is used to encode this information as follows:

disp<15:14>	Mnemonic	Predicted Target <15:0>	Prediction Stack Action
00	JMP	PC+{4*disp< 13:0>}	—
01	JSR	PC+{4*disp<13:0>}	Push PC
10	RET	Prediction stack	Pop
11	JSR_COROUTINE	Prediction stack	Pop, push PC

TABLE 14.2   Integer Control Instructions

Mnemonic	Operands	Operation
BEQ	Ra.rq, disp.al	Branch if Rav = 0
BGE	Ra.rq, disp.al	Branch if Rav ≥ 0
BGT	Ra.rq, disp.al	Branch if Rav > 0
BLBC	Ra.rq, disp.al	Branch if Ra LSB = 0
BLBS	Ra.rq, disp.al	Branch if Ra LSB = 1
BLE	Ra.rq, disp.al	Branch if Rav ≤ 0
BLT	Ra.rq, disp.al	Branch if Rav < 0
BNE	Ra.rq, disp.al	Branch if Rav ≠ 0
BR	Ra.wq, disp.al	Unconditional branch
BSR	Ra.wq, disp.al	Branch to subroutine
JMP	Ra.wq,(Rb.ab),hint	Jump
JSR	Ra.wq,(Rb.ab),hint	Jump to subroutine
RET	Ra.wq,(Rb.ab),hint	Return from subroutine
JSR_COROUTINE	Ra.wq,(Rb.ab),hint	Jump to subr. return

Hint is an encoding to possible branch prediction logic.

The four different opcodes set different bit patterns in disp<15:14>, and the hint operand sets disp<13:0>.

The updated PC value (updated after the instruction is fetched) is stored in register Ra, and then the PC is loaded with the target virtual address, supplied from register Rb. The low 2 bits of Rb are ignored. Fields Ra and Rb may specify the same register; the target address calculation using the old value of Rb is done before the new value is assigned.

The Alpha architecture specifies three types of branching hints in instructions [Sits 93]:

1. Architected static branch prediction rule: forward conditional branches are predicted not-taken, and backward ones taken. To the extent that compilers and hardware implementations follow this rule, programs can run more quickly with little hardware cost. This hint does not preclude doing dynamic branch prediction in an implementation, but it may reduce the need to do so.

2. Describes computed jump targets. Otherwise unused instruction bits are defined to give the low bits of the most likely target, using the same target calculation as unconditional branches. The 14 bits provided are enough to specify the instruction offset within a page, which is often enough to start a fastest-level instruction cache fetch many cycles before the actual target value is known.

3. Describes subroutine and coroutine returns. By marking each branch and jump as "call," "return," or "neither," the architecture provides in implementation enough information to maintain a small stack of likely subroutine return addresses. This implementation stack can be used to prefetch subroutine returns quickly.

The conditional move instructions and the branching hints eliminate some branches and speed up the remaining ones without compromising multiple instruction issue.

Example   Given a jump to subroutine instruction JSR R26, (R27), 0x4123;

The hint is the hexadecimal value 4123. The most significant 2 bits of the hint 01 specify that the computed jump is a call to a subroutine (JSR), and not a return or a regular jump. The remaining 14 bits of the hint (00 0001 0010 0011) are an offset of the most likely jump target (the most likely value to be in R27 at run time). The actual calculation using these bits is exactly as the PC-relative calculation: multiply by 4 (shift 2 bits left) and add to the updated PC.

Integer arithmetic instructions

The integer arithmetic instructions of Alpha (a total of 20) are listed in Table 14.3.

TABLE 14.3   Integer Arithmetic Instructions

Mnemonic	Operation
ADDL	Add longword
ADDQ	Add quadword
S4ADDL	Scaled add longword by 4
S8ADDL	Scaled add longword by 8
S4ADDQ	Scaled add quadword by 4
S8ADDQ	Scaled add quadword by 8
CMPEQ	Compare signed quadword =
CMPLT	Compare signed quadword <
CMPLE	Compare signed quadword < or =
CMPULT	Compare unsigned quadword <
CMPULE	Compare unsigned quadword < or =
MULL	Multiply longword
MULQ	Multiply quadword
UMULH	Unsigned multiply quadword high
SUBL	Subtract longword
SUBQ	Subtract quadword
S4SUBL	Scaled subtract longword by 4
S8SUBL	Scaled subtract longword by 8
S4SUBQ	Scaled subtract quadword by 4
S8SUBQ	Scaled subtract quadword by 8

The operands for the integer arithmetic instructions may be of two forms:

```
Ra.rq, Rb.rq, Rc.wq
```

or

```
Ra.rq, #b.ib, Rc.wq
```

The ADD instructions work in general as follows:

```
Rc ← Rav + Rbv
```

For a longword (32-bit) addition ADDL, the high-order 32 bits of Ra and Rb are ignored, and the sum is sign-extended:

```
Rc ← SE [(Rav + Rbv)<31:0>]
```

In the scaled add instructions, register Ra is shifted left by 2 bits (for S4), or by 3 bits (for S8), and then added to Rb or a literal with the sum going to Rc. For a longword addition, the upper 32 bits of Ra and Rb are ignored.

In the compare instructions (CMPxx, CMPUxx), the value in register Ra is compared to the value in register Rb, or to a literal #b. If the specified relationship is true, the value one is stored into Rc. Otherwise, Rc is cleared.

Example   CMPLE R1, R2, R5 ; compare signed quadword less than or equal.

```
Rav = 50, Rbv = 100, Rav < Rbv; relationship LE (less or equal)
is true
therefore Rc ← 1
```

In the multiplication instruction MULQ Rav is multiplied by Rbv and the 64-bit product is stored in Rc. In the longword multiply, the high 32 bits of Ra and Rb are ignored, and the sign-extended 32-bit product is stored in Rc.

```
Thus, for MULQ:           Rc ← Rav * Rbv
For MULL:                 Rc ← SE [(Rav * Rbv) <31:0>]
```

If an overflow occurs in MULQ, the least significant 64 bits of the product are written into Rc. The upper 64 bits of the 128-bit product can be generated using the UMULH (unsigned multiply quadword high) instruction, in which only the upper 64 bits of the 128-bit unsigned product of Ra and Rb (or a literal) are stored into Rc.

```
For UMULH:    Rc ← Unsigned (Rav * Rbv) <127:64>
```

In the subtract instructions, the value of Rb or a literal is subtracted from the value of Ra and the result is stored in Rc. As in ADD, the upper 32 bits of Ra and Rb are ignored in a longword subtraction, and the result is sign-extended.

```
For SUBQ:                 Rc ← Rav − Rbv
For SUBL:                 Rc ← SE [(Rav − Rbc) <31:0>]
```

The scaled subtract instructions work in complete analogy with the scaled add instructions. For instance, in the S8SUBQ instruction, Ra is left-shifted by 3 bits before Rbv is subtracted from it, with the 64-bit difference going to Rc.

The Alpha architecture does not feature an integer divide instruction, however, a floating-point divide is available.

Logical and shift instructions

The Alpha logical instructions perform quadword Boolean operations. The conditional move instructions (not featured in most other systems) in this group performs conditional transfers from register to register without a branch. The shift instructions perform left and right logical shift and right arithmetic shift. These instructions (a total of 17) are summarized in Table 14.4.

The operand specification for the above instructions is

```
Ra.rq, Rb.rq, Rc.wq
```

or

```
Ra.rq, #b.ib, Rc.wq
```

TABLE 14.4   Logical and Shift Instructions

Mnemonic	Operation
AND	Logical AND
BIC	Logical AND with complement
BIS	Logical OR
EQV	Logical equivalence (XORNOT)
ORNOT	Logical OR with complement
XOR	Exclusive OR
CMOVxx	Conditional move integer
SLL	Shift left logical
SRA	Shift right arithmetic
SRL	Shift right logical

The logical operations are performed as follows:

Mnemonic	Operation
AND	Rc ← Rav AND Rbv
BIC	Rc ← Rav AND (NOT Rbv)
BIS	Rc ← Rav OR Rbv
ORNOT	Rc ← Rav OR (NOT Rbv)
XOR	Rc ← Rav XOR Rbv
EQV	Rc ← Rav XOR (NOT Rbv)

The pure complement function NOT can be performed by doing the ORNOT with Ra = R31 (always zero).

```
Thus: ORNOT R31, Rb, Rb;      0 OR (NOT Rbv) = NOT Rbv → Rb.
```

The OR mnemonic may be used instead of BIS. It has the same meaning and it is accepted by the assembler.

The CMOVxx instruction (xx is the condition mnemonic) works by testing Rav for the specified condition. If the condition is true, the content of Rb is transferred to Rc. There are eight conditions featured on CMOVxx (thus there are eight CMOVxx instructions):

Condition (xx)	CMOVE If Register Ra:
EQ	Equal to zero
GE	Greater than or equal to zero
GT	Greater than zero
LBC	Low bit clear
LBS	Low bit set
LE	Less than or equal to zero
LT	Less than zero
NE	Not equal to zero

In the shift instructions register Ra is shifted left or right 0 to 63 bits. The bit count (by how many bits shifted) is either in register Rb (in Rbv<5:0>) or given as a literal #b. The shifted result is stored in Rc. In the logical shift, zero bits are propagated into the vacated bit positions. In the arithmetic shift, the sign bit (Rav<63>) is propagated into the vacated bit positions.

Byte-manipulation instructions

The Alpha architecture features five types of byte-manipulation instructions (a total of 24 instructions) within registers. This is an unusual feature compared to other systems. The byte-manipulation instructions can be used with the load and store unaligned instructions to manipulate short unaligned strings of bytes. The five types of byte-manipulation instructions will be discussed next one after the other, along with their particular options. All of the byte-manipulation instructions have the same operand specifications as the logical and shift instructions described above.

1. *Compare byte*.   This group contains a single instruction CMP-BGE (compare byte greater or equal). It does eight parallel unsigned byte comparisons between corresponding bytes of Rav and Rbv, storing the eight results in the low 8 bits of Rc. The high 56 bits of Rc are set to zero. Bit 0 of Rc corresponds to byte 0 of Ra and Rb, bit 1 of Rc corresponds to byte 1, and so on. A result bit is set in Rc if the corresponding byte of Rav is greater or equal to that of Rbv (unsigned). The compare byte instruction allows character-string search and compare to be done 8 bytes at a time.

2. *Extract byte*.   This group features seven options for the EXTxx instruction:

EXTxx Option	Extract –
BL	Byte low
WL	Word low
LL	Longword low
QL	Quadword low
WH	Word high
LH	Longword high
QH	Quadword high

The EXTxL shifts register Ra right by 0 to 7 bytes, inserts zeros into vacated bit positions, and then extracts 1, 2, 4, or 8 bytes into register Rc. Instruction EXTxH shifts register Ra left by 0 to 7 bytes, inserts zeros into vacated bit positions, and then extracts 2, 4, or 8 bytes into register Rc. The number of bytes to shift is specified by the 3 LSBs of Rb (Rbv<2:0>) or by an immediate value of a literal #b. The number of bytes to extract is specified in the function code. Remaining bytes are filled with zeros. A single EXTxL instruction can perform byte or word loads (the load instructions work with longwords and quadwords only), pulling the datum out of a quadword and placing it in the low end of a register with high-order zeros. A pair of EXTxL/EXTxH instructions can perform unaligned loads, pulling the two parts of an unaligned datum out of two quadwords and placing the parts in result registers, where they are ready for combining into the full datum by a simple OR.

Example Assume a quadword data item HGFE DCBA (a string of byte characters) is stored unaligned in two memory quadword locations: X(R11) and X + 7(R11). The 3 LSBs of X(R11) are 101(or 5). The value in X(R11) is CBAx xxxx, and the value in X + 7(R11) is xxxH GFED (x is a "don't care" byte). The following is a small program placing the complete quadword datum into register R1:

```
LDQ_U R1, X(R11);   R1 ← CBAx xxxx
LDQ_U R2, X + 7(R11);   R2 ← xxxH GFED
LDA R3, X(R11);   R3 <2:0> ← 101 ( = 5)
EXTQL R1, R3, R1;   R1 ← 0000 0CBA
EXTQH R2, R3, R2;   R2 ← HGFE D000
BIS R2, R1, R1;   R1 ← HGFE DCBA
```

3. *Byte insert.*   The byte insert instruction INSxx has seven options:

INSxx Option	Insert
BL	Byte low
WL	Word low
LL	Longword low
QL	Quadword low
WH	Word high
LH	Longword high
QH	Quadword high

INSxL and INSxH shift bytes from register Ra and insert them into a field of zeros, storing the result in register Rc. Register Rbv<2:0> or a literal #b select the shift amount (0 to 7), and the function code selects the maximum field width: 1, 2, 4, or 8 bytes. The instructions can generate a byte, word, longword, or quadword datum that is spread across two registers at an arbitrary byte alignment.

4. *Byte mask.*   The byte mask MSKxx instruction has seven options same as the byte insert and byte extract instructions (MSKBL, MSKWL, MSKLL, MSKQL, MSKWH, MSKLH, MSKQH). MSKxL and MSKxH set selected bytes of register Ra to zero, storing the result in register Rc. Register Rbv<2:0> or a literal selects the starting position of the field of zero bytes, and the function code selects the maxi-

mum width: 1, 2, 4, or 8 bytes. The instructions generate a byte, word, longword, or quadword field of zeros that can spread across two registers at an arbitrary byte alignment. The INSxx and MSKxx instructions position new data and zero out old data in registers for storing bytes, words, and unaligned data.

5. *Zero bytes.* This group contains two instructions:

```
ZAP                         Zero bytes
ZAPNOT                      Zero bytes not
```

These instructions set selected bytes of register Ra to zero, and store the result in register Rc. Register Rbv<7:0> or a literal selects the bytes to be zeroed; bit 0 of Rb corresponds to byte 0 of Ra, bit 1 of Rb corresponds to byte 1 of Ra, and so on. A result byte is set to zero if the corresponding bit of Rb is a one for ZAP and a zero for ZAPNOT. The ZAP instructions allow zeroing of arbitrary patterns of bytes in a register.

The CMPBGE and ZAP instructions allow very fast implementations of the C language string routines, among other uses.

Floating-point load and store instructions

The floating-point load and store instructions (a total of eight) move floating-point data between memory and floating-point registers. The instructions are summarized in Table 14.5.

In the load instructions Fa is the destination register. The memory address is computed by adding a sign-extended displacement to Rbv for both load and store instructions. Register Fa serves as the source register in the store instructions.

Floating-point control instructions

There are six floating-point branch instructions. These instructions test the value of a floating-point register Fa, and conditionally change the value of the PC. The operand format for all six instructions is

TABLE 14.5   Floating-point Load and Store Instructions

Mnemonic	Operands	Operation
LDF	Fa.wf, disp.ab(Rb.ab)	Load F_floating
LDG	Fa.wg, disp.ab(Rb.ab)	Load G_floating
LDS	Fa.ws, disp.ab(Rb.ab)	Load S_floating
LDT	Fa.wt, disp.ab(Rb.ab)	Load T_floating
STF	Fa.rf, disp.ab(Rb.ab)	Store F_floating
STG	Fa.rg, disp.ab(Rb.ab)	Store G_floating
STS	Fa.rs, disp.ab(Rb.ab)	Store S_floating
STT	Fa.rt, disp.ab(Rb.ab)	Store T_floating

TABLE 14.6   Floating-point Branch Instructions

Mnemonic	Operation
FBEQ	Floating branch equal
FBGE	Floating branch > or equal
FBGT	Floating branch >
FBLE	Floating branch < or equal
FBLT	Floating branch <
FBNE	Floating branch not equal

```
Fa.rq, disp.al
```

The instructions are summarized in Table 14.6.

These instructions use PC-relative addressing. If the specified condition for Fav with respect to zero is true, the displacement value, shifted left by 2 bits, is added to PC, to form a new branch target address.

Floating-point operate instructions

This is the largest group of instructions featured by the Alpha architecture for a total of 47. It can be subdivided into four categories:

1. Arithmetic   16
2. Convert   13
3. Compare   7
4. Miscellaneous 11

The floating-point arithmetic instructions are summarized in Table 14.7.

The operands for all floating-point arithmetic instructions are defined as follows:

```
Fa.rx, Fb.rx, Fc.wx
```

The four arithmetic operations are performed as follows:

```
Addition:        Fc ← Fav + Fbv
Subtraction:     Fc ← Fav − Fbv
Multiplication:  Fc ← Fav * Fbv
Division:        Fc ← Fav/Fbv
```

The floating-point convert operations are summarized in Table 14.8.

In all the above operations Fb contains the datum to be converted, and Fc is the destination where the converted datum is stored.

TABLE 14.7   Floating-point Arithmetic Instructions

Mnemonic	Operation
ADDF	Add F_floating
ADDG	Add G_floating
ADDS	Add S_floating
ADDT	Add T_floating
SUBF	Subtract F_floating
SUBG	Subtract G_floating
SUBS	Subtract S_floating
SUBT	Subtract T_floating
MULF	Multiply F_floating
MULG	Multiply G_floating
MULS	Multiply S_floating
MULT	Multiply T_floating
DIVF	Divide F_floating
DIVG	Divide G_floating
DIVS	Divide S_floating
DIVT	Divide T_floating

There are two types of floating-point compare instructions:

1. CMPGxx Compare G_floating, operands: Fa.rg, Fb.rg, Fc.wq
where xx may take the options:

```
EQ    equal
LE    less than or equal
LT    less than
```

for a total of three instructions.

2. CMPTxx Compare T_floating, operands: Fa.rx, Fb.rx, Fc.wq
where xx may take the options EQ, LE, LT as for CMPGxx, and another option

TABLE 14.8   Floating-point Convert Operations

Mnemonic	Operands	Operation
CVTLQ	Fb.rq, Fc.wx	Convert longword to quadword
CVTQL	Fb.rq, Fc.wx	Convert quadword to longword
CVTDG	Fb.rx, Fc.wx	Convert D_ to G_floating
CVTGD	Fb.rx, Fc.wx	Convert G_ to D_floating
CVTGF	Fb.rx, Fc.wx	Convert G_ to F_floating
CVTGQ	Fb.rx, Fc.wq	Convert G_floating to quadword
CVTQF	Fb.rq, Fc.wx	Convert quadword to F_floating
CVTQG	Fb.rq, Fc.wx	Convert quadword to G_floating
CVTQS	Fb.rq, Fc.wx	Convert quadword to S_floating
CVTQT	Fb.rq, Fc.wx	Convert quadword to T_floating
CVTTQ	Fb.rx, Fc.wq	Convert T-floating to quadword
CVTTS	Fb.rx, Fc.wq	Convert T_ to S_floating
CVTST	Fb.rx, Fc.wx	Convert S_ to T_floating

```
UN     unordered
```

for a total of four instructions (total seven compare instructions).

In all the floating-point compare instructions the operands in Fa and Fb are compared. If the specified relationship is true, a nonzero floating-point value (0.5 for CMPGxx, 2.0 for CMPTxx) is written into Fc. Otherwise, a true zero is written into Fc.

The floating-point miscellaneous operate instructions (a total of 11) are summarized in Table 14.9.

The xx in FCMOVxx stands for six possible conditions:

```
EQ     equal to zero
GE     greater than or equal to zero
GT     greater than zero
LE     less than or equal to zero
LT     less than zero
NE     not equal to zero
```

featuring six floating-point conditional move instructions.

For the CPYS and CPYSN instructions, the sign bit of Fa is fetched (and complemented in the case of CPYSN) and concatenated with the exponent and fraction bits from Fb; the result is stored in Fc. For CPYSE, the sign and exponent bits from Fa are fetched and concatenated with the fraction bits from Fb; the result is stored in Fc. These instructions can be used to generate special operations not available in the regular instruction set, such as:

```
Register move             CPYS  Fx,Fx,Fy
Floating-point negation   CPYSN Fx,Fx,Fy
```

In the FCMOVxx instructions register Fa is tested and compared with zero. If the specified relationship xx is true, Fbv is moved to Fc. Otherwise, no operation is performed.

In the MF_FPCR and MT_FPCR instructions, the content of FPCR is moved to Fa, and Fav is moved to FPCR, respectively. The same register Fa must be specified in all three fields.

TABLE 14.9   Floating-point Miscellaneous Operate Instructions

Mnemonic	Operands	Operation
CPYS	Fa.rq, Fb.rq, Fc.wq	Copy sign
CPYSE	Fa.rq, Fb.rq, Fc.wq	Copy sign and exponent
CPYSN	Fa.rq, Fb.rq, Fc.wq	Copy sign negate
FCMOVxx	Fa.rq, Fb.rq, Fc.wq	Floating conditional move
MF_FPCR	Fa.rq, Fa.rq, Fa.wq	Move from FPCR
MT_FPCR	Fa.rq, Fa.rq, Fa.wq	Move to FPCR

FPCR is floating-point control register.

TABLE 14.10   Miscellaneous Instructions

Mnemonic	Operands	Operation
CALL_PAL	fnc.ir	Call PAL routine
FETCH	0(Rb.ab)	Prefetch data
FETCH_M	0(Rb.ab)	Prefetch data, modify intent
MB	—	Memory barrier
WMB	—	Write memory barrier
RPCC	Ra.wq	Read process cycle counter
TRAPB	—	Trap barrier

Miscellaneous instructions

There are seven miscellaneous instructions, listed in Table 14.10.

The CALL_PAL instruction causes a trap to PALcode (discussed later in this section).

In the FETCH and FETCH_M instructions an aligned 512-byte block of data is specified by Rbv. An implementation may optionally attempt to move all or part of this block (or a larger surrounding block) of data to a faster-access part of the memory hierarchy, in anticipation of subsequent load or store instructions that may access that data. The FETCH instruction is a hint to the implementation that may allow faster execution. The FETCH_M instruction gives the additional hint that modifications (store instructions) to some or all of the data block are anticipated.

The MB instruction is required only in multiprocessor systems. It facilitates synchronization between dependent iterations of loops, scheduled to run on different processors [Ston 93, Tabk 90a].

In the RPCC instruction the content of the process cycle counter (PCC) is written into Ra. The PCC is used for timing intervals in each processor in a multiprocessor system.

The TRAPB instruction allows software to guarantee that in a pipelined implementation, all previous arithmetic instructions will complete without incurring any arithmetic traps before any instructions after the TRAPB are issued.

Privileged architecture library (PAL) code

The PALcode [Sits 92b] provides a mechanism to implement the following functions without resorting to a microcoded machine:

1. Instructions that require complex sequencing as an atomic operation

2. Instructions that require VAX-style interlocked memory accesses

3. Privileged instructions

4. Memory management control

5. Context swapping

6. Interrupt and exception dispatching

7. Power-up initialization and booting

8. Console functions

9. Emulation of instructions with no hardware support

PAL functions are implemented in Alpha architecture in standard machine code, resident in main memory. PALcode environment differs from the normal environment in the following ways:

1. There is complete control of the machine state allowing all functions of the machine to be controlled.

2. Interrupts are disabled, allowing the system to provide multi-instruction sequences as atomic operations.

3. Implementation-specific hardware functions are enabled, allowing access to low-level system hardware.

4. Instruction stream memory management traps are prevented, allowing PALcode to implement memory management functions such as translation buffer (TB) fills.

PALcode uses the Alpha instruction set for most of its operations. There are five opcodes reserved to implement additional PALcode functions: PALRES 0, 1, 2, 3, 4. These instructions produce an illegal instruction trap if executed outside the PALcode environment. PALcode is allocated space in the physical memory.

The Alpha architecture is intended to be implemented on a series of high-performance microprocessor products. The first Alpha AXP implementations are described in the subsequent section.

14.3   Alpha AXP Implementations

The first implementation of the Alpha architecture is the 21064 microprocessor chip [DECM 92, Dobb 92, McLe 93]. The 21064 is fabricated in a 0.75-micron CMOS technology utilizing three levels of metalization and optimized for 3.3-V operation. The die size is 16.8×13.9 mm and it contains 1.68 million transistors. Its initial operating frequency is within the 150- to 200-MHz interval. Power dissipation at 200 MHz is 30 W. The processor is a two-issue superscalar. The chip includes a dual-cache, 8-kbyte instruction and 8-kbyte data. It also includes a four-entry, 32 bytes-per-entry write buffer, a pipelined 64-bit integer execution unit with a 32×64 register file, and a pipelined FPU with a 32×64 register file of its own. The pin interface includes integral support for an external secondary cache of 128 kbytes up to 8 Mbytes. The internal caches are direct-mapped. All caches have 32 bytes/line. The internal data cache is a write-through, read allocate, physical cache. The chip package is a 431-pin pin grid array (PGA) with 140 pins dedicated to power supply voltage and ground.

A block diagram of the 21064 is shown in Fig. 14.7. It features the following main subsystems:

- IBOX—issues instructions (two at a time), maintains the integer pipeline, and performs PC calculations. It decodes two instructions in parallel and checks availability of resources. There is no out-of-order issue. There will be an issue if appropriate resources are available. If resources are not available for the first instruction, there will be no issue. The IBOX contains branch prediction logic, instruction translator buffers (ITBs), interrupt logic, and performance counters (issues, nonissues, total cycles, pipe dry, pipe freeze, cache misses). There are two ITBs:

  1. Small-page ITB, eight-entry, fully associative, contains recently used instruction stream page table entries (PTEs) for 8-kbyte pages.
  2. Large-page ITB, four-entry, fully associative, for 512×8 kbyte pages (4 Mbytes).

- EBOX—integer execution unit. It contains a 64-bit adder, logic box, barrel shifter, bypassers, integer multiplier, 32×64 IRF with four read and two write ports.

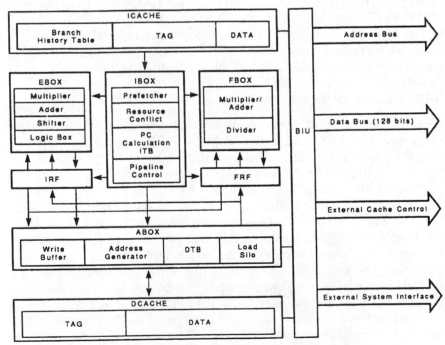

Figure 14.7 Block diagram of the 21064. (*Courtesy of Digital Equipment Corp.*)

- ABOX—address generation unit. It contains address translation datapath, load silo, data cache interface, internal processor registers (IPRs), and the BIU, and a 32-entry, fully associative data translation buffer (DTB). The load silo is a memory reference pipeline that can accept a new load or store instruction every cycle until a data cache fill is required. The BIU has an external 128-bit data bus.

- FBOX—the FPU. It contains in addition to the operation units a 32 × 64 floating-point register file (FRF), and a user-accessible floating-point control register (FPCR).

- ICACHE—instruction cache. 8 kbytes, direct-mapped, physical-addressed, 32 bytes/line.

- DCACHE—data cache. 8 kbytes, direct-mapped, physical-addressed, 32 bytes/line, write-through, read allocate.

An example of an external interface interconnection of the 21064 is shown in Fig. 14.8 [Dobb 92]. It is designed to directly support an off-chip secondary cache (also called backup cache, or B-cache) that can range from 128 kbytes to 8 Mbytes and can be constructed from ordinary SRAMs. The interface is designed to allow all cache policy decisions to be controlled by logic external to the CPU chip. There are 3 control bits associated with each B-cache line: valid (V), shared (S), and dirty (D). The chip completes a B-cache read as long as valid is true. A write is processed by the CPU only if valid is true and shared is false. When a write is performed, the dirty bit is set to true. In all other cases,

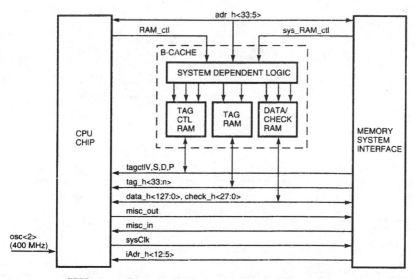

Figure 14.8 CPU external interface. (*Courtesy of Digital Equipment Corp., © IEEE.*)

the chip defers to an external state machine to complete the transaction. This state machine operates synchronously with the sysCLK output (see Fig. 14.8) of the chip, which is a mode-controlled submultiple of the CPU clock rate ranging from divide by 2 to divide by 8. It is also possible to operate without a B-cache. As shown in Fig. 14.8, the external cache is connected between the CPU chip and the memory system interface. The cache access begins with the address delivered on the adr_h lines and results in ctl, tag, data, and check bits appearing at the chip receivers within the prescribed access time. In 128-bit mode, B-cache accesses require two external data cycles to transfer the 32-byte $(256 = 2 \times 128$ bits) cache line across the 16-byte pin bus. In 64-bit mode, it is four cycles. This yields a maximum B-cache read bandwidth of 1.2 Gbytes/sec and a write bandwidth of 711 Mbytes/sec. Internal cache lines can be invalidated at the rate of one line per cycle using the dedicated invalidate address pins iAdr_h<12:5>.

The 21064 IU and FPU pipelines are illustrated in Fig. 14.9 [Dobb 92]. The integer pipeline is seven stages deep. The first four stages are associated with instruction fetching, decoding, and scoreboard checking of operands for possible data dependency (see Chaps. 5 and 6). Pipeline stages 0 through 3 can be stalled. Beyond stage 3, however, all pipeline stages advance every cycle. Most ALU operations complete in cycle 4 (A1). Primary cache accesses complete in cycle 6 (WR), so cache delay is three cycles. The instruction stream is based on autonomous prefetching in cycles 0 and 1 with the final resolution of

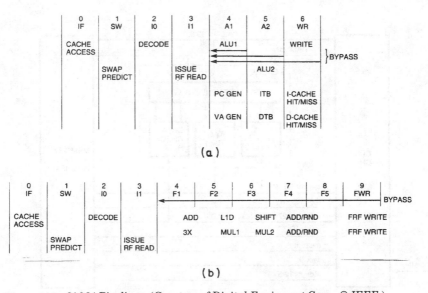

Figure 14.9   21064 Pipelines. (*Courtesy of Digital Equipment Corp., © IEEE.*)

ICACHE hit not occurring until cycle 5. The prefetcher includes a branch history table and a subroutine return stack. The architecture provides a convention for compilers to predict branch decisions and destination addresses, including those for register indirect jumps. The penalty for branch mispredict is four cycles.

The FPU pipeline is 10 stages deep. It is identical and mostly shared with the IU pipeline in stages 0 through 3. All operations, 32- and 64-bit, have the same timing (except divide). Divide is handled by a non-pipelined, single bit per cycle, dedicated divide unit. In cycle 4 (F1), the register file data is formatted to fraction, exponent, and sign. In the first stage adder exponent difference is calculated and a 3xmultiplicand is generated for multiplies. In addition, a predictive leading 1 or 0 detector using the input operands is initiated for use in result normalization. In cycles 5 (F2) and 6 (F3), for add/subtract, alignment or normalization shift are performed. For both single- and double-precision multiplication, the multiply is done in a radix-8 pipelined array multiplier. In cycles 7 (F4) and 8 (F5), the final addition and rounding are performed in parallel and the final result is selected and driven back to the register file in cycle 9 (FWR). With an allowed bypass of the register write data, floating-point delay is six cycles.

The superscalar dual issue of instructions is restricted to the following pairs:

Any load/store in parallel with any operate

An integer operate in parallel with a floating-point operate

A floating-point operate and a floating-point branch

An integer operate and an integer branch

The system supports pages of 8 kbytes, 64 kbytes, 512 kbytes, and 4 Mbytes.

A number of subsequent Alpha AXP implementation microprocessors have been produced by DEC.

The 21064A is a 0.5-micron, three-metal-layer, CMOS-5, 2.5-million-transistor, 431-pin PGA microprocessor, running at frequencies ranging from 225 to 275 MHz. It has double the on-chip cache than the 21064: 16 kbytes instruction, 16 kbytes data, for a total of 32 kbytes.

The 21066 is a highly integrated implementation of Alpha, whose on-chip functions include an I/O controller, IU, FPU, memory controller, graphics accelerator, instruction and data caches (8 kbytes each, as on the 21064), and an external cache controller [McKi 94]. The 21066 is a 0.68-micron, three-metal-layer, CMOS-4, 287-pin PGA microprocessor, running at 166 MHz. The 21068 is a lower-frequency version of the 21066, running at 66 MHz.

14.4   Concluding Comment

The first implementation of the Alpha architecture, the 21064, running at a frequency of 150 to 200 MHz, is listed in the October 1992 *Guinness Book of Records* as the world's fastest single-chip microprocessor [Sits 93]. The 21064A runs at 275 MHz, and there are subsequent implementations which will run at even higher frequencies (over 300 MHz). How long DEC will retain this status, considering the developing competition, remains to be seen.

The Alpha architecture is certainly new for the most part. The Alpha architecture was developed for a new 64-bit RISC-type processor family, while its predecessor at DEC, the VAX architecture supported a CISC-type 32-bit family of superminicomputers. On the other hand, one cannot say that the Alpha architecture starts a completely new page. The DEC VAX, and even the 16-bit PDP, are still alive within the Alpha architecture through a number of features:

1. Although the Alpha is a 64-bit architecture, the terminology for data types has its roots in the 16-bit PDP family. A "word" is still 16-bit, as it was in the PDP and the VAX. A 32-bit data item is a "longword," as in the VAX. As can be seen from the subsequent chapters in this part, most RISC-type systems use the term "word" for 32-bit data, and "halfword" for 16-bit data.

2. The VAX floating-point format is kept on as an optional data type in the Alpha architecture, featured in parallel with the IEEE standard, which is adopted exclusively on all other modern systems, RISC and CISC.

3. Some of the old PDP instruction mnemonics, such as BIS instead of OR, are still kept with the Alpha architecture.

Nevertheless, the Alpha architecture is a pioneer in 64-bit architectures (the only other 64-bit system is the MIPS R4000, described in Chap. 16). Most of its features are innovative and will probably appear in other future systems.

DEC designers are currently working on future extensions of the Alpha architecture and future, more powerful implementations [McLe 93, Sits 93]. The cache will be doubled in size, and the superscalar issue will be increased to 4 in future implementations. While initial implementations use only 43 bits of virtual address, they check the remaining 21 bits (of 64), so that software can run unmodified on later implementations that use all 64 bits. Similarly, while initial implementations use only 34 bits of physical address, the architected PTE formats and page size choices allow growth to 48 bits. By expanding into a 16-bit PTE field that is not currently used by mapping hardware, a further 16 bits of physical address growth (for a total of 64 bits) can be achieved, if needed.

While initial implementations use only 8-kbyte pages, the design accommodates limited growth to 64-kbyte pages. Beyond that, page table granularity hints allow groups of 8, 64, or 512 pages to be treated as a single large page, thus effectively extending the page size range by a factor of over 1000 [Sits 93]. Each architected PTE format also has 1 bit reserved for future expansion.

There are a number of areas of instruction set flexibility designed into the architecture. Four of the 6-bit opcodes are nominally reserved for adding integer and floating-point aligned octaword (128-bit) load/store instructions. Nine more 6-bit opcodes remain for other expansions. In addition, the function field contains more room for further expansion.

One of the most notable intended applications of the Alpha AXP microprocessors is in the Cray T3D massively parallel processor (MPP) system [Cray 93]. The Cray T3D MPP is composed of numerous processing element nodes, each node containing two processing elements (PE). The nodes are interconnected by a bidirectional three-dimensional torus network. The system can be configured to contain 32, 64, 128, 256, 1024, or 2048 PEs. A simplified diagram of T3D node interconnections is shown in Fig. 14.10, and a diagram of a PE node is

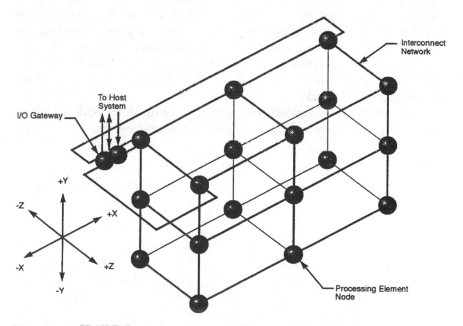

Figure 14.10 CRAY T3D system components. (*Courtesy of Cray Research Corp.*)

Processing Element Node

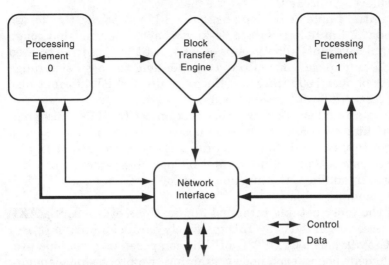

Figure 14.11 Processing element node. (*Courtesy of Cray Research Corp.*)

shown in Fig. 14.11. Each PE consists of an Alpha AXP microprocessor (21064 or other implementations), local memory (two or four megawords; a word on the T3D is 64 bits wide), and support circuitry (mainly interface). The Cray T3D system is managed by a host computer. All T3D applications are compiled by the host computer, which may be any Cray computer system that has an I/O subsystem model E (IOS-E). Host systems include the Cray Y-MP E, Y-MP M90, and Y-MP C90 series computer systems.

We are going to see indeed new, more powerful Alpha AXP architecture implementations and applications in the near future.

15

The PowerPC Family

15.1 Introduction

As mentioned in Chap. 6, the development of the first RISC-type computer started in the mid-seventies at IBM with the experimental IBM 801 system [Radi 83], although the term "RISC" was coined at Berkeley in 1980. The goal of the IBM 801 designers was to create a new system, oriented toward the pervasive use of HLLs. The system was designed to execute an instruction at almost every clock cycle ("almost," because there are a few instructions, such as load or store, which access memory, multiply, divide, that cannot be executed in a single cycle). To achieve the above (see also Chap. 6), a primitive instruction set, completely hardwired, was adopted. These are certainly RISC characteristics.

Subsequently, in the mid-eighties, a commercial RISC-type processor, the ROMP (Research Office products division Microprocessor), was announced by IBM [Tabk 87, Chap. 4; Tabk 90b, Chap. 12]. Compared to the 801, it had a smaller percentage of instructions executing in a single cycle and 65 percent of its instructions were 16 bits long, while the others were 32-bits. In contrast, all 801 instructions were uniformly 32 bits long. Thus, the ROMP appeared to be somewhat "less of a RISC," compared to the original 801. The ROMP was used in the IBM RT 6150 and RT 6151 workstations.

The IBM next RISC-type system, announced in 1990, is the RISC System (RS)/6000 or RS/6000 [BaWh 90, GrOe 90]. Its architecture, which constitutes the basis of the RS/6000 design, is denoted by IBM as POWER (performance optimization with enhanced RISC). The RS/6000 will be described later in this chapter in Sec. 15.4.

In 1991, IBM got together with Motorola and Apple to form a new alliance with a goal to seize the lead in the desktop computer industry by developing a new and powerful family of RISC-type microprocessors. This new family is called the PowerPC family. This alliance puts together the vast experience of the three leading companies in hardware and software. It cuts significantly development costs and avoids useless duplication of effort.

The first PowerPC implementation is the PowerPC 601 microprocessor (also called MPC 601 by Motorola, and PPC 601 by IBM), to be described in Sec. 15.3. It is to be followed by subsequent implementations such as MPC 603, 604, 620, and others in the near future. All these new systems are based on the PowerPC architecture [DiOH 94, PoPC 93], derived from the IBM POWER architecture. The PowerPC architecture will be described in Sec. 15.2.

15.2   The PowerPC Architecture

The PowerPC architecture can be viewed as consisting of the following three layers:

1. User instruction set architecture—includes user-level registers, programming model, data types, addressing modes, and the base user-level instruction set (excluding a few user-level memory-control instructions).

2. Virtual environment architecture—describes the semantics of the memory model that can be assumed by software processes and includes descriptions of the cache model, cache-control instructions, address aliasing, and other related issues.

3. Operating environment architecture—includes the structure of the memory management model, supervisor-level registers, privileged instructions, and the exception model.

Systems based on the PowerPC architecture operate in two basic modes of operation: user and supervisor modes, similar to the M68000 family (see Part 3).

Register set and programming model

The PowerPC register set and programming model is shown in Fig. 15.1 [PoPC 93]. it is subdivided into two main parts, corresponding to the two modes of operation:

User programming model

Supervisor programming model

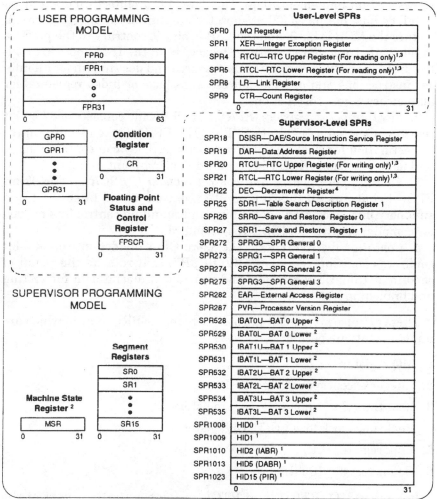

Figure 15.1   Programming model—registers. (*Courtesy of Motorola, Inc.*)

Programs executing in the supervisor mode can access all registers. Programs executing in the user mode can access registers in the user programming model only.

The user programming model includes the following registers:

1. *General-purpose registers* (*GPR*).   The GPR file consists of 32 GPRs denoted as GPR0 to GPR31. The GPRs are 32 bits wide in 32-bit implementations, and 64 bits wide in 64-bit implementations (Fig. 15.1

shows a 32-bit implementation). The PowerPC architecture is extended to full 64 bits, but has a 32-bit subset. Note that in the PowerPC architecture the MSB is bit 0, and the LSB is bit 31, contrary to the practice in most other microprocessors (a "legacy" from the IBM 360/370 architecture). The GPRs serve as the data source and destination for all integer instructions and provide addresses (as base or index registers) for all memory access instructions.

2. *Floating-point registers (FPR)*.   The FPR file consists of 32 64-bit registers denoted as FPR0 to FPR31. The FPRs, serve as data source or destination for floating-point instructions. They can contain single- or double-precision IEEE 754 standard floating-point data.

3. *Floating-point status and control register (FPSCR)*.   The 32-bit FPSCR contains all floating-point exception signal bits, exception summery bits, exception enable bits, and rounding control bits needed for compliance with the IEEE 754 standard.

4. *Condition register (CR)*.   The 32-bit CR is divided into eight 4-bit separate condition registers, CR0 to CR7, that reflects the result of certain arithmetic operations and provides a mechanism for testing and branching.

There are also six special-purpose registers (SPR), some of which are implementation-specific:

1. MQ register

2. Integer exception register (XER)

3. Real-time clock upper (RTCU) register

4. Real-time clock lower (RTCL) register

5. Link register (LR)

6. Count register (CTR)

Registers MQ, RTCU, and RTCL are not actually a part of the PowerPC architecture. They are holdovers from the IBM POWER architecture implemented in the PowerPC 601 for compatibility. The PowerPC architecture specifies a 64-bit time-base (TB) register instead of the RTC.

The supervisor programming model includes:

Machine state register (MSR). The MSR defines the state of the processor.

Segment registers (SR). The 32-bit 16 SRs are present only in 32-bit PowerPC implementations.

The remaining registers are SPRs, some of which are implementation dependent.

The 32-bit condition register (CR) reflects the results of certain operations and provides a mechanism for testing and branching. The CR is subdivided into eight 4-bit separate condition registers CR0 to CR7, shown in Fig. 15.2.

The CR condition registers can be set in the following ways:

1. Specified fields in the CR can be set by a move instruction from a GPR (**mtcrf** or **mcrfs**).

2. Specified fields of the CR can be moved from one CRi field to another with the **mcrf** instruction.

3. A specified field of the CR can be set by a move (**mcrxr**) from the XER register.

4. CR logical instructions can be used to operate on specified bits in the CR.

5. CR0 can be the implicit result of an integer operation.

6. CR1 can be the implicit result of a floating-point operation.

7. A specified CR field can be the explicit result of either an integer or floating-point compare operation.

Setting CR0 and CR1 is an optional implicit result, controlled by opcode.

Instructions are provided to test individual CR bits. The CR is cleared by a hard reset.

The CRi bit settings for compare operations are

Bit 0, less than, LT for integer, FL for floating-point

Bit 1, greater than, GT for integer, FG for floating-point

Bit 2, equal, EQ for integer, FE for floating-point

Bit 3, summary overflow, SO for integer, unordered, FU (or UO) for floating-point

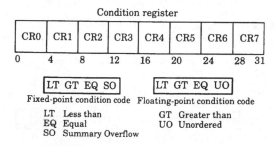

Figure 15.2 Condition register and condition code formats. (*Courtesy of IBM, Inc.*)

The CR0 field has special setting interpretations, if set from arithmetic, rather than compare instructions:

Bit 0, negative (LT), set when the result is negative

Bit 1, positive (GT), set when the result is positive

Bit 2, zero (EQ), set when the result is zero

Bit 3, summary overflow (SO), set on overflow

The bits of the MSR, shown in Fig. 15.3, are used as follows:

Bit 16, external interrupt enable (EE)—external interrupts enabled when EE = 1, disabled when EE = 0

Bit 17, privilege level (PR)—the processor can only execute user-level instructions when PR = 1, and all instructions when PR = 0

Bit 18, floating-point available (FP)—floating-point available for processor execution when FP = 1, and unavailable when FP = 0

Bit 19, machine check enable (ME)—machine check exceptions enabled when ME = 1, disabled when ME = 0

Bits 20 and 23, floating-point exception mode 0 and 1 (FE0 and FE1)—2 encoded bits:

FE0	FE1	Mode
0	0	Floating-point exceptions disabled
0	1	Floating-point imprecise nonrecoverable
1	0	Floating-point imprecise recoverable
1	1	Floating-point precise mode

Bit 21, single-step trace enable (SE)—single-step operation when SE = 1, normal execution when SE = 0.

Bit 25, exception prefix (EP)—specifies an exception vector offset:

- When EP = 0, exceptions are vectored to the physical address 000n nnnn (hex)
- When EP = 1, exceptions are vectored to the physical address FFFn nnnn (hex)

Figure 15.3 Machine State Register (MSR). (*Courtesy of Motorola, Inc.*)

Bit 26, instruction address translation (IT)—instruction address translation enabled when IT = 1, disabled when IT = 0

Bit 27, data address translation (DT)—data address translation enabled when DT = 1, disabled when DT = 0.

Other bits of the MSR are reserved.

Data types

The PowerPC architecture recognizes the following integer data types:

Operand	Size (bytes)	Size (bits)	Address bits 28 to 31 if aligned
Byte	1	8	xxxx
Halfword	2	16	xxx0
Word	4	32	xx00
Doubleword	8	64	x000

Best performance is always attained when data operands are aligned. The PowerPC architecture supports both big-endian and little-endian byte ordering (see Chap. 3). The default byte ordering is big-endian (true for the MPC 601 so far). Byte ordering can be changed to little-endian by setting a bit in the MSR register.

PowerPC architecture implements the IEEE 754-1985 standard single- and double-precision floating-point formats, described in Chap. 3.

Addressing modes

As in all RISC-type systems, all operations are register-to-register, using the following two modes:

1. Register direct; the operand is in a GPR or FPR
2. Immediate; the operand is a part of the instruction

An EA to memory is needed in two classes of instructions:

Load and store

Branch

The appropriate addressing modes for the above are
For load and store instructions:

1. Register indirect; a GPR register contains the address of the operand in memory.

2. Register indirect with immediate index; an immediate displacement d (called in PowerPC architecture d operand or immediate index) is added to the value in a GPR to form an EA.

3. Register indirect with index; the contents of two GPR registers, one serving as register indirect and the other as index, is added to form the EA. Any GPR can serve as a register indirect or index register.

For branch instructions:

1. Immediate addressing; the branch address is a part of the instruction. It is sign-extended to form a target EA. This mode is generally called direct or absolute (see Chap. 3).

2. Link register indirect; the target EA is in the link register.

3. Count register indirect; the target EA is in the count register.

Instruction formats

Some of the main instruction formats of the PowerPC architecture are shown in Fig. 15.4 [PoPC 93]. All PowerPC instructions are 32 bits long. The format in Fig. 15.4(a) is used for most operations. The upper 6 bits (0 to 5) contain the opcode. The opcode is extended into the subopcode field (bits 22 to 30). The two register source operands are in fields A (bits 11 to 15) and B (bits 16 to 20). The destination operand is in field D (bits 6 to 10). Control bit OE (bit 21) enables overflow detection when set (OE = 1). Control bit Rc (bit 31), called the record bit, updates CR if set (Rc = 1). In the PowerPC assembly notation of instructions, the order of the operands is usually the same as in the instruction format.

Example

```
add rD, rA, rB; rD ← (rA) + (rB)
```

Any GPR (integer operation) and any FPR (floating-point operation) can serve as rA, rB, or rD in this example.

Operations with an immediate operand have the format shown in Fig. 15.4(b). The SIMM field (bits 16 to 31) provides for the storage of a 16-bit immediate operand.

Example

```
addi rD, rA, SIMM; add immediate, rD ← (rA) + EXTS(SIMM)
```

where EXTS(SIMM) is a sign-extended SIMM field.

The format in Fig. 15.4(b) can also be used by load and store instructions using the register indirect with immediate (displacement) addressing mode. In this case field A stores the register indirect encoding, while field D is used for the destination register for load, and for the source register for store instructions.

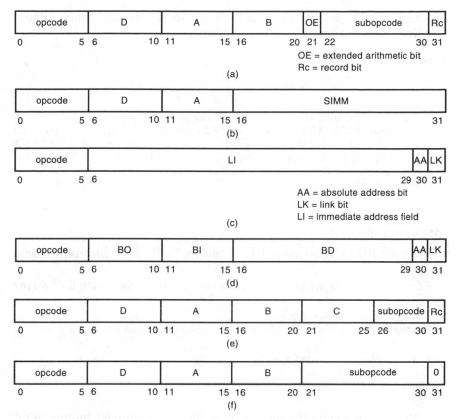

Figure 15.4 PowerPC main instruction formats. (*a*) Register-to-register operations; OE = extended arithmetic; Rc = record bit; (*b*) Immediate operations, load/store reg. indirect and immediate; (*c*) Branch. AA = absolute address bit, LI = immediate address field; LK = link bit; (*d*) Branch conditional; (*e*) Combined arithmetic (floating-point multiply and add); (*f*) Load/store reg. indirect and index.

The format in Fig. 15.4(*c*) is used for branch instructions. The LI field is the immediate address field. If the absolute address bit AA (bit 30) is clear (AA = 0), the LI is shifted 2 bits left, filling the 2 lower bits with zeros, and added (sign-extended) to the instruction address to form the branch target address. If AA = 1, the shifted as above and signed extended LI field constitutes the branch target address (direct, or PowerPC immediate addressing mode). If the link bit LK (bit 31) is set (LK = 1), then the EA of the instruction following the branch instruction is placed into the LR.

The branch conditional instruction format is illustrated in Fig. 15.4(*d*). The BO field specifies the conditions under which the branch is taken. The BI field specifies the bit in the CR to be used as the condition of the branch. The BD field is used to form the branch target address. The AA and LK bits are used in a manner similar to that of the unconditional branch, shown in Fig. 15.4(*c*).

The PowerPC architecture features composite instructions, involving floating-point multiply and add, for instance.

Example

fmadd frD, frA, frC, frB; frD ← (frA)*(frC) + (frB)

The format for such an instruction is shown in Fig. 15.4(*e*).

The format in Fig. 15.4(*f*), similar to the format in 15.4(*a*), is used for load and store instructions using the register indirect with index addressing mode. The A field is used for the register indirect, and the B field is used for the index register. As before, the D field serves as a destination for load and as a source for store instructions.

Instruction set

The PowerPC instruction set is listed in Table 15.1 in alphabetical order.

Some of the instructions, such as integer arithmetic, have four forms of operation.

Example Take, for instance, the **add** instruction. Its three additional forms are:

add.   add with CR update
addo   add with overflow update
addo.   add with overflow and CR updated

Most floating-point instructions have just the additional **"mnemonic."** option.

Example

fadd   floating-point add
fadd.   floating-point add with CR update

The branch **b** and branch conditional **bc** have four different options. The three additional options are (in addition to **b**):

ba branch absolute, AA = 1, LK = 0, the branch target address is the sum of the shifted LI field and the address of the branch instruction

bl branch then link, AA = 0, LK = 1, the LI field is the target address, and the address of the next instruction is placed in LR

bla branch absolute then link, AA = 1, LK = 1, a combined effect of ba and bl.

The same options are also in effect for the branch conditional (bc) instruction: bca, bcl, and bcla.

TABLE 15.1   PowerPC Instruction Set

Mnemonic	Operation
1. add	add
2. addc	add carrying
3. adde	add extended
4. addi	add immediate
5. addic	add immediate carrying
6. addic.	add immediate carrying and record
7. addis	add immediate shifted
8. addme	add to minus one extended
9. addze	add to zero extended
10. and	and
11. andc	and with complement
12. andi.	and immediate
13. andis.	and immediate shifted
14. b	branch
15. bc	branch conditional
16. bcctr	branch conditional to CTR
17. bclr	branch conditional to LR
18. cmp	compare
19. cmpi	compare immediate
20. cmpl	compare logical
21. cmpli	compare logical immediate
22. cntlzw	count leading zeros word
23. crand	CR AND
24. crandc	CR AND with complement
25. creqv	CR equivalent
26. crnand	CR NAND
27. crnor	CR NOR
28. cror	CR OR
29. crorc	CR OR with complement
30. crxor	CR XOR
31. dcbf	data cache block flush
32. dcbi	data cache block invalidate
33. dcbst	data cache block store
34. dcbt	data cache block touch
35. dcbtst	data cache block touch for store
36. dcbz	data cache block set to zero
37. divs	divide short
38. divw	divide word
39. divwu	divide word unsigned
40. eciwx	external control input word indexed
41. ecowx	external control output word indexed
42. eieio	enforce in-order execution of I/O
43. eqv	equivalent
44. extsb	extend sign byte
45. extsh	extend sign halfword
46. fabs	floating-point absolute value
47. fadd	floating-point add
48. fcmpo	floating-point compare ordered
49. fcmpu	floating-point compare unordered
50. fctiw	floating-point convert to integer word
51. fctiwz	floating-point convert to integer word with round to zero

TABLE 15.1   PowerPC Instruction Set *(Cont.)*

Mnemonic	Operation
52. fdiv	floating-point divide (single-precision)
53. fmadd	floating-point multiply-add (single-precision)
54. fmr	floating-point move register
55. fmsub	floating-point multiply-subtract (single-precision)
56. fmul	floating-point multiply (single-precision)
57. fnabs	floating-point negative absolute value
58. fneg	floating-point negate
59. fnmadd	floating-point negative multiply-add (single precision)
60. fnmsub	floating-point neg. multiply-subtract (single-precision)
61. frsp	floating-point round to single-precision
62. fsub	floating-point subtract (single-precision)
63. icbi	instruction cache block invalidate
64. isync	instruction synchronize
65. lbz	load byte and zero
66. lbzu	load byte and zero with update
67. lbzux	load byte and zero with update indexed
68. lbzx	load byte and zero indexed
69. lfd	load floating-point double-precision
70. lfdu	load floating-point double-precision with update
71. lfdux	load floating-point double-precision with update indexed
72. lfdx	load floating-point double-precision indexed
73. lfs	load floating-point single-precision
74. lfsu	load floating-point single-precision with update
75. lfsux	load floating-point single-precision with update indexed
76. lfsx	load floating-point single-precision indexed
77. lha	load halfword algebraic
78. lhau	load halfword algebraic with update
79. lhaux	load halfword algebraic with update indexed
80. lhax	load halfword algebraic indexed
81. lhbrx	load halfword byte-reverse indexed
82. lhz	load halfword and zero
83. lhzu	load half-word and zero with update
84. lhzux	load halfword and zero with update indexed
85. lhzx	load halfword and zero indexed
86. lmw	load multiple word
87. lswi	load string word immediate
88. lswx	load string word indexed
89. lwarx	load word and reverse indexed
90. lwbrx	load word byte-reverse indexed
91. lwz	load word and zero
92. lwzu	load word and zero with update
93. lwzux	load word and zero with update indexed
94. lwzx	load word and zero indexed
95. mcrf	move CR field
96. mcrfs	move to CR from FPSCR
97. mcrxr	move to CR from XER
98. mfcr	move from CR
99. mffs	move from FPSCR
100. mfmsr	move from MSR
101. mfspr	move from SPR
102. mfsr	move from SR

TABLE 15.1   PowerPC Instruction Set *(Cont.)*

Mnemonic	Operation
103. mfsrin	move from SR indirect
104. mtcrf	move to CR fields
105. mtfsb0	move to FPSCR bit 0
106. mtfsb1	move to FPSCR bit 1
107. mtfsf	move to FPSCR fields
108. mtfsfi	move to FPSCR field immediate
109. mtmsr	move to MSR
110. mtspr	move to SPR
111. mtsr	move to SR
112. mtsrin	move to SR indirect
113. mulhw	multiply high word
114. mulhwu	multiply high word unsigned
115. mullw	multiply low
116. mulli	multiply low immediate
117. nand	NAND
118. neg	negate
119. nor	NOR
120. or	OR
121. orc	OR with complement
122. ori	OR immediate
123. oris	OR immediate shifted
124. rfi	return from interrupt
125. rlwimi	rotate left word immediate then mask insert
126. rlwinm	rotate left word immediate then AND with mask
127. rlwnm	rotate left word then AND with mask
128. sc	system call
129. slw	shift left word
130. sraw	shift right algebraic word
131. srawi	shift right algebraic word immediate
132. srw	shift right word
133. stb	store byte
134. stbu	store byte with update
135. stbux	store byte with update indexed
136. stbx	store byte indexed
137. stfd	store floating-point double-precision
138. stfdu	store floating-point double-precision with update
139. stfdux	store floating-point double-precision with update indexed
140. stfdx	store floating-point double-precision indexed
141. stfs	store floating-point single-precision
142. stfsu	store floating-point single-precision with update
143. stfsux	store floating-point single-precision with update indexed
144. stfsx	store floating-point single-precision indexed
145. sth	store halfword
146. sthbrx	store halfword byte-reverse indexed
147. sthu	store halfword with update
148. sthux	store halfword with update indexed
149. sthx	store halfword indexed
150. stmw	store multiple word
151. stswi	store string word immediate
152. stswx	store string word indexed
153. stw	store word

TABLE 15.1   PowerPC Instruction Set *(Cont.)*

Mnemonic	Operation
154. stwbrx	store word byte-reverse indexed
155. stwcx.	store word conditional indexed (CR0 field affected)
156. stwu	store word with update
157. stwux	store word with update indexed
158. stwx	store word indexed
159. subf	subtract from
160. subfc	subtract from carrying
161. subfe	subtract from extended
162. subfic	subtract from immediate carrying
163. subfme	subtract from minus one extended
164. subfze	subtract from zero extended
165. sync	synchronize
166. tlbie	TLB invalidate entry
167. tw	trap word
168. twi	trap word immediate
169. xor	XOR
170. xori	XOR immediate
171. xoris	XOR immediate shifted

Some instructions, such as **andi.** or **andis.,** come with the CR update option only.

PowerPC instructions are subdivided into the following categories [PoPC 93]:

1. Integer instructions. There are four integer instruction types.
   a. Integer arithmetic, including add, addi, addis, addic, subf, subfic, addc, subfc, adde, subfe, addme, subfme, addze, subfze, neg, mulli, mullw, mulhw, mulhwu, divw, divwu
   b. Integer compare, including cmp, cmpi, cmpl, cmpli
   c. Integer logical, including and, andi, andis., or, ori, oris, xor, xori, xoris, nand, nor, eqv, andc, orc, extsb, extsh, cntizw
   d. Integer rotate and shift, including rlwinm, rlwnm, rlwimi, slw, srw, srawi, sraw

2. Floating-point instructions. There are six floating-point instruction types.
   a. Floating-point arithmetic, including all arithmetic instructions starting with "f" (see Table 15.1).
   b. Floating-point Multiply-Add, including all combined multiply-add and multiply-subtract instructions starting with "f" (see Table 15.1).
   c. Floating-point rounding and conversion, including frsp, fctiw, fctiwz
   d. Floating-point compare, including fcmpu and fcmpo

      e. Floating-point status and CR, including mffs, mcrfs, mtfsfi, mtfsf, mtfsb0, mtfsb1.

      f. Floating-point move, including fmr

3. Load and store instructions. There are eight load and store instruction types

      a. Integer load, including lbz, lbzx, lbzu, lbzux, lhz, lhzx, lhzu, lhzux, lha, lhax, lhau, lhaux, lwz, lwzx, lwzu, lwzux

      b. Integer store, including stb, stbx, stbu, stbux, sth, sthx, sthu, sthux, stw, stwx, stwu, stwux

      c. Integer load and store with byte reversal, including lhbrx, lwbrx, sthbrx, stwbrx

      d. Integer load and store multiple, including lmw and stmw

      e. Integer move string, including lswi, lswx, lscbx, stswi, stswx

      f. Memory synchronization, including eieio, isync, lwarx, stwcx., sync

      g. Floating-point load, including lfs, lfsx, lfsu, lfsux, lfd, lfdx, lfdu, lfdux

      h. Floating-point store, including stfs, stfx, stfsu, stfsux, stfd, stfdx, stfdu, stfdux

4. Flow control instructions, including b, bc, bclr, bcctr, crand, cror, crxor, crnand, crnor, creqv, crandc, crorc, mcrf, sc, rfi, twi, tw

5. Processor control instructions, including mtspr, mfspr, mtcrf, mcrxr, mfcr, mtmsr, mfmsr

6. Memory control instructions, including dcbi, dcbt, dcbtst, dcbz, dcbst, dcbf, mtsr, mtsrin, mfsr, mfsrin, tlbie

7. External control instructions, including eciwx and ecowx

Examples

```
subf    rD, rA, rB; (rB) - (rA) → rD
mullw   rD, rA, rB; low-order 32 bits of (rA)*(rB) → rD
mulhw   rD, rA, rB; high-order 32 bits of (rA)*(rB) → rD
divw    rD, rA, rB; low-order 32 bits of (rA)/(rB) → rD
```

    To compute the remainder, remainder = dividend − (quotient) * (divisor):

```
divw    rD, rA, rB; quotient → rD
mullw   rD, rD, rB; (quotient) * (divisor) → rD
subf    rD, rD, rA; (rA) - (rD) = remainder → rD
srw     rA, rS, rB; shift rS right by a number of bits specified by
        rB (bits 26 to 31),
        zeros on left, shifted result → rA
stwx    rS, rA, rB; (rS) → (EA), EA = (rA) + (rB)
fmr     frD, frB; frD ← (frB)
```

Exceptions

The PowerPC architecture supports four types of exceptions:

1. Synchronous, precise. Caused by an instruction. All instruction-caused exceptions are handled precisely; that is, the machine state at the time the exception occurs is known and can be completely restored.

2. Synchronous, imprecise. Involves imprecise floating-point exceptions. May or may not be restartable.

3. Asynchronous, precise. Involves maskable external interrupt exceptions.

4. Asynchronous, imprecise. Involves nonmaskable imprecise exceptions such as system reset and machine check. When exceptions occur, information, such as the instruction that should be executed after control is returned to the original program and the content of the MSR, is saved in the save/restore registers SRR0 and SRR1 (SPR26 and SPR27), program control passes from user to supervisor level, and software begins execution of the exception handler.

15.3   The PowerPC 601

The PowerPC 601, also called MPC 601, is the first microprocessor implementation of the PowerPC architecture. It is a 66-MHz, 2.8-million transistor, 0.6-micron, four-layer metal CMOS, packaged in 304-pin ceramic quad flat pack microprocessor [AlBe 93, BAMo 93, Moor 93, PaSi 93, PoPC 93]. Its power dissipation is 9 W at 3.6 V, 66 MHz operation. It is a three-issue superscalar system (see Chap. 5). The MPC 601 block diagram is shown in Figure 15.5.

The instruction unit (Fig. 15.5) prefetches instructions from the cache and places them in the instruction queue, illustrated in Fig. 15.6. The instruction queue has space for eight instructions in slots Q0 to Q7. The eight instructions can be loaded into the queue in one cycle (there are $32 * 8 = 256$ bits leading from the cache to the instruction unit). Instructions move from the top of the queue (Q7) toward the bottom (Q0) and a full range of shift amounts through the queue is supported. The instruction queue (Q4 to Q7) provides buffering to reduce the need to access the cache. Some initial decoding is performed in the lower half (Q0 to Q3). Only three instructions at a time can be issued to the execution units. Some instructions can be issued out of order from any of the bottom four queue entries.

The MPC 601 has three independent execution units:

1. Integer unit, IU. It is a 32-bit unit executing all integer and memory access instructions (including those required for floating-point registers). The IU contains an ALU, a multiplier, a divider, the integer

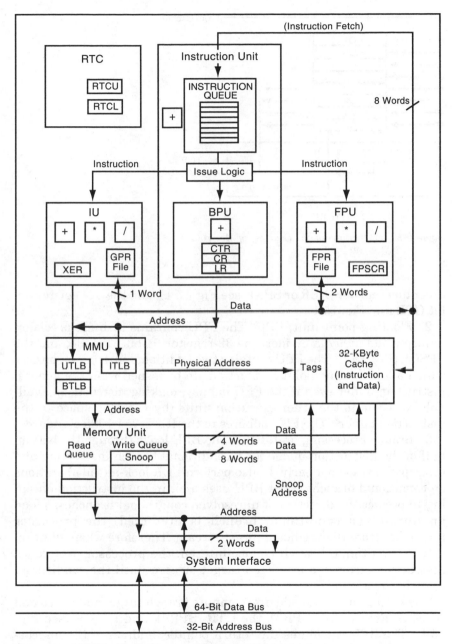

Figure 15.5 MPC601 block diagram. (*Courtesy of Motorola, Inc.*)

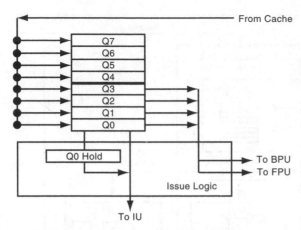

Figure 15.6   Instruction queue. (*Courtesy of Motorola, Inc.*)

exception register (XER or SPR1, see Fig. 15.1), and the 32-register 32-bit GPR file (Fig. 15.1).

2. Floating-point unit, FPU. The FPU contains a single-precision multiply-add array, a divider, the 32-register 64-bit FPR file, and the FPSCR (Fig. 15.1). The FPU contains two additional instruction buffers which allow floating-point instructions to be issued from the general instruction buffer even if the FPU is busy, making instructions available for issue to the other execution units (by vacating space in the instruction buffer). The FPU adheres to the IEEE 754-1985 standard.

3. Branch processing unit, BPU. The BPU looks through the bottom half of the instruction queue for a conditional branch instruction and attempts to resolve it early. It also performs CR look-ahead operations on conditional branches. The BPU uses a bit in the instruction encoding to predict the direction of unresolved conditional branches. When an unresolved conditional branch is predicted, the processor prefetches from the predicted target stream. Therefore when an unresolved conditional branch is encountered, the processor prefetches instructions from the predicted target stream until the conditional branch is resolved.

The BPU contains an adder to compute branch target addresses and three SPRs: the LR (SPR8), the CTR (SPR9), and the CR (see Fig. 15.1). The BPU calculates the return pointer (address of the instruction following the subroutine call) for subroutine calls and saves it in the LR. The LR may also contain the branch target address for the branch conditional to LR (bclr) instruction. The CTR contains the branch target address for the branch conditional to CTR (bcctr) instruction. Because the BPU uses dedicated registers rather than

GPRs or FPRs, execution of branch instructions is independent from execution of integer and floating-point instructions.

Three simultaneously issued instructions can execute simultaneously only if they are of the appropriate type to be issued to the three independent execution units, that is, if one instruction is integer, one floating-point, and one branch. If two back-to-back integer instructions are encountered, they will be executed in sequence through the IU, although branch and floating-point instructions may be dispatched from behind.

The MPC 601 features four types (physically three pipelines) of pipelines of different depth [Moor 93]:

1. Branch pipeline, two stages:

    Fetch

    Dispatch, decode, execute, predict; all in one cycle

2. Integer instructions pipeline, four stages:

    Fetch

    Dispatch, decode

    Execute

    Write back

3. Load/Store instructions pipeline, five stages:

    Fetch

    Dispatch, decode

    Address generation

    Cache

    Write back

    Stages fetch, dispatch, decode, address generation (or execute), and write back, are physically the same on pipelines 2 and 3. The cache stage is used by pipeline 3 only.

4. Floating-point pipeline, six stages:

    Fetch

    Dispatch

    Decode

    Execute 1

    Execute 2

    Write back

The load/store pipeline needs an extra cycle for cache access (more cycles will be needed in case of a cache miss; see also Chaps. 4 and 5). The floating-point pipeline also needs extra cycles for some operations, particularly those of double-precision.

The MPC 601 MMU, shown in Fig. 15.7, supports up to 4 petabytes (2^{52} bytes) of virtual memory and 4 Gbytes (2^{32} bytes) of physical mem-

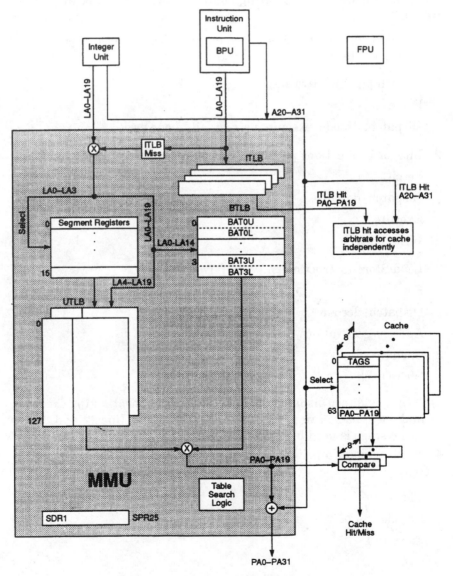

Figure 15.7 MMU block diagram. (*Courtesy of Motorola, Inc.*)

ory. The MMU also controls access privileges for these spaces on block and page granularities. The MPC has three types of TLBs (see Chap. 4):

1. Unified TLB (UTLB), 256-entry, two-way set-associative, for 4-kbyte pages.
2. Block TLB (BTLB), four-entry, fully associative, for blocks (optionally 128 kbytes to 8 Mbytes).
3. Instruction TLB (ITLB), four-entry, fully associative, most recently used instruction address translations.

The BPU generates all instruction and data addresses. After a logical address (LA) is generated, its upper order bits, LA0 to LA19 (Fig. 15.7), are translated by the MMU into physical address bits PA0 to PA19. Simultaneously, the lower address bits, A20 to A31, are directed to the on-chip cache (discussed later in this Sec.) to point to the appropriate set and the data item accessed. After translating the address (or rather its upper 20 bits), the MMU passes the upper 20 bits to the cache for tag match lookup. The untranslated lower bits are concatenated with the translated upper bits to form a physical address accessing the external memory.

The sixteen 32-bit segment registers (SR), shown in Figs. 15.1 and 15.7, control both the page and I/O controller interface address translation mechanism. One of the SRs is selected by the four highest-order address bits LA0 to LA3. The contents of the selected SR are concatenated with the low-order 28 bits of the logical address to generate an interim 52-bit virtual address. Page address translation corresponds to the conversion of this virtual address into the 32-bit physical address. In most cases, the physical address for the page resides in the UTLB. If there is a miss in the UTLB, the processor automatically searches the page tables in memory.

Block address translation done by the BTLB, occurs in parallel with page translation and is similar to it, except that there are fewer upper order LA bits to be translated, since blocks are much larger than 4-kbyte pages.

For instruction accesses, the MMU first performs a lookup in the four-entry ITLB. In case of a miss in ITLB, UTLB, and BTLB are looked up.

The MPC 601 memory unit (Fig. 15.8) contains read and write queues that buffer operations between the external interface and the cache. These operations result from load and store instructions for which a cache miss occurred, and operations required to maintain cache coherency, such as dirty line copybacks. As shown in Fig. 15.8, the read queue contains two elements and the write queue contains three. Each element of the write queue can contain eight words (256 bits or 32 bytes). Each element of the read queue contains 32 address bits and control information bits.

The MPC 601 has a 32-kbyte unified (code and data in the same cache), eight-way set-associative, 64 bytes/line cache, illustrated in Fig. 15.9. Each line is divided into two eight-word sectors, each of which can be snooped, loaded, flushed, or invalidated independently.

The size of the cache being 2^{15} bytes, line size 2^6 bytes, we have

$2^{15}/2^6 = 2^9$ lines in the cache, and

$2^9/2^3 = 2^6$ sets in the cache.

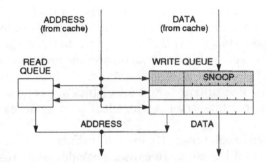

Figure 15.8 Memory unit. (*Courtesy of Motorola, Inc.*)

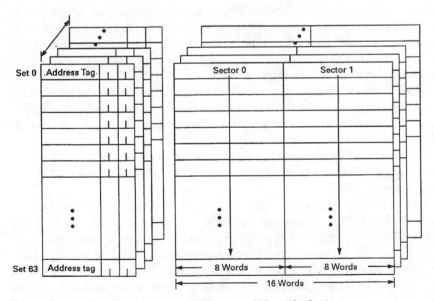

Figure 15.9 Cache unit organization. (*Courtesy of Motorola, Inc.*)

In forming the address accessing the cache, we need 6 bits to identify each byte in a line, and 6 bits to identify the set (total of 12 bits also constituting the internal page offset in a 4-kbyte page), leaving 20 bits for the TAG (see Chap. 4).

PowerPC implementations can control access modes on a page or block basis, featuring:

Write-back/write-through options

Cache-inhibited mode

Memory coherency

The cache coherency is handled by the modified, exclusive, shared, invalid (MESI) protocol [DiAl 92], featuring the following four states:

1. Modified; the cache line was modified and it holds the only valid data for this address.

2. Exclusive; the cache line holds data identical to the data at this address in memory. No other cache has this data.

3. Shared; the cache line holds data identical to the data at this address in memory and perhaps in one other cache.

4. Invalid; the data in the cache line is invalid.

Each sector has 2 bits that store the encoding for the above four states. An LRU algorithm is used for cache line replacement.

The cache organization is specific to the MPC 601 implementation. Caches of future PowerPC implementations may have different parameters and organization. For instance, other PowerPC implementations feature a dual cache, which yields better performance for pipeline handling (see Chaps. 5 and 6), particularly for superscalar systems.

The MPC 601 implements the following classes of exceptions:

1. Asynchronous imprecise for machine check and system reset

2. Asynchronous precise for external interrupts

3. Synchronous precise for instruction-caused exceptions

The external signals of the MPC 601 are shown in Fig. 15.10. PowerPC has adopted a set of interface signals similar to that of the Motorola MC88110 (see Chap. 19), but not identical. The 64-bit external data bus should be particularly noted.

15.4 The IBM RS/6000

The IBM RS/6000 is a predecessor of the PowerPC architecture. The PowerPC architecture is taken mostly from the IBM POWER architec-

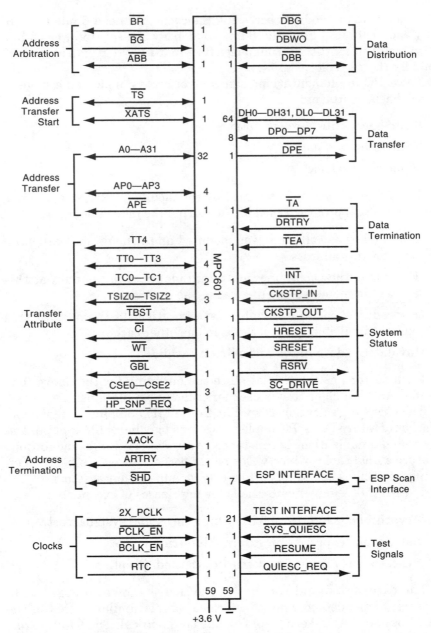

Figure 15.10 MPC601 signal groups. (*Courtesy of Motorola, Inc.*)

ture, on which the RS/6000 is based. Most of the POWER instructions are actually implemented by the PowerPC architecture. About 34 POWER architecture instructions were not implemented by PowerPC [PoPC 93]. A block diagram of the RS/6000 system is shown in Fig. 15.11 [BaWh 90, GrOe 90].

The primary RS/6000 subsystems, shown in Fig. 15.11, are

Instruction cache unit—ICU

Fixed-point unit—FXU

Floating-point unit—FPU

Data cache unit—DCU

Storage control unit—SCU

Initial program load—IPL

Read-only storage—ROS

I/O channel controller—IOCC

Serial link adapter—SLA

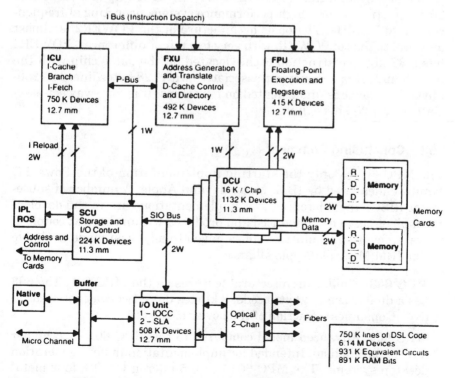

Figure 15.11   RS/6000 block diagram. (*Courtesy of IBM Corp.*)

System I/O bus—SIO bus

Description language code—DSL code

The bus size is expressed in Fig. 15.11 in units of W (word = 32 bits). Each block in Fig. 15.11 represents a separate chip (except blocks designated as memory cards, native I/O, and IPL ROS). If we count the FPU chip, we can say that the RS/6000 CPU is realized on three chips. The ICU contains an 8-kbyte Icache and a branch processor. The branch processor processes the incoming instruction stream from the Icache and feeds a steady flow of instructions to the FXU and the FPU. The RS/6000 is a four-issue superscalar, however, only a specific combination of four instructions can be executed simultaneously (one branch, one integer, one floating-point, and one CR instruction, for instance). The branch processor provides all of the branching, interrupt, and condition code functions within the system.

There exists also a single-chip version of the RS/6000. The RS/6000 instruction pipeline has five stages: fetch, dispatch, decode, execute, write back; somewhat different, but close to the MPC 601 integer pipeline. IBM has followed up the POWER line of system development with a new system called POWER2. It is an eight-chip (a total of over 20 million transistors) high-performance system, operating at frequencies up to 72 MHz. The basic architecture of the POWER2 is almost identical to that of POWER, with very few minor differences. POWER2 has a 32-kbyte instruction cache, located on the same chip with the branch unit. It is two-way set-associative with 128 bytes/line. The 256-kbyte data cache is implemented on four chips. It is four-way set-associative with 256 bytes/line.

15.5 Concluding Comment

The MPC 601 is only the starting implementation of the PowerPC family, endeavored by IBM, Motorola, and Apple. A number of subsequent RISC-type microprocessor implementations are in the development pipeline, to be announced and produced within the next couple of years. A few representative names have already been made available by the IBM/Motorola/Apple alliance:

MPC 603. Similar architectural features as the MPC 601. The 603 has a dual cache (8 kbytes code, 8 kbytes data, two-way set-associative). Consumes a fraction of the power of the 601.

MPC 604. Advanced model compared to the MPC 601 with higher level of parallelism. Intended for implementation in next-generation desktop systems. The MPC 604 is a 0.5 micron CMOS, four metal layer, 3.6 million transistors chip. It is a four-issue superscalar with an eight-entry instruction buffer and six execution units. It has a dual on-chip cache; 16-kbyte Icache, 16-kbyte Dcache (total 32 kbytes). The caches are four-way set-associative with 32 bytes/line.

MPC 620. A more advanced, fully 64-bit system (64-bit data, 64-bit address). Higher level of parallelism. Implemented with a dual cache and as a higher-issue superscalar.

The 603, 604, and 620 are implementations of the pure PowerPC architecture, whereas 601 is a "bridge" chip implementing a superset of the IBM POWER and PowerPC architectures.

Many more MPC 6xx-family systems are undoubtedly under development. How will they compare with other top RISC performers, such as the DEC Alpha (see Chap. 14), only future experience will tell.

16

The Sun SPARC Family

16.1 Introduction

The SPARC architecture was initiated by Sun Microsystems, Inc. in Mountain View, CA. Before announcing the SPARC, Sun Microsystems produced a very popular family of M68000-based Sun workstations. One of the things that differentiates SPARC from other RISC-type systems is that Sun does not have a history of preceding microprocessors, RISC or CISC, so it had no software compatibility constraints to worry about. SPARC designers could start from a clean slate. Another thing that differentiates Sun from other RISC microprocessor manufacturers is that it does not manufacture the chips it designs. Sun Microsystems designed the SPARC architecture, and then produced the workstations implementing the SPARC microprocessors (the famous and very popular SPARCstations). The actual manufacturing of microprocessors implementing the SPARC architecture is licensed out to a number of chip manufacturing companies in the United States, Europe, and Japan (the situation may somewhat change in the future with Sun starting to market the SPARC microprocessors as well). In fact, the very first marketed implementation of the SPARC was manufactured by the Japanese Fujitsu Microelectronics, Inc. The latest top level SPARC implementation is the SuperSPARC; a joint venture of Sun Microsystems and Texas Instruments (TI) in Houston, Texas.

The name SPARC stands for scalable processor architecture [Garn 88, KlWi 88, Paul 94, Sprc 90, Tabk 90b]. The concept of scalability, as seen by the creators of SPARC, is the wide spectrum of its possible price/performance implementations, ranging from microcomputers to supercomputers [Garn 88]. The scalability of the SPARC can also be interpreted in the number of CPU registers that can be used in various versions of products, implementing the SPARC architecture. The SPARC architecture follows the Berkeley RISC design philosophy [Kate 85, PaSe 82, Patt 85, Tabk 90b] by stressing of the importance of the relatively large CPU register file and by implementing similar register window features (see also Chap. 6). This will be further elaborated in the next section describing the SPARC architecture.

16.2   SPARC Architecture

Register file

SPARC architecture features a comparatively large CPU register file of over 100 registers. As in the Berkeley RISC, any procedure running on the SPARC can access only 32 registers, denoted r0 to r31. Eight of the registers (r0 to r7) are global, accessible by all procedures. The other 24 registers are the window registers, assigned to each procedure, with an overlap of eight registers between procedures. The 24 window registers are subdivided into three groups of eight registers each, as illustrated in Fig. 16.1, for a sequence of three nested procedures.

The group subdivision of the window registers is

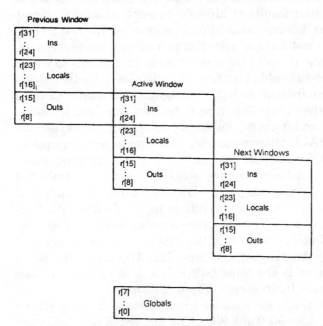

Figure 16.1 Three overlapping windows and globals. (*Copyright Sun Microsystems, Inc., 1987*)

r31 to r24 ins, contain parameters passed to the procedure by the calling procedure

r23 to r16 locals, contain local parameters of the procedure

r15 to r8 outs, contain parameters passed to the called procedure

As can be seen in Fig. 16.1, the outs registers of the calling procedure, are physically the ins registers of the called procedure. The calling procedure passes parameters to the called procedure through its outs registers, which are the ins registers of the called procedure. The register window of the currently running procedure, called the active window, is pointed to by the current window pointer (CWP) in the processor state register (PSR).

The number of windows (NWINDOWS) that can be used in different versions of the SPARC ranges from 2 to 32, for a total number of general-purpose IU registers (including the eight globals) ranging from 48 to 548, respectively. Most current SPARC implementation microprocessors feature eight windows for a total of 136 registers. Implemented windows are contiguously numbered 0 to (NWINDOWS − 1). An example of an eight-window implementation, where the windows are circularly interconnected, is shown in Fig. 16.2.

The CPU contains a 32-bit control register called window invalid mask (WIM). Each bit of WIM, $w_i (i = 0, 1, ..., 31)$, corresponds to one of the possible 32 windows (even if less than 32 are implemented). If $w_i = 1$, window i is considered to be invalid, and a trap condition exists. The CPU's program counter (PC) is a separate register, not included in the general-purpose register file. SPARC implementations may have several PCs containing addresses of subsequent instructions.

Some of the SPARC IU registers have specially designated tasks. The r0 is hardwired to a zero value, as it is in many other RISC-type systems (see also Chaps. 4 and 6). A CALL instruction writes its own address into the outs register r15. The CWP is decremented with a SAVE instruction on a procedure call and incremented by a RESTORE instruction on a procedure return. Procedures can also be called without changing the window.

Suppose that in the case of NWINDOWS = 8 (Fig. 16.2), window 0 is the currently running active window. In this case, CWP = 0. Since window 0 is the last free window, when the procedure, using window 0 calls another procedure, a window overflow occurs. A new register window wraps around to overwrite the previously used window 7, whose contents must be saved in the memory by software. After a return, and when the register file was out of windows, we have a window underflow. Software must restore previously used register windows in this case. A window overflow trap is caused by the overflow. The overflow

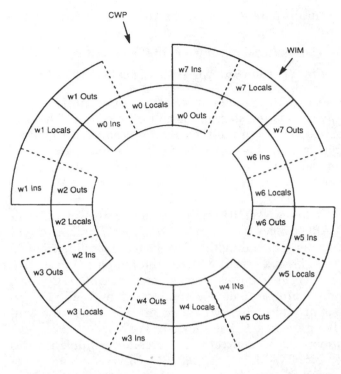

Figure 16.2 Circular stack of window registers. (*Copyright Sun Microsystems, Inc., 1987*)

trap handler uses the locals of window 7 for pointers into the memory where the overflowed window is stored. Window 7 is invalidated during the trap handling by setting bit w_7 of the WIM register.

Data types

SPARC architecture recognizes the following data types, in Fig. 16.3:

- Integer
  Signed, unsigned byte 8 bits
  Signed, unsigned halfword 16 bits
  Signed, unsigned word 32 bits
  Doubleword 64 bits
- Floating-point (IEEE 754-1985 standard)
  Single-precision 32 bits
  Double-precision 64 bits
  Quad-precision exponent: 15 bits, mantissa: 63 bits

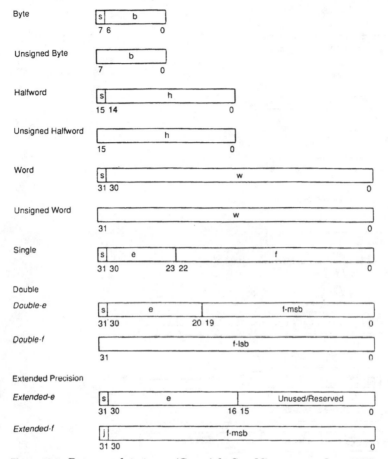

Figure 16.3 Processor data types. (*Copyright Sun Microsystems, Inc., 1987*)

The halfwords are aligned on 2-byte address boundaries (even addresses), words on 4-byte boundaries (addresses divisible by 4), and doublewords on 8-byte boundaries (addresses divisible by 8). The storage of the data types in memory is illustrated in Fig. 16.4. A big-endian byte ordering is used.

Instruction formats

The SPARC instruction formats are shown in Fig. 16.5. There are three basic instruction format types:

1. CALL

2. Branch instructions

3. Operate instructions (register-to-register).

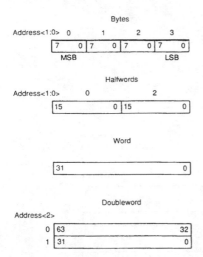

Figure 16.4 Address conventions. (*Copyright Sun Microsystems, Inc., 1987*)

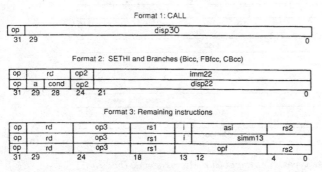

Figure 16.5 SPARC instruction formats. (*Copyright Sun Microsystems, Inc., 1987*)

As can be seen in Fig. 16.5, the operations instructions implement three-operand addressing, and all formats are of a single word length (32 bits). The fields in the instructions have the following designation: op bits 31,30 in all formats. They are interpreted as follows:

op	Instructions
01	Call
00	Bicc, FBfcc, CBccc, SETHI
10 or 11	All other instructions

op2 bits 24 to 22 in format 2. Selects the instruction as follows:

op2	Instruction
000	UNIMP
010	Bicc
100	SETHI
110	FBfcc
111	CBccc

rd   bits 29 to 25 in formats 2 and 3. Selects the source register for store instructions and the destination register for all other instructions.

a   bit 29 in format 2. Annul bit. Changes the behavior of the instruction encountered immediately after a control transfer.

cond   bits 28 to 25 in format 2. Selects the condition code for conditional branches.

imm22   bits 21 to 0 in format 2. A 22-bit constant value used by the SETHI instruction.

disp22   bits 21 to 0 in format 2. A 22-bit sign-extended word displacement for branch instructions.

disp30   bits 29 to 0 in format 1. A 30-bit sign-extended word displacement for PC-relative call instructions.

op3   bits 24 to 19 in format 3. Opcode extension.

i   bit 13 in format 3. Selects the type of the second ALU operand for non-floating-point operation instructions.

i = 0: the second operand is in a register rs2.

i = 1: the second operand is sign-extended simm13.

asi   bits 12 to 5 in format 3. An 8-bit address space identifier generated by load and store alternate instructions.

rs1   bits 18 to 14 in format 3. Selects the first source operand register.

rs2   bits 4 to 0 in format 3. Selects the second source operand register.

simm13   bits 12 to 0 in format 3. A sign-extended 13-bit immediate value.

opf   13 to 5 in format 3. Identifies a floating-point operate instruction.

Addressing modes

Besides the standard register direct and immediate addressing modes, there are only three addressing modes for memory access:

1. Register indirect with displacement; register + signed 13-bit constant

2. Register indirect indexed; register1 + register2

3. PC-relative; used in CALL instructions with a 30-bit displacement

Instruction set

The SPARC architecture features the following types of instructions:

1. Load/store
2. Arithmetic/logical/shift
3. Control transfer
4. Read/write control registers
5. Floating-point operate
6. Coprocessor operate (not needed on latest highly integrated implementations)

The SPARC instruction set is summarized in Table 16.1.

The load and store instructions (the only ones to access memory) generate a 32-bit byte address. In addition to the address, the processor always generates a 9-bit address space identifier (asi), interpreted as follows:

asi (decimal)	Assignment
0–7	Implementation-definable
8	User instruction space
9	Supervisor instruction space
10	User data space
11	Supervisor data space
12–255	Implementation-definable

Examples

```
LDSB    addr,rd; load signed byte from memory at address addr into
        register rd.
LD      addr,rd; load word from addr into rd.
```

Most of the arithmetic-logical instructions have dual versions (such as ADD and ADDcc) that modify the integer condition codes (icc) as a side effect.

Examples

```
ADD     rs1,rs2,rd; (rs1) + (rs2) → rd, no icc modification
ADDcc   rs1,rs2,rd; (rs1) + (rs2) → rd, icc modified
```

In some SPARC assemblers, registers are denoted as follows:

%0 to %31   All general registers r0 to r31

%g0 to %g7   Global registers, same as %0 to %7

TABLE 16.1   SPARC Instruction Set

Opcode	Name
LDSB (LDSBA*)	Load Signed Byte (from Alternate space)
LDSH (LDSHA*)	Load Signed Halfword (from Alternate space)
LDUB (LDUBA*)	Load Unsigned Byte (from Alternate space)
LDUH (LDUHA*)	Load Unsigned Halfword (from Alternate space)
LD (LDA*)	Load Word (from Alternate space)
LDD (LDDA*)	Load Doubleword (from Alternate space)
LDF	Load Floating-point
LDDF	Load Double Floating-point
LDFSR	Load Floating-point State Register
LDC	Load Coprocessor
LDDC	Load Double Coprocessor
LDCSR	Load Coprocessor State Register
STB (STBA*)	Store Bytes (into Alternate space)
STH (STHA*)	Store Halfword (into Alternate space)
ST (STA*)	Store Word (into Alternate space)
STD (STDA*)	Store Doubleword (into Alternate space)
STF	Store Floating-point
STDF	Store Double Floating-point
STFSR	Store Floating-point State Register
STDFQ*	Store Double Floating-point Queue
STC	Store Coprocessor
STDC	Store Double Coprocessor
STCSR	Store Coprocessor State Register
STDCQ*	Store Double Coprocessor Queue
LDSTUB (LDSTUBA*)	Atomic Load-Store Unsigned Byte (in Alternate space)
SWAP (SWAPA*)	Swap r Register with Memory (in Alternate space)
ADD (ADDcc)	Add (and modify icc)
ADDX (ADDXcc)	Add with Carry (and modify icc)
TADDcc (TADDccTV)	Tagged Add and modify icc (and Trap on overflow)
SUB (SUBcc)	Subtract (and modify icc)
SUBX (SUBXcc)	Subtract with Carry (and modify icc)
TSUBcc (TSUBccTV)	Tagged Subtract and modify icc (and Trap on overflow)
MULScc	Multiply Step and modify icc
AND (ANDcc)	And (and modify icc)
ANDN (ANDNcc)	And Not (and modify icc)
OR (ORcc)	Inclusive-Or (and modify icc)
ORN (ORNcc)	Inclusive-Or Not (and modify icc)
XOR (XORcc)	Exclusive-Or (and modify icc)
XNOR (XNORcc)	Exclusive-Nor (and modify icc)
SLL	Shift Left Logical
SRL	Shift Right Logical
SRA	Shift Right Arithmetic
SETHI	Set High 22 bits of r register
SAVE	Save caller's window
RESTORE	Restore caller's window
Bicc	Branch on integer condition codes
FBfcc	Branch on floating-point condition codes
CBccc	Branch on coprocessor condition codes
CALL	Call
JMPL	Jump and Link
RETT*	Return from Trap

TABLE 16.1   SPARC Instruction Set   *(Continued)*

Opcode	Name
Ticc	Trap on integer condition codes
RDY	Read Y register
RDPSR*	Read Processor State Register
RDWIM*	Read Window invalid Mask register
RDTBR*	Read Trap Base Register
WRY	Write Y register
WRPSR*	Write Processor State Register
WRWIM*	Write Window Invalid Mask register
WRTBR*	Write Trap Base Register
UNIMP	Unimplemented instruction
IFLUSH	Instruction cache Flush
FPop	Floating-point Operate: FiTO(s,d,x), F(s,d,x)TOi FsTOd, FsTOx, FdTOs, FdTOx, FxTOs, FxTOd, FMOVs, FNEGs, FABs, FSQRT(s,d,x), FADD(s,d,x), FSUB(s,d,x), FMUL(s,d,x), FDIV(s,d,x), FCMP(s,d,x), FCMPE(s,d,x)
CPop	Coprocessor operate

*Privileged instruction.

SOURCE: Copyright Sun Microsystems, Inc., 1987.

%o0 to %o7   Outs registers, same as %8 to %15

%l0 to %l7   Local registers, same as %16 to %23

%i0 to %i7   Ins registers, same as %24 to %31

Examples

```
SUB %12,%11,%15; (r12) - (r11) → r15.
AND %5,%3,%1; (r5) AND (r3) → r1.
```

The SETHI (set high 22 bits of rd) instruction writes a 22-bit constant from the instruction into the high-order bits of the destination register rd. It clears the low-order 10 bits of rd and does not change the condition codes. It can be used to construct a 32-bit constant using two instructions. It is used in the following form (format 2): SETHI const22,rd; const22 is a 22-bit constant number.

There are five types of control transfer instructions:

1. Conditional branch (Bicc, FBfcc, CBccc)

2. Jump and link

3. Call (CALL)

4. Trap (Ticc)

5. Return from trap (RETT)

Each of the above can be further categorized according to whether it is (1) PC-relative or register-indirect, or (2) delayed or nondelayed.

A PC-relative control transfer computes its target address by adding the (shifted) sign-extended immediate displacement to the PC value. A control transfer instruction is delayed if it transfers control to the target address after a one-instruction delay. This is essentially the delayed branch feature, described in Chap. 6. The a (annul) bit in format 2 influences the execution of the delay instruction, that follows the branch instruction. The annul operation is as follows:

a	Type of Branch	Delay Instruction Executed?
1	Unconditional	No
	Conditional, taken	Yes
	Conditional, not taken	No
0	Unconditional	Yes
	Conditional, taken	Yes
	Conditional, not taken	Yes

A procedure that requires a register window is invoked by executing both a CALL (or JMPL) and a SAVE instruction. A procedure that does not need a register window, a so-called leaf routine, is invoked by executing only a CALL (or a JMPL). Leaf routines can use only the outs registers. The CALL instruction stores PC, which points to the CALL itself, into register r15 (an outs register). The JMPL (jump and link) instruction stores PC into the specified register.

Example JMPL addr,rd; PC stored in rd.

The SAVE instruction is similar to an ADD instruction, except that it also decrements the CWP by one, causing the active window to become the previous window, thereby saving the caller's window.

Example SAVE rs1,rs2,rd; the operands rs1 and rs2 are from the previous (old) window and rd is in the new window, addressed by the new value in the CWP field.

A procedure that uses a register window returns by executing both a RESTORE and a JMPL instruction. A leaf procedure returns by executing a JMPL only. The RESTORE instruction (also similar to an ADD) increments the CWP by one, causing the previous window to become the active window, thereby restoring the caller's window. Also, the source registers for the addition are from the current window while the result is written into the previous window. Both SAVE and RESTORE compare the new CWP against the WIM to check for window overflow or underflow.

The SPARC architecture features the following multiprocessor operation support instructions:

SWAP—exchanges the content of an IU register with a word from memory, while preventing other memory accesses from intervening. SWAP is an atomic instruction (see Chap. 3).

LDSTUB—atomic load and store unsigned byte, reads a byte from memory into an IU register and then rewrites the same byte in memory to all 1s, while precluding intervening accesses. Can be used to construct semaphores (see Chap. 3).

The floating-point instructions, featured by the SPARC architecture, which were mentioned in Table 16.1, are listed in more detail in Table 16.2.

16.3   The SuperSPARC

The SuperSPARC is a 3.1-million transistor, 0.8-micron, three-layer metal BiCMOS, 293 ceramic pin grid array (PGA) microprocessor, manufactured by TI, in cooperation with Sun Microsystems [BlKr 92, SSPC 92]. The processor chip contains an IU, FPU, MMU, and a dual cache (20 kbytes code, 16 kbytes data, total: 36 kbytes). A block diagram of the SuperSPARC is shown in Fig. 16.6.

The SuperSPARC is a three-issue superscalar system. The SuperSPARC can issue and execute three instructions every cycle subject to the following constraints:

TABLE 16.2   Floating-point SPARC Instructions

Mnemonic	Operation
FiTOs, d, q	Convert integer to floating-point s, d, q
F(s, d, q)TOi	Convert floating-point s, d, q to integer
FsTOd, q	Convert floating-point single to d, q
FdTOs, q	Convert floating-point double to s, q
FqTOs, d	Convert floating-point quad to s, d
FMOVs	Move floating-point single
FNEGs	Negate floating-point single
FSQRTs, d, q	Floating-point square root s, d, q
FADDs, d, q	Floating-point add s, d, q
FSUBs, d, q	Floating-point subtract s, d, q
FMULs, d, q	Floating-point multiply s, d, q
FDIVs, d, q	Floating-point divide s, d, q
FsMULd	Floating-point multiply single produce double
FdMULq	Floating-point multiply double produce quad
FCMPs, d, q	Floating-point compare s, d, q
FCMPEs, d, q	Floating-point compare no exceptions s, d, q

where s = single-precision
      d = double-precision
      q = quad or extended-precision

Note: Instructions FsMULd and FdMULq are new on the SuperSPARC and have not been featured in earlier SPARC implementations.

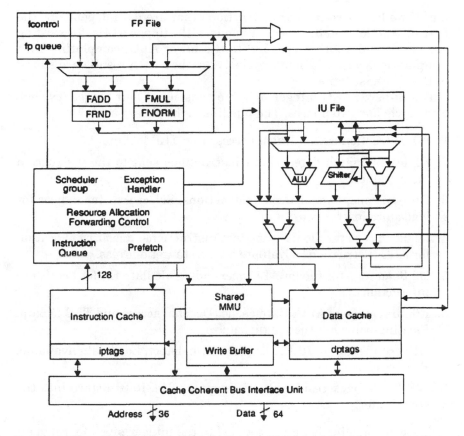

Figure 16.6 SuperSPARC functional block diagram. (*Courtesy of Sun Microsystems / Texas Instruments from SuperSPARC data sheets.*)

1. Maximum of two integer results
2. Maximum of one data memory reference
3. Maximum of one floating-point arithmetic instruction
4. Terminate group of instructions after each control transfer

Data dependencies (see Chap. 5) are solved on the SuperSPARC by

1. Cascading dependent instructions in the same group
2. Forwarding dependent instructions in consecutive groups

The block diagram in Fig. 16.6 shows the details of the structure of the IU. There are four ALUs and a shifter. The lower leftmost ALU is used for address computations; its output is forwarded to the MMU and the data cache. The other three ALUs are used for data processing. If we have two integer instructions that are data-dependent, the first instruction can be executed in one of the upper ALUs, forwarding the result and another operand to the lower ALU, completing both computations within a single cycle. After that, both results are stored in the IU register file.

The SuperSPARC integer pipeline consists of four stages (cycles). Each cycle has two phases. The pipeline eight phases are

1. F0 Instruction cache (Icache) access and TLB lookup.

2. F1 Icache match detect. Four instructions sent to the instruction queue (iqueue).

3. D0 Issue one, two, or three instructions. Select register indices for load/store instructions.

4. D1 Read register file for load/store instructions. Resource allocation for ALU instructions. Evaluate branch target address.

5. D2 Read register file for ALU operands. Calculate EA for load/store instructions.

6. E0 First stage of ALU. Data cache (Dcache) access and TLB lookup. Floating-point instruction dispatch.

7. E1 Second stage of ALU. Dcache match detect. Load data available. Resolve exceptions.

8. WB Write back result into the register file. Retire store into the store buffer.

The FPU pipeline is tightly coupled to the integer pipeline. An operation may be started every cycle; the delay of most floating-point operations is three cycles. In the E0 phase, one floating-point arithmetic instruction is selected for execution and its operands are read during E1. Two stages of execution delay are required for the double-precision FPU adder and FPU multiplier. The first cycle of the adder examines exponents, aligns mantissas, and produces a result. The first cycle of the multiplier computes and adds partial products. Independent second stages round and normalize the result of the respective units. Forwarding paths are provided to chain results of one FPU operation into the source of a subsequent operation.

Figure 16.7 illustrates an example of pipelined execution of a set of ALU operations, with a load instruction in between. Up to four instructions can be fetched during the (F0, F1) cycle, but only up to three instructions can be issued as a group (GRP) during the D0 phase. Forwarding of results between subsequent groups of instructions, is shown by arrows in Fig. 16.7.

Pipelined execution of a set of instructions, which includes a conditional branch, is shown in Fig. 16.8. A taken branch case in Fig.

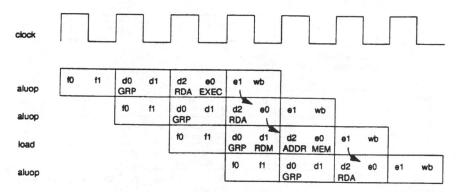

Figure 16.7 Pipelined execution of load with forwarding. ADDR = calculate EA; EXEC = execute; GRP = issue group of instructions; MEM = Dcache access; RDA = read register file for ALU operands; RDM = read register file for memory access. (*Courtesy of Sun Microsystems, Inc. from SuperSPARC data sheets.*)

16.8(*a*), and a nontaken case in Fig. 16.8(*b*). The original sequential instructions are denoted as S1 and S2, and the target instructions as T1, T2, T3, and T4. The delay instruction, placed after the conditional branch instruction (BNE in this example), is denoted by DI, while C1 refers to the certainty instruction stream. The SuperSPARC processor can group the compare (CMP) and the conditional branch instruction (BNE), to speed execution. The processor statically predicts that all branches are taken. When a control transfer instruction relative to the PC is issued, its DI is fetched concurrently. During the D1 phase the target address (TA) is computed. As the branch instruction enters phase D2, the target instruction stream is fetched (FT). The fetch completes as the DI advances to phase D1 and the compare and branch instructions enter phase E0. The compare instruction computes new integer condition codes in phase E0 and the branch direction is resolved. When a branch is taken, all sequential path instructions (S1 and on; grouped together with the DI) are invalidated (squash S1), as shown in Fig. 16.8(*a*). When a branch is not taken (untaken), sequential path instructions (SA+) remain valid and the target instructions (T1 and on) fetched are discarded. This scheme does not introduce a pipeline bubble (stall) for either branch path. The PC and prefetch PC values for both directions are precomputed. The SuperSPARC branch implementation can execute nontaken branches somewhat more efficiently than taken branches.

The SuperSPARC implements a precise exception model. At any given time there can be up to nine instructions in the IU pipeline and four more in the floating-point queue. Exceptions and the instructions that caused them propagate through the IU pipeline. They are resolved in the execute stage before their results can modify visible state in the register file, control registers, or memory.

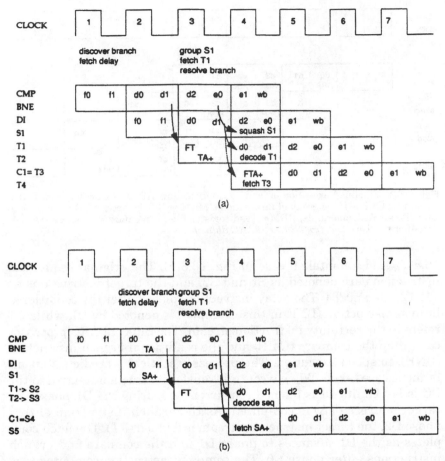

Figure 16.8 (a) Taken conditional branch. (b) Untaken conditional branch. (*Courtesy of Sun Microsystems/TI from SuperSPARC data sheets.*)

The FPU consists of a floating-point controller (FPC), two independent pipelines FADD and FMUL, a floating-point queue, and a 32-bit 32-register floating-point file (FP file; see Fig. 16.6). The FP file is organized as sixteen 64-bit double words to optimize double-precision performance. Each 32-bit word of the FP file can be accessed separately, however. The FP file has three read and two write ports. The FPC is tightly coupled to the IU pipeline and is capable of executing a floating-point memory event and a floating-point operation in the same cycle. The FPC also handles floating-point exceptions.

There are two types of floating-point instructions:

1. FPOPs—floating-point operations, such as add, multiply, convert, and so on.

2. FPEVENTs—floating-point events, such as load/store to/from float-ing-point registers, load/store to/from floating-point status register, store floating-point queue, integer multiply, and integer divide. FPEVENTs are executed by the FPU but do not enter the FPU queue.

The FPU pipeline consists of four stages:

1. FRD—decode and read
2. FM/FA—execute multiply or add
3. FN/FR—normalization and rounding
4. FWB—write-back to FP file

All floating-point instructions are issued by the IU in the E0 phase of the IU pipeline. Once issued, the floating-point instructions proceed through the FPU pipeline. The FPU pipeline stalls the IU pipeline in a few situations: the FPU queue will become full after several long delay FPU arithmetic instructions are encountered in close proximity. Forwarding paths are provided to chain the result of one FPOP to source operands of a subsequent FPOP without stalling for an FP register write. Similarly, an FPEVENT load can forward data to FPOP source operands and an FPOP result can forward data to an FPEVENT store.

All FPU instructions start in order and complete in order. They are executed in one of the two independent units: FADD and FMUL. These instructions are held in an FIFO queue that holds up to four entries. Each entry can hold a 32-bit FPOP instruction and a 32-bit FPOP address.

An FPU exception remains pending until another FPOP or FPEVENT is requested by the IU. It will be reported to the subsequent floating-point instruction at that time.

The SuperSPARC has a dual cache: 20 kbyte Icache, and 16 kbyte Dcache for a total of 36-kbyte on-chip primary cache. There is support for an external, second-level 1-mbyte cache. Figure 16.9 shows the SuperSPARC MMU and cache organization. This organization uses the term "set" in a different manner than in the computer literature, as presented in Chap. 5. In the SuperSPARC a set is a block of data of 4 kbytes, the page size in this system. The instruction cache contains five such sets for a total of five pages or 20 kbytes. The data cache contains four such sets for a total of four pages or 16 kbytes. Based on this, the instruction cache is said to be five-way set-associative, and the data cache is four-way set-associative. Both caches use a pseudo-LRU replacement algorithm. The line size on the instruction cache is 64 bytes, and the line size on the data cache is 32 bytes. The instruction

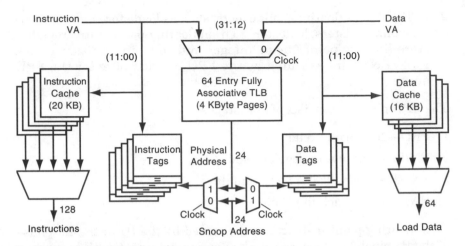

Figure 16.9 Cache/MMU organization. VA = virtual address. (*Courtesy of Sun Microsystems / Texas Instruments from SuperSPARC data sheets.*)

cache is accessed by a 128-bit fetch path, allowing the fetching of 4 instructions simultaneously. The data cache is accessed by a 64-bit path, allowing transmission of double-precision floating-point data in one bus cycle. The hit rate for SPEC 92 was reported by Sun to be 98 percent for the instruction cache and 90 percent for the data cache. The caches are physically addressed. The SuperSPARC TLB has 64 entries and it is fully associative.

The following instructions, not available on the early SPARC implementations, are featured by the SuperSPARC:

UMUL (UMULcc): unsigned integer multiply (and modify icc)

SMUL (SMULcc): signed integer multiply (and modify icc)

UDIV (UDIVcc): unsigned integer divide (and modify icc)

SDIV (SDIVcc): signed integer divide (and modify icc)

RDASR: read ancillary state register

WRASR: write ancillary state register

STBAR: store barrier

FsMULd: floating-point multiply single produce double

FdMULq: floating-point multiply double produce quad

16.4   Earlier SPARC Implementations

The very first SPARC architecture implementations included a CPU chip with a single-ALU IU, a register file (usually 136 registers from the start), PCs, PSR, and a few other control and status registers. The FPU, MMU, and cache had to be realized on separate chips. Some of the first manufacturers to implement the SPARC architecture were Fujitsu Microelectronics, Inc. (Japan), Cypress Semiconductor (Ross Technology), Bipolar Integrated Technology, Inc. (BIT), LSI Logic Corporation, and Texas Instruments (TI), the manufacturer of the SuperSPARC.

The original Cypress SPARC CPU was labeled CY7C601. Cypress followed it up by a subsequent model called HyperSPARC, or CY7C620. The HyperSPARC is a two-issue superscalar, with over 1 million transistors, containing on-chip IU and FPU, operating at 55.5 MHz (later versions are promised to run at up to 100 MHz), with a 64-bit wide data bus. The MMU and cache have to be configured outside of the CPU chip.

TI also features a lower-level, scalar, single-pipeline microprocessor, called MicroSPARC. The MicroSPARC is an 0.8-micron, 800-000 transistor, 5 V microprocessor, consuming 3.5 W at 50 MHz. It contains an IU, FPU, and a modest on-chip dual cache of 4 kbytes code and 2 kbytes data. A more recent version, MicroSPARC II, will operate within the frequency interval of 40 to 100 MHz.

16.5 Concluding Comment

There exist numerous implementations of the SPARC architecture, manufactured by a number of companies all over the world. Many more are currently being developed, and undoubtedly new SPARC-architecture more powerful microprocessors will be announced in the near future. It is quite possible that some new versions of the SuperSPARC will be higher-issue superscalar systems, operating at higher frequencies.

Chapter

17

The MIPS Rx000 Family

17.1 Introduction

Similarly to Sun Microsystems, MIPS Computer Systems, Inc., is a creator of an original RISC-type architecture. Similarly to Sun, MIPS performs the architectural design and licenses the actual chip manufacturing to other companies. MIPS manufactures, however, workstations and other computer systems based on these chips (again, similarly to Sun). MIPS RISC-type chips have been adopted as CPUs by notable computer systems manufacturers, such as DEC in its DECstation systems, Silicon Graphics in its Iris series of workstations, and many others. Silicon Graphics apparently developed such a strong taste for MIPS microprocessors, that it eventually purchased MIPS Computer Systems.

The MIPS system originated at Stanford University in the early eighties [Hein 93, Henn 82, Henn 84, HePa 90, Przy 84, Tabk 90b]. The MIPS acronym stands for *m*icroprocessor without *i*nterlocked *p*ipeline *s*tages. The reason for the above name is that MIPS pipeline hazards are handled by software (by the MIPS compiler), as opposed to hardware pipeline interlocks, practiced on other systems (see also Chap. 5).

Some of the MIPS developers at Stanford, together with other professionals, founded the MIPS Computer Systems company, which continued the development of the MIPS into a successful commercial product. MIPS developers have designed the RISC-type systems R2000, R3000, R6000, R4000, and R4400, which can be denoted in general as the Rx000 family, to be described in the following sections in this chapter. All of the above are implementations of the MIPS architecture, to be described in the next section.

381

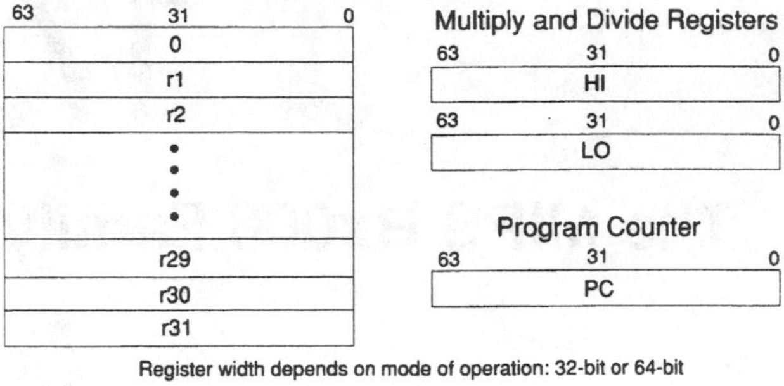

General-Purpose Registers

Figure 17.1   CPU registers. (*Courtesy of MIPS Computer Systems, Inc.*)

17.2   MIPS Architecture

Register file

The MIPS user-visible CPU integer registers are shown in Fig. 17.1. There are 32 general-purpose registers r0 to r31, a PC, and two registers that hold the results of integer multiply and divide operations:

HI—multiply and divide register higher result (remainder for divide)

LO—multiply and divide register lower result (quotient for divide)

All registers shown in Fig. 17.1 are 64-bit for 64-bit microprocessor implementations. The same registers are 32-bit for 32-bit implementations.

Two of the general-purpose registers have a special function:

r0 is hardwired to a zero value, as in most RISC-type systems (see Chaps. 5 and 6).

r31 is the link register for jump and link instructions. It should not be used explicitly by other instructions.

All other registers, r1 through r30, are free to be used by the programmer in any way.

The floating-point register file in the FPU is shown in Fig. 17.2. It consists of:

1. Thirty-two floating-point general-purpose registers FGR0 to FGR31. As shown in Fig. 17.2, the register configuration depends on the value of the FR bit in the status register (SR; contains the operating mode, interrupt enabling, and the diagnostic states of the processor).

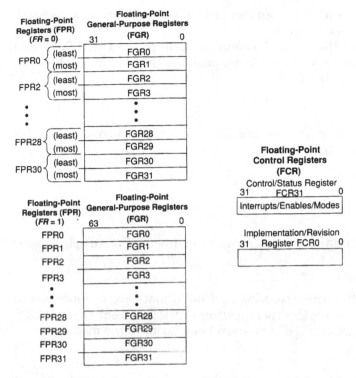

Figure 17.2 FPU registers. (*Courtesy of MIPS Computer Systems, Inc.*)

If FR = 1, the FGRs are 64 bits wide, and if FR = 0, the FGRs are 32 bits wide.

2. Control/status register FCR31. The 32-bit FCR31 contains control and status data. It controls the arithmetic rounding mode and the enabling of user-mode traps. It also identifies exceptions that occurred in the most recently executed instructions.

3. Implementation/revision register FCR0. The 32-bit FCR0 specifies the implementation and revision number of the FPU.

Data types

The MIPS architecture recognizes the following data types:

Byte—8 bits

Halfword—16 bits

Word—32 bits

Doubleword—64 bits

Byte ordering within a word can be configured either as big-endian or little-endian (see Chap. 3).

The MIPS floating-point formats conform to the IEEE 754-1985 standard, featuring the 32-bit single-precision and 64-bit double-precision (see Chap. 3) [IEEE 85].

Addressing modes

The MIPS architecture features the following addressing modes (defined in Chap. 3):

Register

Immediate

PC-relative

Register indirect (base register), with or without offset (displacement)

Instruction formats

The three MIPS architecture instruction formats are summarized in Fig. 17.3, along with the specification of the different fields. MIPS instructions must be aligned on word boundaries in the memory.

Instruction set

MIPS architecture defines four coprocessors, denoted CP0 through CP3. CP0, incorporated on the CPU chip, supports the virtual memory

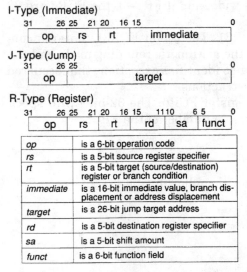

Figure 17.3 CPU instruction formats. (*Courtesy of MIPS Computer Systems, Inc.*)

system and exception handling. CP1 is the FPU coprocessor, incorporated on the CPU chip in the latest implementations. CP2 is reserved for future definition by MIPS, and CP3 is used to provide certain extensions to the MIPS instruction set architecture (ISA). The basic instruction set (ISA), common to all MIPS Rx000 microprocessors is listed in Table 17.1. Table 17.2 lists instructions which are extensions to the ISA in latest implementations.

The on-chip coprocessor CP0 special instructions are listed in Table 17.3. The FPU instructions are listed in Table 17.4.

Examples

```
Load word: LW rt,offset(rs);
```

```
The I-type format (Fig. 17.3) is used. The rt field represents the
destination CPU register. The rs field points to the base register.
The immediate field contains a 16-bit offset (displacement sign-
extended to 32 bits), added to the content of rs to form the memory
EA, whose content is loaded into rt.
```

```
Subtract:    SUB rd,rs,rt; rd ← (rs) − (rt), R-type format
Multiply:    MULT rs,rt; R-type format (rd not used), (rs) * (rt) →
             (HI,LO) 64-bit product
Divide:      DIV rs,rt; R-type format (rd not used), (rs)/(rt),
             quotient → LO, remainder → HI
Jump:        J target; J-type format
```

The 26-bit target field is shifted left 2 bits, combined (on the left) with the 4 most significant bits of the current PC value, to form the jump target address. The delayed branch feature (see Chap. 6) is implemented on the scalar implementations of the MIPS.

```
Jump and link: JAL target;
```

Used as a subroutine call. Works exactly the same way as the jump, except that the address of the instruction after the delay slot, (PC) + 8, is placed in the link register r31.

```
Branch on equal: BEQ rs,rt,offset; I-type format.
```

The contents of registers rs and rt are compared. If equal, the program branches to a new target address. The new target address is computed from the sum of the address of the instruction in the delay slot (following BEQ) and the 16-bit offset, shifted left 2 bits and sign-extended to 32 bits.

17.3 The MIPS R4000 and R4400

The MIPS R4000 is a 64-bit, 1.3-million transistor, 1-micron CMOS, 447-pin microprocessor. It is a two-issue superpipelined system, with an external frequency of 50 MHz, and the internal pipeline running at

TABLE 17.1   CPU Instruction Set (ISA)

OP	Description	OP	Description
	Load and Store Instructions		Multiply and Divide Instructions
LB	Load Byte		
LBU	Load Byte Unsigned	MULT	Multiply
LH	Load Halfword	MULTU	Multiply Unsigned
LHU	Load Halfword Unsigned	DIV	Divide
LW	Load Word	DIVU	Divide Unsigned
LWL	Load Word Left	MFHI	Move From HI
LWR	Load Word Right	MTHI	Move To HI
SB	Store Byte	MFLO	Move From LO
SH	Store Halfword	MTLO	Move To LO
SW	Store Word		
SWL	Store Word Left		Jump and Branch Instructions
SWR	Store Word Right		
		J	Jump
	Arithmetic Instructions	JAL	Jump And Link
	(ALU Immediate)	JR	Jump Register
		JALR	Jump And Link Register
ADDI	Add Immediate	BEQ	Branch on Equal
ADDIU	Add Immediate Unsigned	BNE	Branch on Not Equal
SLTI	Set on Less Than Immediate	BLEZ	Branch on Less than or Equal to Zero
SLTIU	Set on Less Than Immediate Unsigned	BGTZ	Branch on Greater Than Zero
ANDI	AND Immediate	BLTZ	Branch on Less Than Zero
ORI	OR Immediate	BGEZ	Branch on Greater than or Equal to Zero
XORI	Exclusive OR Immediate		
LUI	Load Upper Immediate	BLTZAL	Branch on Less Than Zero And Link
	Arithmetic Instructions (3-Operand, R-type)	BGEZAL	Branch on Greater than or Equal to Zero And Link
ADD	Add		
ADDU	Add Unsigned		Coprocessor Instructions
SUB	Subtract		
SUBU	Subtract Unsigned	LWCz	Load Word to Coprocessor z
SLT	Set on Less Than	SWCz	Store Word from Coprocessor z
SLTU	Set on Less Than Unsigned		
AND	AND	MTCz	Move to Coprocessor z
OR	OR	MFCz	Move from Coprocessor z
XOR	Exclusive OR	CTCz	Move Control to Coprocessor z
NOR	NOR	CFCz	Move Control From Coprocessor z
	Shift Instructions	COPz	Coprocessor Operation z
SLL	Shift Left Logical	BCzT	Branch on Coprocessor z True
SRL	Shift Right Logical	BCzF	Branch on Coprocessor z False
SRA	Shift Right Arithmetic		
SLLV	Shift Left Logical Variable		Special Instructions
SRLV	Shift Right Logical Variable		
SRAV	Shift Right Arithmetic Variable	SYSCALL	System Call
		BREAK	Break

SOURCE: Courtesy of MIPS Computer Systems, Inc.

TABLE 17.2   Extensions to the ISA

OP	Description	OP	Description
	Load and Store Instructions		**Multiply and Divide Instructions (*Cont.*)**
LD	Load Doubleword		
LDL	Load Doubleword Left	DDIV	Doubleword Divide
LDR	Load Doubleword Right	DDIVU	Doubleword Divide Unsigned
LL	Load Linked		**Jump and Branch Instructions**
LLD	Load Linked Doubleword		
LWU	Load Word Unsigned	BEQL	Branch on Equal Likely
SC	Store Conditional	BNEL	Branch on Not Equal Likely
SCD	Store Conditional Doubleword	BLEZL	Branch on Less than or Equal to Zero Likely
SD	Store Doubleword		
SDL	Store Doubleword Left	BGTZL	Branch on Greater Than Zero Likely
SDR	Store Doubleword Right		
SYNC	Sync	BLTZL	Branch on Less Than Zero Likely
	Arithmetic Instructions (ALU Immediate)	BGEZL	Branch on Greater Than or Equal to Zero Likely
DADDI	Doubleword Add Immediate	BLTZALL	Branch on Less Than Zero And Link Likely
DADDIU	Doubleword Add Immediate Unsigned	BGEZALL	Branch on Greater Than or Equal to Zero And Link Likely
	Arithmetic Instructions (3-Operand, R-type)	BCzTL	Branch on Coprocessor z True Likely
DADD	Doubleword Add	BCzFL	Branch on Coprocessor z False Likely
DADDU	Doubleword Add Unsigned		
DSUB	Doubleword Subtract		**Exception Instructions**
DSUBU	Doubleword Subtract Unsigned	TGE	Trap if Greater Than or Equal
	Shift Instructions	TGEU	Trap if Greater Than or Equal Unsigned
DSLL	Doubleword Shift Left Logical	TLT	Trap if Less Than
DSRL	Doubleword Shift Right Logical	TLTU	Trap if Less Than Unsigned
DSRA	Doubleword Shift Right Arithmetic	TEQ	Trap if Equal
		TNE	Trap if Not Equal
DSLLV	Doubleword Shift Left Logical Variable	TGEI	Trap if Greater Than or Equal Immediate
DSRLV	Doubleword Shift Right Logical Variable	TGEIU	Trap if Greater Than or Equal Immediate Unsigned
DSRAV	Doubleword Shift Right Arithmetic Variable	TLTI	Trap if Less Than Immediate
DSLL32	Doubleword Shift Left Logical + 32	TLTIU	Trap if Less Than Immediate Unsigned
DSRL32	Doubleword Shift Right Logical + 32	TEQI	Trap if Equal Immediate
DSRA32	Doubleword Shift Right Arithmetic + 32	TNEI	Trap if Not Equal Immediate
	Multiply and Divide Instructions		**Coprocessor Instructions**
		DMFCz	Doubleword Move From Coprocessor z
DMULT	Doubleword Multiply	DMTCz	Doubleword Move To Coprocessor z
DMULTU	Doubleword Multiply Unsigned	LDCz	Load Double Coprocessor z
		SDCz	Store Double Coprocessor z

SOURCE: Courtesy of MIPS Computer Systems, Inc.

TABLE 17.3   CP0 Instructions

OP	Description
DMFC0	Doubleword Move From CP0
DMTC0	Doubleword Move To CP0
MTC0	Move to CP0
MFC0	Move from CP0
TLBR	Read Indexed TLB Entry
TLBWI	Write Indexed TLB Entry
TLBWR	Write Random TLB Entry
TLBP	Probe TLB for Matching Entry
ERET	Exception Return

SOURCE: Courtesy of MIPS Computer Systems, Inc.

a double frequency of 100 MHz. It has an on-chip FPU and a dual cache of 8 kbytes code and 8 kbytes data, for a total of 16 kbytes. Although the R4000 is a full-fledged 64-bit system, its instructions are uniformly 32-bit, as specified in Sec. 17.2.

There is also a more advanced version of the R4000: the R4400. The R4400 is architecturally identical to the R4000. The R4400 is a 0.6-micron CMOS, containing 2.2 million transistors, with a double dual cache of 16 kbytes code and 16 kbytes data, for a total of 32 kbytes. Its external frequency is 75 MHz, and the internal pipeline runs at 150 MHz.

In addition to the above, there is also an intermediate version, the R4200, produced by MIPS and NEC (Nippon Electric Company, Japan). The R4200 combines the integer and the floating-point processing in the same functional area on the chip. It has a dual nonsymmetric cache: 16 kbytes code, 8 kbytes data, for a total of 24 kbytes. Its external frequency is 40 MHz (internal 80 MHz) with a power dissipation of 1.5 W. It has a simpler five-stage pipeline, as opposed to the R4000 or R4400 eight-stage pipeline (to be discussed in detail later in this section). The R4200 is manufactured by NEC and it is strongly targeted for embedded designs.

Another implementation of the same architecture is the Integrated Device Technology (IDT; Santa Clara, CA) and Toshiba (Irvine, CA) R4600, also called Orion. It has essentially the same features as the R4400, except that it has a shorter pipeline and its caches are two-way set-associative (other R4x00 systems are direct-mapped). According to IDT's announcement, the R4600 outperforms the R4400 by 35 percent for integer, and by 8 percent for floating-point operations. Its power dissipation is typically 4 W.

A block diagram of the R4000 is shown in Fig. 17.4. One can clearly see in it the IU, FPU, CP0, and the dual cache.

TABLE 17.4   FPU Instruction Summary

OP	Description
	Load/Store/Move Instructions
LWC1	Load word to FPU
SWC1	Store word from FPU
LDC1	Load doubleword to FPU
SDC1	Store doubleword from FPU
MTC1	Move word to FPU
MFC1	Move word from FPU
CTC1	Move control word to FPU
CFC1	Move control word from FPU
DMTC1	Doubleword move to FPU
DMFC1	Doubleword move from FPU
	Conversion Instructions
CVT.S.fmt	Floating-point Convert to Single FP
CVT.D.fmt	Floating-point Convert to Double FP
CVT.W.fmt	Floating-point Convert to Single Fixed Point
ROUND.w.fmt	Floating-point Round
TRUNC.w.fmt	Floating-point Truncate
CEIL.w.fmt	Floating-point Ceiling
FLOOR.w.fmt	Floating-point Floor
	Computational Instructions
ADD.fmt	Floating-point Add
SUB.fmt	Floating-point Subtract
MUL.fmt	Floating-point Multiply
DIV.fmt	Floating-point Divide
ABS.fmt	Floating-point Absolute value
MOV.fmt	Floating-point Move
NEG.fmt	Floating-point Negate
SQRT.fmt	Floating-point Square Root
	Compare Instructions
C.cond.fmt	Floating-point Compare
	Branch on FP Condition
BC1T	Branch on FPU True
BC1F	Branch on FPU False
BC1TL	Branch on FPU True Likely
BC1FL	Branch on FPU False Likely
.fmt	format specifier
.cond	condition specifier

SOURCE: Courtesy of MIPS Computer Systems, Inc.

The R4000 eight-stage instruction pipeline is shown in Figs. 17.5 and 17.6. Figure 17.5 shows the eight pipeline stages for eight consecutive instructions, while Fig. 17.6 illustrates the details of the pipeline operation. One can see in Fig. 17.5 the issue of two consecutive instructions, one every half master clock cycle, in a two-issue super-pipelined operation (see Chap. 5).

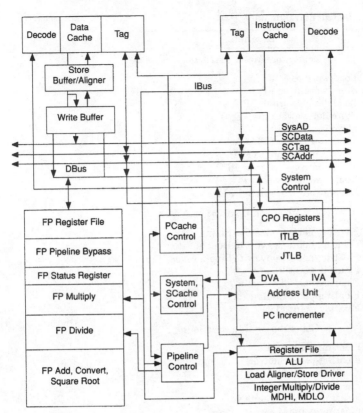

Figure 17.4 R4000 internal block diagram. (*Courtesy of MIPS Computer Systems, Inc.*)

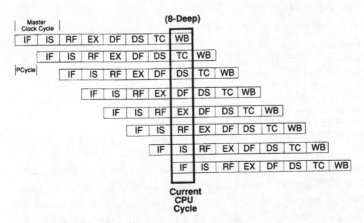

Figure 17.5 R4000 pipeline and instruction overlapping. (*Courtesy of MIPS Computer Systems, Inc.*)

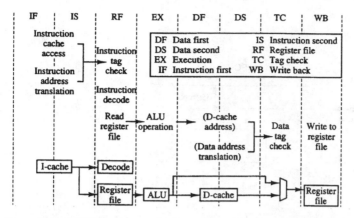

| IF | IS | RF | EX | DF | DS | TC | WB |

Instruction cache access

Instruction address translation

Instruction tag check

Instruction decode

DF	Data first	IS	Instruction second
DS	Data second	RF	Register file
EX	Execution	TC	Tag check
IF	Instruction first	WB	Write back

Read register file — ALU operation — (D-cache address) — Data tag check — Write to register file

(Data address translation)

I-cache → Decode

Register file → ALU → D-cache → Register file

Figure 17.6 R4000 hardware pipeline stages. (*Courtesy of MIPS Computer Systems, Inc.*)

The R4000 pipeline stages are the following [Hein 93; MR40 91]:

1. IF—instruction fetch, first half. An instruction address is selected by the branch logic and the instruction cache fetch begins. The instruction TLB (ITLB) begins the virtual to physical address translation.

2. IS—instruction fetch, second half. The instruction cache fetch and the ITLB translation are completed.

3. RF—register fetch. The instruction decoder decodes the instruction and checks for interlock conditions. The instruction cache tag is checked against the page frame number obtained from the ITLB. Any required operands are fetched from the register file.

4. EX—execution. For register-to-register instructions, the ALU performs the arithmetic or logical operation. For load and store instructions, the ALU calculates the data virtual address. For branch instructions, the ALU determines whether the branch condition is true and calculates the virtual branch target address.

5. DF—data fetch, first half. For load and store instructions, the data cache fetch and the data address translation begin. For branch instructions, the target address translation and TLB update begin. Register-to-register instructions perform no operations during the DF, DX, and TC stages.

6. DS—data fetch, second half. For load and store instructions, the data cache fetch and data address translation are completed. The shifter aligns the data to the word or double word boundary. For branch instructions, the target address translation and TLB update are completed.

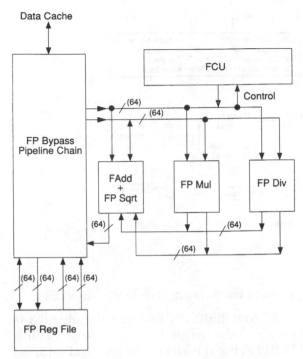

Figure 17.7 FPU functional block diagram. (*Courtesy of MIPS Computer Systems, Inc.*)

7. TC—tag check. For load and store instructions, the cache performs the tag check. The physical address from the TLB is checked against the cache tag to determine if there is a hit or a miss.

8. WB—write back. For register-to-register instructions, the result is written back to the register file. Load, store, and branch instructions perform no operation during this stage.

The above pipeline has a branch delay of three cycles and a load delay of two cycles. Any pipeline interrupts are handled by hardware interlocks. In that respect, the design of the R4000 deviates from the original Stanford MIPS.

The R4000 FPU is shown in Fig. 17.7 [MR40 91]. Like the IU, the FPU uses a load/store oriented instruction set, with single-cycle load and store operations. Floating-point operations are started in a single cycle and their execution is overlapped with other fixed-point or floating-point operations. The FPU conforms to the IEEE 754-1985 standard [IEEE 85]. The FPU has three operation units adder, multiplier, and divider, subject to the following constraints:

Adder: allows one clock cycle overlap between each newly issued instruction and the instruction being completed.

Multiplier: allows up to two pipelined MUL.[S,D] instructions to be processed as long as:

1. Two idle cycles are required after MUL.S.
2. Three idle cycles are required after MUL.D.

Divider: handles only one nonoverlapped divide instruction in its pipeline at any one time.

Both R4000 caches are direct-mapped, 8 kbytes each, organized with either a four-word (16-byte) or eight-word (32-byte) line. The write policy for the data cache is write-back (see Chap. 4). Each line of the Icache has associated with it in addition to a 24-bit address tag, a valid bit, and a parity bit. Each line of the data cache has associated with it in addition to a 24-bit address tag, a two-bit line state indicator, and a write-back bit. The write-back bit indicates when the cache line contains modified data that must be written back to memory or to the secondary cache. There are four states defined for the primary on-chip cache:

1. Invalid—The cache line does not contain valid information.
2. Shared—The line contains valid information and may be present in another cache. The line may or may not be consistent with memory, and may or may not be owned.
3. Clean exclusive—The line contains valid information and is not present in any other cache. The line is consistent with memory and is not owned.
4. Dirty exclusive—The line contains valid information and is not present in any other cache. The line is inconsistent with memory and is owned by the processor.

The above states are particularly used in protocols for maintaining cache coherency in multiprocessing systems [Hwan 93].

The R4000 contains on-chip logic to interface to an optional external secondary cache. It can be configured either as a unified or as a dual cache. The interface is designed to support the following secondary cache parameters: direct mapping, write-back policy for data, with optional line sizes of 4-word (16-byte), 8-word (32-byte), 16-word (64-byte), or 32-word (128-byte). The secondary cache line has associated with it, in addition to a 19-bit address tag, a 3-bit primary cache index, and a 3-bit line state indicator (five states).

The primary cache index provides a pointer to the virtual address of primary cache lines that may contain data from the secondary cache line. The five secondary cache states are:

1. Invalid—The line does not contain valid information.

2. Shared—The line contains valid information and may be present in another cache. The line may or may not be consistent with memory, and is not owned.

3. Dirty shared—The line contains valid information and may be present in another cache. The line is inconsistent with memory and is owned.

4. Clean exclusive—The line contains valid information and is not present in any other cache. The line is consistent with memory and is not owned.

5. Dirty exclusive—The line contains valid information and is not present in any other cache. The line is inconsistent with memory and is owned.

The primary cache shared state corresponds to the secondary cache states shared and dirty shared.

The secondary cache may be configured to have a total size of 128 kbytes to 4 Mbytes. There are 128 data lines between the CPU chip and the secondary cache.

The R4000 virtual address space is defined for two modes:

1. The 32-bit mode: 32-bit address, maximal user process size is 2 Gbytes = 2^{31} bytes.

2. The 64-bit mode: 64-bit address, maximal user process size is 1 terabyte = 2^{40} bytes.

The 36-bit physical address space is 64 Gbytes = 2^{36} bytes.
The R4000 functions in three operating modes:

1. User, regular operation for nonsupervisory programs.

2. Supervisor, intermediate mode which can be used to more easily build a secure OS.

3. Kernel, most privileged mode, analogous to "supervisor" mode in other systems.

The R4000 TLB has 48 entries. Each entry maps two consecutive pages, having the effect of a 96-entry TLB.

There are three types of the R4000 processor:

1. R4000SC, a high-performance uniprocessor, 447-pin, secondary cache control

2. R4000MC, large cache-coherent CPU for multiprocessors, 447-pin, secondary cache control

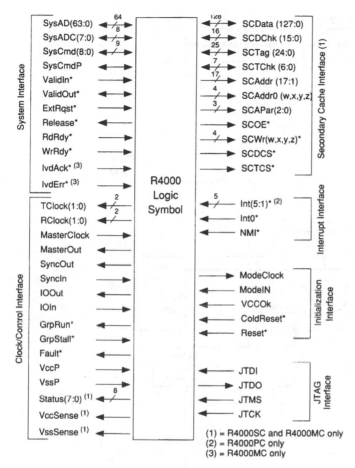

Figure 17.8   R4000 logic symbol diagram. (*Courtesy of MIPS Computer Systems, Inc.*)

3. R4000PC, for cost-sensitive systems, 179-pin, no support for secondary cache

The pinout diagram of the R4000 is shown in Fig. 17.8. Of particular note are the 128 data lines for secondary cache interface on the right (SCData), and 64 address and data lines (multiplexed), interfacing to the system bus (SysAD).

17.4   Earlier MIPS Implementations

Early MIPS implementation microprocessors (R2000, R3000, R6000) contained just an integer CPU and the CP0 on chip. The FPU was implemented on auxiliary chips (R2010, R3010, R6010). Some of the

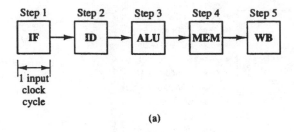

(a)

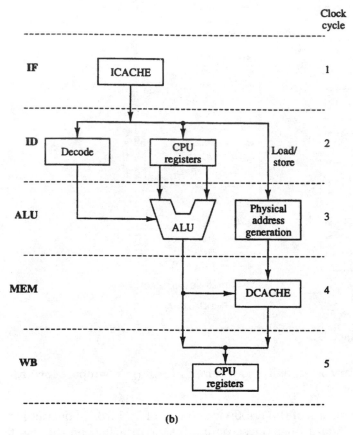

(b)

Figure 17.9   MIPS R2000/R3000 pipeline. (*a*) Functional representation of a simplified five-stage instruction pipeline. (*b*) Simplified physical representation of five-stage instruction pipeline. (Here we assume the existence of an Icache and a Dcache on-chip. *Note:* stages IF and ALU may involve TLB look-up. (*Courtesy of MIPS Computer Systems, Inc.*)

primary MIPS manufacturers are Integrated Device Technology (IDT), Inc. (Santa Clara, CA), LSI Logic Corporation (Milpitas, CA), and Performance Semiconductor Corporation (Sunnyvale, CA). The on-chip logic was designed to interface to a dual external cache from the beginning.

The R2000/R3000 pipeline (essentially the same for R6000), illustrated in Fig. 17.9, has five stages:

1. IF—instruction fetch. Get instruction physical address from the TLB.

2. ID—instruction decode. Complete instruction fetch from external instruction cache. Read operands from the CPU register file while decoding the instruction.

3. ALU—arithmetic logic unit. Perform the required operation on the instruction operands.

4. MEM—access memory. Access external data cache for load or store instructions.

5. WB—write back. Store ALU result or value loaded from the data cache into the register file.

17.5 Concluding Comment

Systems implementing the MIPS architecture microprocessors have been extensively implemented both by MIPS Computer Systems and other companies in a wide range of workstations and other computing systems. Among some latest implementations one can mention the use of the R4400 in an up to 24 CPUs multiprocessor of the Nile series, produced by Pyramid Technology Corporation (San Jose, CA), which was one of the first manufacturers of commercial RISC-type systems in the early eighties [Tabk 90b]. Silicon Graphics (Mountain View, CA), which later purchased MIPS, produced a large variety of MIPS architecture microprocessor based graphics workstations of the Iris series. Of particular note is one of the latest models, an R4000-based workstation Crimson. Many more MIPS applications exist and will be announced in the near future.

Chapter

18

The Intel i860 Family

18.1 Introduction

The Intel i860 family microprocessors (also known as 80860) are the first RISC-type products of Intel. For many years Intel was and still is a leader in the microprocessor industry, primarily known for its x86 CISC-type family, described in detail in Part 2. The i860 RISC was first announced in 1989 [Atki 91, KoMa 89]. It featured on-chip FPU, dual cache, and a graphics unit (first microprocessor with such a feature). The latest model, i860 XP, has an on-chip dual cache of a total of 32 kbytes; 16 kbytes instruction, 16 kbytes data. Since its inception, the i860 was implemented in a number of notable computing systems, such as the Alliant 2800 multiprocessor and Intel's latest massively parallel processing systems, such as the Paragon.

The i860 architecture will be presented in the next section, followed by details of its implementation. As is the case with other Intel microprocessors, the i860 is manufactured by Intel.

18.2 i860 Architecture

Register set

The i860 CPU registers and data paths are illustrated in Fig. 18.1. There are two separate sets of general-purpose registers:

1. Integer register file: thirty-two 32-bit registers, r0 to r31. Register r0 is permanently hardwired to a value of zero, as it is practiced in most RISC-type systems.

2. Floating-point register file: thirty-two 32-bit registers, f0 to f31. Registers f0 and f1 are permanently hardwired to a zero value. When accessing 64-bit floating-point values, an even-odd pair of registers can be used such as (f2, f3). When accessing 128-bit values, an aligned set

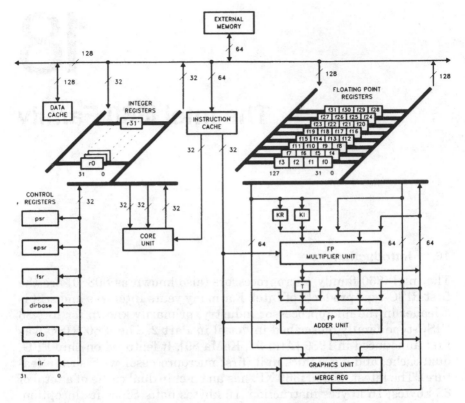

Figure 18.1 Registers and data paths. (*Courtesy of Intel Corporation*)

of four registers (f0, f1, f2, f3), (f4, f5, f6, f8),..., or (f28, f29, f30, f31), can be used. The instruction must designate the lowest register number of the set of registers containing 64- or 128-bit values. The register with the lowest number contains the least significant value.

The 32-bit processor status register (**psr**), shown in Fig. 18.2, contains the state information for the current process. The psr bits and fields are the following:

Bit 0, BR—break read and bit 1, BW—break write enable a data access trap when the operand address matches the address in the data breakpoint register (**db**), and a read or write, respectively, occurs.

Bit 2, CC—condition code, set by various instructions according to tests they perform. The branch on condition code (bc) instructions tests its value.

Bit 3, LCC—loop condition code, set and tested by the branch on LCC and add (bla) instruction.

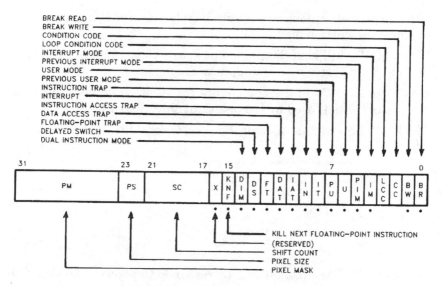

*Can be changed only from supervisor level.

Figure 18.2 Processor status register. (*Courtesy of Intel Corporation*)

Bit 4, IM—interrupt mode, enables external interrupts if set (IM = 1); disables interrupts if clear (IM = 0).

Bit 5, PIM—previous interrupt mode and bit 7, PU—previous user mode, save the corresponding status bits (IM and U) on a trap, because those status bits are changed when a trap occurs. They are restored into their corresponding status bits when returning from a trap handler with a branch indirect instruction when a trap flag is set in the psr.

Bit 6, U—user mode, set when the processor is executing in the user mode; cleared when executing in the supervisor mode.

Bit 8, IT—instruction trap, bit 9, IN—interrupt, Bit 10, IAT—instruction access trap, bit 11, DAT—data access trap, and bit 12, FT—floating-point trap are trap flags. They are set when the corresponding trap occurs. The trap handler routine examines these bits to determine which conditions have caused the trap.

Bit 13, DS—delayed switch, set if a trap occurs during the instruction cycle before the dual instruction mode is entered or exited.

Bit 14, DIM—dual instruction mode, set when a trap occurs if the processor is executing in dual-instruction mode (when the processor fetches two instructions on each clock cycle); cleared if it is executing in single-instruction mode.

Bit 15, KNF—kill next floating-point instruction, when set, the next floating-point instruction is suppressed.

Bits 17 to 21, SC—shift count, store the 5-bit shift count (by how many bits shifted) used by the last right shift instruction.

Bits 22,23, PS—pixel size, control pixel size according to the encoding:

PS	Pixel size, bits	Pixel size, bytes
00	8	1
01	16	2
10	32	4
11	Undefined	Undefined

Bits 24 to 31, PM—pixel mask, corresponds to pixels to be updated by the pixel store instruction (**pst.d**). If a bit of PM is set, then **pst.d** stores the corresponding pixel.

The KR, KI, and T registers are special-purpose registers used by the dual-operation floating-point instructions which initiate both an adder and a multiplier unit operation. These registers can store values from one dual-operation instruction and supply them as inputs to subsequent dual-operation instructions.

The MERGE register is used only by the vector-integer instructions. It accumulates (or merges) the results of multiple addition operations that use as operands the color intensity values from pixels or distance values from a buffer. The accumulated results can then be stored in one 64-bit operation.

Data types

The i860 architecture supports integer and floating-point data types. Load and store operations can reference 8-, 16-, 32-, 64-, and 128-bit operands. Integer operations are performed on 32-bit operands. Add and subtract instructions can also operate on 64-bit integers. Arithmetic operations on 8- and 16-bit integers can be performed by sign-extending the values to 32 bits and then using the 32-bit operations. Two's complement is used for signed values. When an 8- or 16-bit integer is stored in a 32-bit register, it is sign-extended to 32 bits.

The i860 supports the IEEE 754-1985 floating-point standard formats of 32 and 64 bits (see Chap. 3). Graphics unit pixels of 8, 16, or 32 bits are supported. Regardless of the pixel size, the i860 always operates on 64 bits worth of pixels at a time.

Addressing modes

The i860 architecture features a few very simple addressing modes.

Register—the operand is in a CPU register.

Immediate—the operand is a part of the instruction.

Offset—absolute address into the first or last 32 kbytes of the logical address space.

Register indirect with offset—EA = (reg) + offset; assembly notation: #const(reg)

Register indirect with index—EA = (reg1) + (reg2); assembly notation: reg1(reg2)

In addition the floating-point load and store instructions may select autoincrement addressing. In this mode (reg2) is replaced by (reg1 + reg2) after performing the load or store. This mode makes stepping through arrays more efficient, because it eliminates one address calculation instruction.

Instruction formats

The i860 architecture instruction formats are illustrated in Fig. 18.3, along with the appropriate notation interpretation.

Instruction set

The i860 instruction set is listed in Table 18.1. It should be noted that the i860 does not feature integer multiply and divide instructions; in fact, it does not have a divide instruction at all. The multiplication operation is performed by the floating-point multiplier. In order to multiply two integers, the **fmlow.p** instruction (see Table 18.1) can be used. There is a floating-point reciprocal value instruction (**frcp.p**) which can be used to perform division via a Newton–Raphson approximation. There is also a reciprocal floating-point square root (**frsqr.p**) instruction.

Examples

```
adds    src1,src2,rdest; add signed, (src1)+(src2) → rdest
subs    src1,src2,rdest; subtract signed, (src1) − (src2) → rdest
ld.x    src1(src2),rdest; load integer from memory EA = (src1)+(src2)
        into rdest register.
```

.x denotes the size of the loaded operand:

.x = **.b,** 8-bit

.x = **.s,** 16-bit

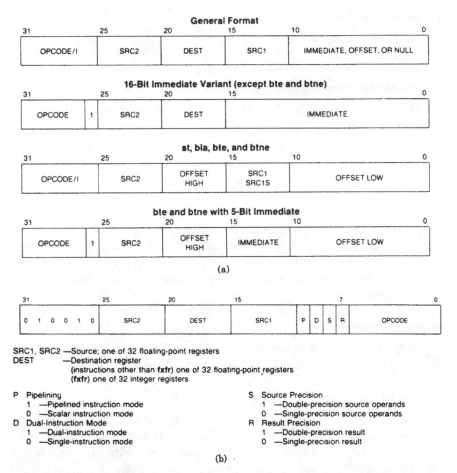

Figure 18.3 80860 instruction formats. (*a*) REG-format variations; (*b*) floating-point instruction encoding. (*Courtesy of Intel Corporation*)

.x = .l, 32-bit

st.x src1,#const(src2); store integer from register src1 into memory EA = (src2) + const.

shl src1,src2,rdest; shift left logical src2 by (src1) bits → rdest.

mov src2,rdest; register-to-register move (src2) → rdest.

The above move instruction is equivalent to: **shl** r0,src2,rdest (r0 is permanently zero).

In most systems, the pipelines are transparent to the programmer and would thus not be a part of the architecture (see discussion in Chap. 3) but of the organization. The i860 has several specific pipeline instructions, giving the user direct control over the pipeline operation.

TABLE 18.1   Instruction Set

Core unit		Floating-point unit	
Mnemonic	Description	Mnemonic	Description
Load and Store Instructions		**Register to Register Move**	
ld.x	Load integer	fxfr	Transfer F-P to integer register
st.x	Store integer		
fld.y	F-P load	**F-P Multiplier Instructions**	
fst.y	F-P store		
pfld.y	Pipelined F-P load	fmul.p	F-P multiply
pst.d	Pixel store	pfmul.p	Pipelined F-P multiply
Register to Register Move		pfmul3.dd	Three-stage pipelined F-P multiply
ixfr	Transfer integer to F-P register	fmlow.p	F-P multiply low
		frcp.p	F-P reciprocal
Integer Arithmetic Instructions		fsqr.p	F-P reciprocal square root
addu	Add unsigned	**F-P Adder Instructions**	
adds	Add signed		
subu	Subtract unsigned	fadd.p	F-P add
subs	Subtract signed	pfadd.p	Pipelined F-P add
Shift Instructions		famov.r	F-P adder move
		pfamov.r	Pipelined F-P adder move
shl	Shift left	fsub.p	F-P subtract
shr	Shift right	pfsub.p	Pipelined F-P subtract
shra	Shift right arithmetic	pfgt.p	Pipelined greater-than compare
shrd	Shift right double	pfeq.p	Pipelined equal compare
Logical Instructions		fix.v	F-P to integer conversion
		pfix.v	Pipelined F-P to integer conversion
and	Logical AND	ftrunc.v	F-P to integer truncation
andh	Logical AND high		
andnot	Logical AND NOT	**Dual-Operation Instructions**	
andnoth	Logical AND NOT high		
or	Logical OR	pfam.p	Pipelined F-P add and multiply
orh	Logical OR high	pfsm.p	Pipelined F-P subtract and multiply
xor	Logical exclusive OR	pfmam.p	Pipelined F-P multiply with add
xorh	Logical exclusive OR high	pfmsm.p	Pipelined F-P multiply with subtract
Control Transfer Instructions		**Long Integer Instructions**	
br	Branch direct		
bri	Branch indirect	fisub.z	Long-integer subtract
bc	Branch on CC	pfisub.z	Pipelined long-integer subtract
bc.t	Branch on CC taken	fiadd.z	Long-integer add
bnc	Branch on not CC	pfiadd.z	Pipelined long-integer add
bnc.t	Branch on not CC taken	**Graphics Instructions**	
bte	Branch if equal		
btne	Branch if not equal	fzchks	16-bit Z-buffer check
bla	Branch on LCC and add	pfzchds	Pipelined 16-bit Z-buffer check
call	Subroutine call		
calli	Indirect subroutine call		
intovr	Software trap on integer overflow		
trap	Software trap		

TABLE 18.1   Instruction Set   *(Continued)*

Core unit		Floating-point unit	
Mnemonic	Description	Mnemonic	Description
Graphics Instructions (*Cont.*)		System Control Instructions	
fzchkl	32-bit Z-buffer check	flush	Cache flush
pfzchkl	Pipelined 32-bit Z-buffer check	ld.c	Load from control register
		st.c	Store to control register
faddp	Add with pixel merge	lock	Begin interlocked sequence
pfaddp	Pipelined add with pixel merge	unlock	End interlocked sequence
		scyc.x	Special bus cycles
faddz	Add with Z merge	Assembler Pseudo-Operations	
pfaddz	Pipelined add with Z merge		
form	OR with MERGE register	Register to Register Move	
pform	Pipelined OR with MERGE register	mov	Integer move
		fmov.r	F-P reg-reg move
I/O Instructions		pfmov.r	Pipelined F-P reg-reg move
		nop	Core no-operation
ldio.x	Load I/O	fnop	F-P no-operation
stio.x	Store I/O	pfle.p	Pipelined F-P less-than or equal
ldint.x	Load interrupt vector		

SOURCE: Courtesy of Intel Corporation.

The pipelines are "pushed" by the user and not by the clock. Therefore, in the i860 the pipelines are a part of the architecture. Another interesting point to be noted is that pipelined loads of the i860 help "hide" memory delays. The i860 pipeline instructions start with a "p" (see Table 18.1), such as **pfmul.p, pfadd.p, pfsub.p,** and others.

Another form of parallelism featured by the i860 architecture is that it can execute both a floating-point and an integer instruction simultaneously. In such a case, the i860 operates as a two-issue superscalar (see Chap. 5). Such parallel execution is called on the i860 dual-instruction mode. When executing in dual-instruction mode, the instruction sequence consists of 64-bit aligned instructions with a floating-point instruction in the lower 32 bits and an integer instruction in the upper 32 bits (the i860 data bus is 64-bit in and out of the chip in all its implementations). Table 18.1 identifies which instructions are executed by the integer (core) unit and which by the FPU. Programmers can specify a dual-instruction mode in one of the following ways:

1. By including in the mnemonic of a floating-point instruction a **d.** prefix

2. By using the assembler directives *.dual* and *.enddual* arranged as follows:

```
.dual
```

code to be executed by the dual-instruction mode

```
.enddual
```

Both of the above specifications cause the D bit of floating-point instructions to be set (see Fig. 18.3), permitting execution in the dual-instruction mode. Special dual-operation floating-point instructions (add and multiply, **pfam.p,** subtract and multiply, **pfsm.p**) use both the multiplier and adder units within the FPU (Fig. 18.1) in parallel to efficiently execute such common tasks as evaluating systems of linear equations, performing the fast Fourier transform (FFT), and graphics transformations.

The instructions pfam.p, pfsm.p, pfmam.p, and pfmsm.p, all operating on operands src1,src2,rdest, initiate both an adder operation and a multiplier operation. Six operands are required, but the instruction format specifies only three operands; therefore, there are special provisions for specifying the operands. These special provisions consist of:

1. Three special registers KR, KI, and T that can store values from one dual-operation instruction and supply them as inputs to subsequent dual-operation instructions.
   a. Registers KR and KI can store the value of src1 and subsequently supply that value to the multiplier pipeline in place of src1.
   b. Register T can store the last-stage result of the multiplier pipeline and subsequently supply that value to the adder pipeline in place of src1.

2. A 4-bit data path control field in the opcode that specifies the operands and loading of the special registers.
   a. Operand 1 of the multiplier can be KR, KI, or src1.
   b. Operand 2 of the multiplier can be src2 or the last-stage result of the adder pipeline.
   c. Operand 1 of the adder can be src1, T, or the last-stage result of the adder pipeline.
   d. Operand 2 of the adder can be src2, the last-stage result of the multiplier pipeline, or the last-stage result of the adder pipeline.

The lower left corner of Fig. 18.1 shows all the possible data paths surrounding the adder and multiplier. The above-mentioned data path control field selects different data paths.

Unconditional branch instructions and the branch and test instruction are never delayed on the i860. Conditional branch instructions can be delayed (see Chap. 6) by the user by adding ".**t**" to the instruction mnemonic. The i860 uses a scoreboarding technique (see Chap. 6) to guarantee proper operation of the code and prevent the use of incorrect data in a pipelined environment.

Memory organization

The i860 memory is byte-addressable with a paged virtual address memory of 4 Gbytes (2^{32} bytes). Data and instructions can be located anywhere in this address space. Normally, data are stored in memory in little-endian format. The i860 also offers the big-endian byte ordering as an option. Data consisting of n bytes are to be aligned on n-byte address boundaries. For instance, 64-bit (8-byte) data are to be aligned on 8-byte boundaries, that is, the 3 LSBs of the address must be zero.

For the 4-kbyte page implementation, the i860 uses the page tables and the same format of virtual address as in the Intel x86 architecture (see Sec. 7.8). The page directory is pointed to by the **dirbase** register (see Fig. 18.1).

18.3 i860 Implementation

The current top implementation of the i860 architecture is the i860 XP. A block diagram of the i860 XP is shown in Fig. 18.4. It is a 2.55-million-transistor, three-layer metal, 0.8-micron CHMOS, 262-pin microprocessor. Its power dissipation is 5 W at 50 MHz. It has an on-chip dual cache of 16 kbytes code and 16 kbytes data, for a total of 32 kbytes [i860 91].

The i860 integer instruction pipeline has four stages [Atki 91], as described in the example in Chap. 5:

1. **F**—fetch instruction from instruction cache
2. **D**—decode instruction and access register file
3. **E**—execute operation
4. **W**—write (store) result in the register file

If the instruction is a branch, stage 3 becomes:

I—access instruction cache for target instruction and decode the instruction in the branch delay slot (next instruction to the branch). If it is a conditional branch, check condition code to determine if taken.

If the instruction is a subroutine call, then during stage 4 (W), write return address to the register file.

For floating-point instructions the execute stage extends for three cycles: E1, E2, and E3. Thus, the floating-point pipeline is six stages deep.

The predecessor of the i860 XP is i860 XR. Their architecture is essentially identical. The i860 XR is a 1-million transistor, 1-micron CHMOS, 168-pin chip. It has a dual cache of 4 kbytes code and 8 kbytes data, for a total of 12 kbytes. Both caches use a random replacement policy (see Chap. 4).

Instructions that are new on the i860 XP: **pfld.q, scyc.x, ldio.x, stio.x, ldint.x** (see Table 18.1).

Beside the **psr** and the **dirbase** system registers, already defined earlier, there are the following 32-bit system registers on the i860 XP (Fig. 18.4, [i860 91]):

epsr—extended psr, contains additional state information for the current process beyond that stored in the **psr**.

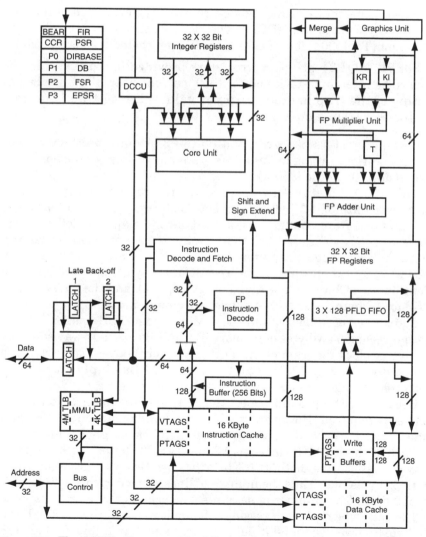

Figure 18.4 The i860XP. (*Courtesy of Intel Corporation*)

db—data breakpoint register, used to generate a trap when the processor makes a data operand access to the address stored in this register.

fir—fault instruction register, contains the address of the trapping instruction, when a trap occurs.

fsr—floating-point status register, contains the floating-point trap and rounding mode status for the current process.

bear—bus error address register, receives the address of the cycle for bus or parity errors.

ccr—concurrency control register, controls the operation of the concurrency control unit (CCU). The CCU feature was licensed from Alliant [Tabk 90a] for efficient compiler generated loop level code.

P0, P1, P2, P3—privileged registers provided for the OS. May be used as interrupt stack pointer (SP), current user SP at the beginning of the trap handler, register values during trap handling, processor ID in a multiprocessor system.

System registers bear, ccr, and P0 to P3 are new on the i860 XP. Eight additional implemented bits were added in register epsr, and one in dirbase on the i860 XP.

The i860 XP features two page sizes: the regular 4 kbytes and a large page size of 4 Mbytes, intended for lengthy software packages and graphics frame buffers. As pointed out in Chap. 4, it follows from some recent research studies that it may be advantageous to feature more than one page size, offered as an option. This design direction was already taken up in some recent microprocessors such as the Pentium (see Chap. 8), the Alpha (see Chap. 14), and the i860 XP. The TLB for 4-kbyte pages has 64 entries encompassing $2^{12} * 2^6 = 2^{18}$ bytes = 256 kbytes of memory, and the 4-Mbyte page TLB has 16 entries, encompassing 64 Mbytes of memory. Both TLBs are four-way set-associative. A block diagram of the TLBs and their address structure, are illustrated in Fig. 18.5(a) and (b). Both TLBs employ a random replacement algorithm because of its simplicity.

Because of the pipeline operation a simultaneous access to both caches is required. However, the TLB can handle only one access at a time. In such a case, data access translation has a higher priority.

Each of the i860 XP caches is 16 kbytes = 2^{14} bytes. Both are four-way set-associative (both i860 XR caches are two-way set-associative) with 32 bytes/line. Thus the number of lines in each cache is $2^{14}/2^5 = 2^9$ = 512 lines. The number of sets in each cache is therefore $2^9/2^2 = 2^7$ = 128 sets. The cache addressing scheme is shown in Fig. 18.6. For a 32-byte line we need a 5-bit field to select a byte in a line, and for 128 sets we need a 7-bit field for set selection, a total of 12 bits, exactly the

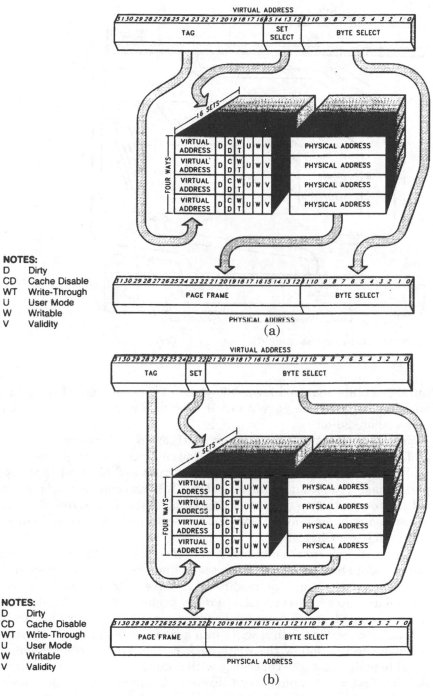

NOTES:
D   Dirty
CD   Cache Disable
WT   Write-Through
U   User Mode
W   Writable
V   Validity

(a)

NOTES:
D   Dirty
CD   Cache Disable
WT   Write-Through
U   User Mode
W   Writable
V   Validity

(b)

Figure 18.5   (a) 4K TLB organization. (b) 4M TLB organization. (*Courtesy of Intel Corporation*)

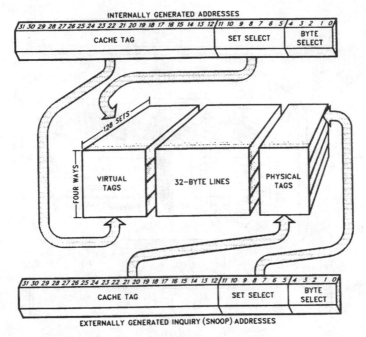

INTERNALLY GENERATED ADDRESSES

EXTERNALLY GENERATED INQUIRY (SNOOP) ADDRESSES

Figure 18.6 Cache address usage. (*Courtesy of Intel Corporation*)

number of bits required for a 4 kbytes $= 2^{12}$ byte page offset. Thus, both the tag and the page number (nonzero part of the page base address) are 20-bit.

Each cache has two sets of tags: virtual tags used for internal access, and physical tags used for snooping (for cache coherency protocols in multiprocessors). The presence of both virtual and physical tags supports aliasing, a situation in which the TLBs associate a single physical address with two or more virtual addresses. Cache line fills are generated only for read misses, not for write misses (a no-write-allocate policy; see Chap. 4).

Figure 18.7 shows the organization of the i860 XP data cache. There are two state bits per physical tag (each line has a tag) and one validity bit per virtual tag. The modified/exclusive/shared/ invalid (MESI) set of four states for the cache coherency protocol, listed in Table 18.2, is implemented. A write-once policy is implemented. A write-through is implemented for the first write into a line, while subsequent writes to the same line follow the write-back policy. The write-once policy helps to maintain cache coherency with a minimal amount of bus traffic. The first write (under write-through) broadcasts to other processors the fact that a line has been modified. The external system can

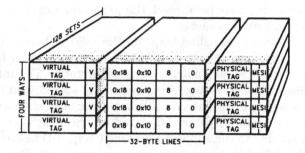

Figure 18.7   Data cache organization. (*Courtesy of Intel Corporation*)

TABLE 18.2   MESI Cache Line States

Cache line state	M modified	E exclusive	S shared	I invalid
This cache line is valid?	Yes	Yes	Yes	No
The memory copy is...	...out of date	...valid	...valid	—
Copies exist in other caches?	No	No	Maybe	Maybe
A write to this line...	...does not go to bus	...does not go to bus	...goes to bus and updates the cache	...goes directly to bus

dynamically change the update policy (write-back, write-through, write-once) of the i860 XP with each cache line.

Figure 18.8 shows the organization of the Icache. Since instructions are not written into, no coherency problem exists, and no state bits are needed. There is only one validity bit per line (common to both virtual and physical tags). Aliasing support for instructions consists not simply of changing the virtual tag, but rather fetching a line whenever a virtual tag miss occurs. If the physical address already exists in the instruction cache, its line and tags are overwritten. So, even though a

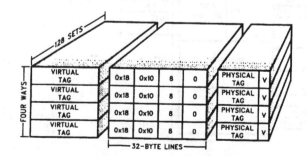

Figure 18.8   Instruction cache organization. (*Courtesy of Intel Corporation*)

physical line may be aliased, the processor never enters the line twice in the instruction cache.

Exceptions are called traps on the i860. When a trap occurs, execution of the current instruction is aborted. Except for bus error and parity error traps, the instruction is restartable. The information needed to restart the interrupted instruction is stored, and the processor executes an appropriate trap handler program in supervisor mode. Floating-point traps are triggered by the next floating-point instruction similar to the Alpha AXP's (see Chap. 14) floating-point traps.

The i860 XP signals are shown in Fig. 18.9. There are 64 external data lines D0 to D63. Of the 32 address lines, only lines A3 to A31 are external. However, there are 8 byte enable lines BE0# to BE7# to identify each byte within a 64-bit double word. Tables 18.3, 18.4, and 18.5

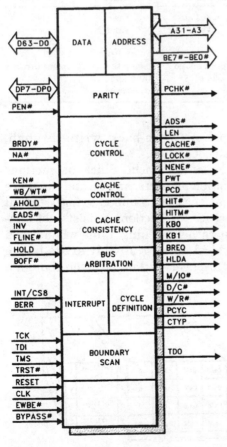

Figure 18.9 Signal grouping. (*Courtesy of Intel Corporation*)

TABLE 18.3   ADS# Initiated Bus Cycle Definitions

M/IO#	D/C#	W/R#	Bus cycle initiated
0	0	0	Interrupt Acknowledge
0	0	1	Special Cycle
0	1	0	I/O Read
0	1	1	I/O Write
1	0	0	Code Read
1	0	1	Reserved
1	1	0	Memory Read
1	1	1	Memory Write

TABLE 18.4   Memory Data Transfer Cycle Types

PCYC	CTYP	W/R#	Data transfer type
0	0	0	Normal read
0	1	0	Pipelined load (**pfld** instruction)
1	0	0	Page directory read
1	1	0	Page table read
0	0	1	Write-through (S-state hit)
0	1	1	Store miss or write-back
1	0	1	Page directory update
1	1	1	Page table update

Note: PCYC and CTYP are defined for memory data transfer cycles (D/C# = 1, M/IO# = 1)

TABLE 18.5   Cycle Length Definition

W/R#	LEN	CACHE#	KEN#	Cycle description	Burst length
0	0	1	—	Noncacheable** 64-bit (or less) read	1
0	0	—	—	Noncacheable 64-bit (or less) read	1
1	0	1	—	64-bit (or less) write	1
—	0	1	—	I/O and Special Cycles	1
0	1	1	—	Noncacheable 128-bit read **(p)fld.q**	2
0	1	—	1	Noncacheable 128-bit read **(p)fld.q**	2
1	1	1	—	128-bit write **fst.q**	2
0	—	0	0	Cache line fill	4
1	—	0	—	Cache write-back	4

Note: **Includes CS8-mode fetches, which may be cached by the processor.

—indicates "don't care" values.

SOURCE: Courtesy of Intel Corporation.

list the operational interpretation of some combinations of i860 XP control and status signals.

18.4 Concluding Comment

The Intel i860 architecture and its microprocessor implementation were discussed in the preceding sections. The i860 is extensively used both in Intel and other manufacturers' applications, particularly in multiprocessors. The i860 XP has a relatively large total on-chip cache of 32 kbytes, second only to the SuperSPARC (36 kbytes total; see Chap. 16). Its performance is competitive with other leading RISC and CISC microprocessors, operating at the same frequencies.

The Motorola M88000 Family

19.1 Introduction

Similarly to Intel, Motorola started its own RISC-type family of micro-processors, the M88000 family, after years of featuring a popular CISC-type family, the M68000 (see Part 3). The first members of the M88000 family are the MC88100 CPU and the MC88200 Cache MMU (CMMU) chips [Alsp 90, Mele 89, Tabk 90b]. They were followed recently by the new generation RISC MC88110, which is also a two-issue superscalar [DiAl 92, M811 91]. The M88000 architecture will be presented in the next section, while the implemented microprocessors will be discussed in the subsequent sections.

19.2 M88000 Architecture

Register file

The M88000 register file (or programming model) is shown in Fig. 19.1. It consists of the following parts:

1. General register file (GRF), consisting of thirty-two 32-bit registers r0 to r31. Register r0 is permanently wired to a zero value, as it is in other RISC-type systems. Register r1 contains the subroutine return pointer. The content of register r1 is not protected; it can be overwritten by software.

2. Extended register file (XRF), consisting of thirty-two 80-bit registers x0 to x31, intended for storing floating-point operands. The operands may be IEEE 754-1985 standard 32-bit single-precision, 64-bit double-precision, or 80-bit extended precision (see Chap. 3). Register x0 is permanently wired to a positive zero value.

All figures in Chap. 19 are courtesy of Motorola, Inc.

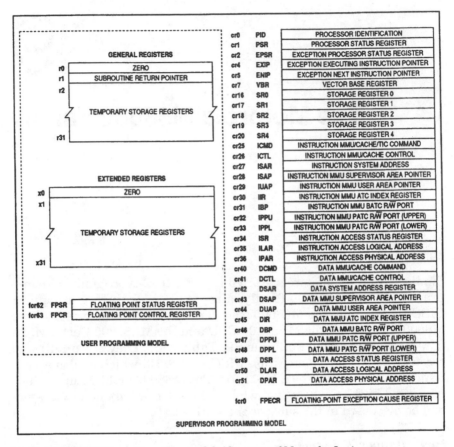

Figure 19.1 M88000 programming model. (*Courtesy of Motorola, Inc.*)

3. Floating-point control register (FPCR), 32-bit.

4. Floating-point status register (FPSR), 32-bit.

5. Control registers cr0 to cr51, 32-bit. In supervisor programming model only.

6. Floating-point exception cause register (FPECR), fcr0, 32-bit. In supervisor programming model only.

Some of the control registers will be discussed later on in the next section dealing with M88000 microprocessor implementation.

Data types

The following data types are recognized as operands by the M88000 architecture:

1. Integer
   a. Byte—8 bits
   b Halfword—16 bits
   c. Word—32 bits
   d Doubleword—64 bits

2. Bit-field. 1 to 32 bits in a 32-bit register
3. Floating-point
   a. Single-precision—32 bits
   b. Double-precision—64 bits
   c. Double extended precision—80 bits

4. Graphics

   a. 32-bit packed nibbles (nibble = 4 bits), bytes, and halfwords
   b. 64-bit packed nibbles, bytes, halfwords, and words

The operand size for each instruction is either explicitly encoded in the instruction or implicitly defined by the instruction operation. Bit fields are defined by width and offset values given in the instruction or in a source register specified by the instruction. Data storage in GRFs and in the XRFs is illustrated in Fig. 19.2(a) and (b). Memory storage alignment of integer and floating-point operands is illustrated in Fig. 19.3. Attempting an incorrectly aligned data transfer will cause a misaligned reference exception, if this exception is not masked. The M88000 architecture supports both big-endian (by default) and little-endian byte ordering [M811 91].

Addressing modes

The M88000 addressing modes are classified into three categories as follows [M811 91]:

1. Computational
   a.   Register (general, extended, control)
   b. Immediate (6-bit, 10-bit, 16-bit)

2. Load/store/exchange
   a. Register indirect with immediate displacement
   b. Register indirect with index
   c. Register indirect with scaled index

3. Flow control
   a. Register indirect
   b. Register indirect with 9-bit vector table index
   c. PC-relative with 16-bit displacement
   d. PC-relative with 26-bit displacement

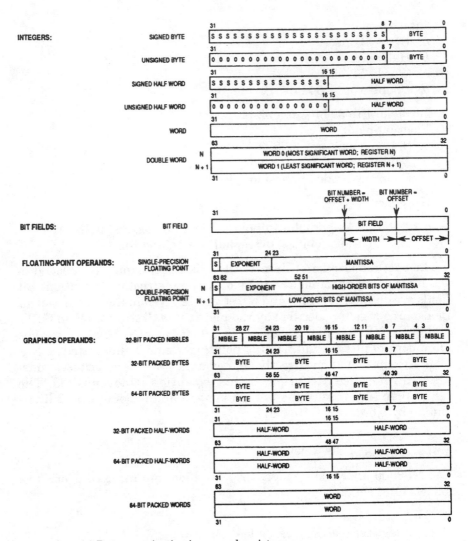

Figure 19.2 (*a*) Data organization in general registers.

Figure 19.2 (*b*) Operands in extended register file. (*Courtesy of Motorola. Inc.*)

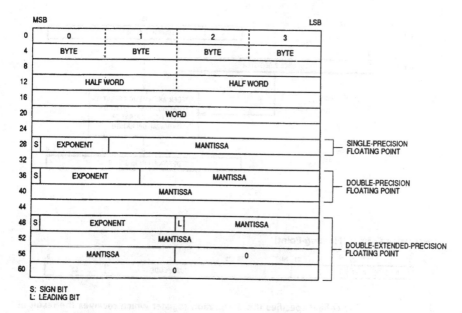

Figure 19.3 Floating-point memory storage alignment. (*Courtesy of Motorola, Inc.*)

Most of the above addressing modes were explained in general in Chap. 3. The only addressing modes that requires further explanation is the register indirect with 9-bit vector table index. This mode is used by trap generation instructions (tb0, tb1, and tcnd). For this addressing mode there is a 9-bit VEC9 field (bits 8 to 0) in the instruction. This field is shifted 3 bits left, filling the 3 LSBs with zeros, and concatenated with the upper 20 bits of the vector base address (which point to the exception vector table) to form a 32-bit address. The reason for the 9-bit VEC9 field is that there are $512 = 2^9$ exceptions featured by the M88000 architecture, and hence 512 vectors. Each vector takes up 8 $= 2^3$ bytes, and therefore the shift left by 3 bits [M811 91].

Instruction formats

The M88000 instruction formats for three-operand register-to-register instructions, for integer and floating-point operations, are illustrated in Fig. 19.4. In other formats the subopcode and S2 fields are superseded by fields for immediate operands. When the PC-relative with a 26-bit displacement mode is used (for br and bsr instructions), there is only a 6-bit opcode field (bits 31 to 26) and then a 26-bit displacement field (bits 25 to 0).

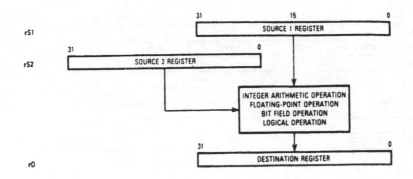

Instruction Format (Floating-Point)

31	26 25	21 20	16 15	5 4	0
1 0 0 0 0 1	D	S1	SUBOPCODE	S2	

D   The D field specifies the destination register which receives the result of the operation.

S1   The S1 field specifies the source 1 operand register. For the **int, nint, flt,** and **trnc** instructions, S1 must be zero.

SUBOPCODE   This field identifies the floating-point instruction (**fadd, fcmp, fdiv, fmul, fsub, int, nint, flt,** and **trnc**).

S2   The S2 field specifies the source 2 operand register.

Instruction Format (Non-Floating-Point)

31	26 25	21 20	16 15	5 4	0
1 1 1 1 0 1	D	S1	SUBOPCODE	S2	

D   The D field specifies the destination register which receives the result of the operation. This field is ignored for instructions that do not generate results.

S1   The S1 field specifies the source 1 operand register. For bit scan and the **rte** instructions, this field is ignored.

SUBOPCODE   This field identifies the non-floating-point instruction (**add, addu, and, cmp, div, divu, ext, extu, ff0, ff1, mak, mul, or, rot, rte, set, sub, subu, trnc,** and **xor**).

S2   The S2 field specifies the source 2 operand register. For the **rte** instruction this field is ignored.

Figure 19.4   Instructions formats for M88000. (*Courtesy of Motorola, Inc.*)

Instruction set

The M88000 instruction set is listed in Table 19.1, subdivided into seven separate instruction groups.

Instruction group	Number of instructions
Integer arithmetic	9
Bit-field	8
Logical	4
Graphics	9
Flow control	13
Load/store/exchange	7
Floating-point	16

The total number of M88000 instructions is 66, a suitable number for a RISC-type system. All instructions are 32 bits long.

Examples

```
add r5,r4,r1; r5 ← (r4) + (r1)
divs r11,r1,r2; r11 ← (r1)/(r2), signed divide, r11 receives the
quotient
```

Floating-point operations permit the mixing of single-, double-, and extended-precision operands.

Examples

```
fmul.dss x1,x4,x8; x1 ← (x4)*(x8)
```

where s = single-precision
       d = double-precision

The double-precision product of the single-precision contents of x4 and x8 is placed into x1.

```
fmul.xds x2,x10,x12; x2 ← (x10)*(x12), where x = extended precision.
```

All combinations of s, d, or x are allowed as sources and destinations.

The M88000 offers the user an optional delayed branch capability, attained by placing ".n" after the instruction mnemonic, such as in jmp.n or br.n (just writing "jmp" or "br" would be a regular, undelayed jump or branch; see Chap. 6). It should be noted, however, that implementation of a delayed branch in superscalar systems introduces extra difficulties both in hardware and software, and for this reason it is not recommended for use in such systems.

TABLE 19.1   MC88110 Instruction Set

Integer Arithmetic Instructions		Flow Control Instructions (*Cont.*)	
Mnemonic	Description	Mnemonic	Description
add	Signed Add	br	Unconditional Branch
addu	Unsigned Add	bsr	Branch to Subroutine
cmo	Integer Compare	illop	Illegal Operation
divs	Signed Divide	jmp	Unconditional Jump
divu	Unsigned Divide	jar	Jump to Subroutine
muls	Signed Multiply	rte	Return from Exception
mulu	Unsigned Multiply	tb0	Trap on Bit Clear
sub	Signed Subtract	tb1	Trap on Bit Set
subu	Unsigned Subtract	tbnd	Trap on Bounds Check
		tcnd	Conditional Trap

Bit-Field Instructions	
Mnemonic	Description
clr	Clear Bit Field
ext	Extract Bit Field
ext u	Unsigned Extract Bit Field
ff0	Find Fist Bit Clear
ff1	Find First Bit Set
mak	Make Bit Field
rot	Rotate Register
set	Set Bit Field

Load/Store/Exchange Instructions	
Mnemonic	Description
ld	Load Register From Memory
lda	Load Address
ldcr	Load from Control Register
st	Store Register to Memory
stcr	Store to Control Register
xcr	Exchange Control Register
xmem	Exchange Register with Memory

Logical Instructions	
Mnemonic	Description
and	And
mask	Logical Mask Immediate
or	Or
xor	Exclusive Or

Floating-Point Instructions	
Mnemonic	Description
fadd	Floating-Point Add
fcmp	Floating-Point Compare
fcmpu	Unordered Floating-Point Compare
fcvt	Convert Floating-Point Precision
fdiv	Floating-Point Divide
fldcr	Load from Floating-Point Control Register
flt	Convert Integer to Floating-Point
fmul	Floating-Point Multiply
fsqrt	Floating-Point Square Root
fstcr	Store to Floating-Point Control Register
fsub	Floating-Point Subtract
fxcr	Exchange Floating-Point Control Register
int	Round Floating-Point to Integer
mov	Register-to-Register Move
nint	Round Floating-Point to Nearest Integer
trnc	Truncate Floating-Point to Integer

Graphics Instructions	
Mnemonic	Description
padd	Pixel Add
padds	Pixel Add and Saturate
pcmp	Pixel Compare
pmul	Pixel Multiply
ppack	Pixel Truncate, Insert, and Pack
prot	Pixel Rotate Left
psub	Pixel Subtract
psubs	Pixel Subtract and Saturate
punpk	Pixel Unpack

Flow Control Instructions	
Mnemonic	Description
bb0	Branch on Bit Clear
bb1	Branch on Bit Set
bcnd	Conditional Branch

SOURCE: Courtesy of Motorola, Inc., © 1989 IEEE.

19.3   The MC88110

The MC88110 is a 1.3-million-transistor, triple-level metal, 0.8-micron CMOS, 299-pin microprocessor. It is a two-issue superscalar system with 10 on-chip operational units, also denoted as execution units [DiAl 92, M811 91]. A block diagram of the MC88110 and the execution units, are illustrated in Fig. 19.5(a) and (b). The architecture defines a major opcode subset denoted as a special function unit (SFU); all floating-point instructions are encoded in SFU1, and all graphics instructions in SFU2 [see Fig. 19.5(b)]. SFU3 to SFU7 are reserved for future extensions. The execution units are

1., 2. Integer

3.   Branch

4.   Bit-field

5.   Multiply (integer and floating-point; Booth algorithm), pipelined

6.   Floating add, pipelined

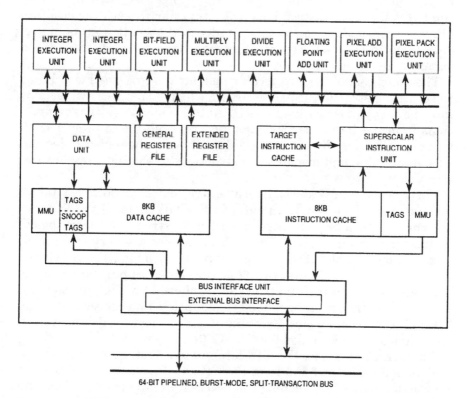

64-BIT PIPELINED, BURST-MODE, SPLIT-TRANSACTION BUS

Figure 19.5   (a) MC88110 block diagram.

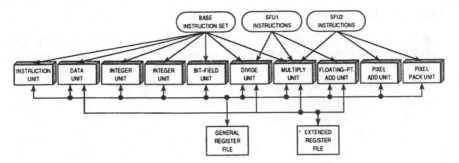

Figure 19.5 (*b*) SFU hardware use. (*Courtesy of Motorola, Inc.*)

7.   Divide (integer and floating-point), iterative, radix-8, 3 bits per clock

8., 9. Graphics

10.   Load/store

The abundance of execution units permits simultaneous execution of a large number of combinations of simultaneously fetched instructions. Even if both instructions are integer arithmetic, or both are graphics instructions, there are two identical execution units to execute them in parallel. This is indeed an important advantage in a superscalar system (see Chap. 5).

The main part of the superscalar instruction unit (see Fig. 19.5) is the central instruction sequencer (CIS), which dispatches instructions into the array of execution units [DiAl 92]. The CIS fetches instructions from the Icache, tracks resource availability and interinstruction dependencies, directs operand flow between the register files and execution units, and dispatches instructions to the individual execution units.

There is also a fully associative, 32-entry, logically addressed, target instruction cache (TIC). Each entry in the TIC contains the first two instructions of a branch target instruction stream, a 31-bit logical address tag, and a valid bit. The 31-bit tag holds a supervisor/user bit and the upper 30 bits of the address of the branch instruction. When a branch instruction occurs, the TIC is accessed in parallel with the decoding of the branch instruction. If there is a TIC hit, the two instructions corresponding to the branch instruction are sent from the TIC to the instruction unit. This eliminates much of the delay involved with changes in instruction flow [M811 91].

On each clock cycle, the CIS fetches two instructions from the instruction cache and two from the TIC. It decodes the appropriate instruction pair while fetching the necessary data operands from the register files (GRF or XRF). If all the required execution units and operands are available, the CIS simultaneously dispatches both instructions to their respective execution units. If the sequencer cannot dispatch both instructions, it tries to dispatch at least the first of the pair. In that case, the second instruction moves into the first issue slot, a new instruction is fetched to replace it, and the CIS tries to issue the new pair on the next clock cycle [DiAl 92].

Of particular interest is the MC88110 load/store execution unit [DiAl 92]. The unit provides a stunt box capability for holding memory references that are waiting for the memory system and allows dynamic reordering of load operations past stalled store operations. The term "stunt box" refers to any device that allows reordering of memory references in the memory access system. The data path from the load/store unit to the register files is 80 bits wide. On each clock cycle, the load/store unit can accept one new load or store instruction from the instruction dispatch unit. The dispatched instruction awaits access to the Dcache in one of the two FIFO queues: the four-deep load buffers for load operations, and the three-deep store reservation station for store operations. Normal instruction dispatch and execution can continue while these instructions await service by the cache or memory system (provided there is no data dependency on load instructions). Store instructions can be dispatched before the store data operand is available. Store instructions wait in the store queue until the instruction computing the required data completes execution. When the operand becomes available, the sequencer directs it into the store reservation station, and the associated store instruction becomes a candidate for access to the data cache. If a store instruction stalls in the reservation station waiting for its operand, subsequently issued load instructions can bypass the store and immediately access the cache. An address comparator detects address hazards and prevents loads from going ahead of stores to the same address, thus loading stale data. This load/store reordering feature allows runtime overlapping of tight loops by permitting loads at the top of a loop to proceed without having to wait for the completion of stores from the bottom of the previous iteration of the loop.

As with stores, the MC88110 provides a reservation station to avoid stalling on conditional branches. The sequencer predicts the branch direction, and instructions down the predicted path execute conditionally, or speculatively, until the branch operand is resolved. The static prediction of the branch direction is based on the opcode of the branch instruction. The branch reservation station provides a place to set aside the branch instruction so that instruction issue can continue while the branch condition can be resolved. If the prediction was incorrect, the system backs up to the branch, undoing all operations along the wrong path, and resumes execution along the correct path [DiAl 92].

The MC88110 has a dual-cache, 8-kbyte instruction, 8 kbytes data, for a total of 16 kbytes. Each cache is physically addressed, two-way set-associative, with 32 bytes/line. Each cache has $2^{13}/2^5 = 2^8 = 256$ lines, that is, $256/2 = 128 = 2^7$ sets. Therefore, when addressing the cache we need 5 bits for the byte in line field, 7 bits for the set field, and 20 bits for the tag (see Chap. 4). This fits exactly the virtual memory addressing for a 4-kbyte $= 2^{12}$ bytes/page, as practiced on the MC88110. Thus, the lowest 12 bits of the virtual address (bits 11 to 0) are the offset into a page, and at the same time the set (bits 11 to 5) and byte (bits 4 to 0) fields for cache access. The upper 20 bits are forwarded to the address translation cache (ATC) for translation into a 20-bit physical page number, to be compared with the 20-bit tags of the appropriate set in the cache (see Chap. 4). The data cache is write-back.

The MC88110 can also be connected to a secondary external cache, controlled by the MC88410 cache controller. The secondary cache can be configured out of 62110 SRAM chips, for a total size of 256 kbytes to 1 Mbyte. The secondary cache is designed to be direct-mapped, write-back, with 32 or 64 bytes/line [DiAl 92].

The MC88110 features two types of ATCs. One is the page ATC (PATC), which is fully-associative and contains 32 entries containing page number translations for 4-kbyte pages. The PATC is automatically maintained by hardware or can be maintained by system software. The other ATC is the block ATC (BATC), which is fully associative, with eight entries, containing address translations for block sizes ranging from 512 kbytes to 64 Mbytes. The term "block" is just an M88000 term for optional larger page sizes [M811 91]. As in the latest implementations of the Motorola M68000 family (see Part 3), the MC88110 features a dual MMU, illustrated in Fig. 19.6, the instruction MMU (IMMU) and the data MMU (DMMU). The IMMU and the instruction cache comprise the instruction memory unit (IMU), and the DMMU with the Dcache comprise the data memory unit (DMU). As already argued in Chaps. 5 and 6, the memory unit duality permits the processor to deal with instruction and data accesses in parallel, thus supporting a more efficient handling of the pipeline. This is particularly important in superscalar systems.

The MC88110 page translation table structure is shown in Fig. 19.7, and the page table lookup is illustrated in Fig. 19.8. The extra S/U bit of the logical address differentiates between the supervisor and user modes, each of which has separate instruction and data areas in memory and separate address pointers (called area descriptors) into their segment tables. As can be seen in Fig. 19.8, the page table structure and the address partitioning of the MC88110 is identical to that of Intel (see Chaps. 7 and 18), except for some different terminology. The "page

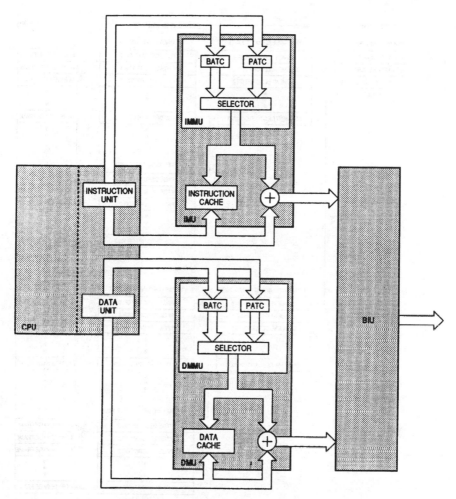

Figure 19.6 MC88110 MMU block diagram. (*Courtesy of Motorola, Inc.*)

table" term is the same on both. What is called "page directory" in Intel, is called "segment table" on the MC88110. All of the above have up to 1024 32-bit entries and may be up to 4 kbytes in size (page size). What is called "PTE" in Intel is called "page descriptor" on the MC88110, and a "PDE" in Intel is a "segment descriptor" on the MC88110. Another difference is that there are four pointers on the MC88110 (and more than one segment table), while there is only one such pointer in Intel (CR3) and one table directory.

The MC88110 processor status register (PSR) is shown in Fig. 19.9. Only a part of its bits are currently implemented.

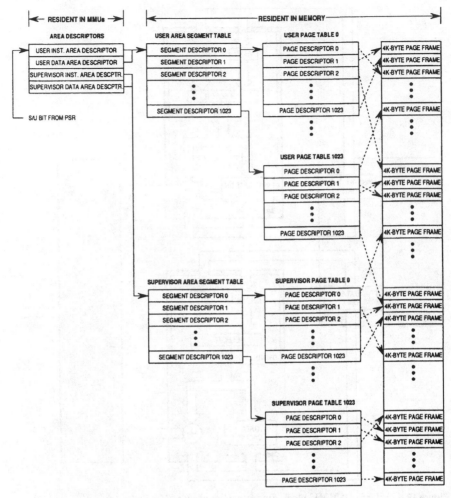

Figure 19.7   Page translation table structure. (*Courtesy of Motorola, Inc.*)

Bit 31, supervisor/user mode (MODE), MODE = 1 supervisor, MODE = 0 user

Bit 30, byte ordering (BO), BO = 1 little-endian, BO = 0 big-endian

Bit 29, serial mode (SER), SER = 1 serial instruction execution, SER = 0 concurrent execution

Bit 28, carry (C), C = 1 carry generated by add or sub instructions, C = 0 carry not generated

Bit 26, signed immediate mode (SGN), SGN = 1 immediate offsets and constants are signed, SGN = 0 immediate offsets and constants are unsigned

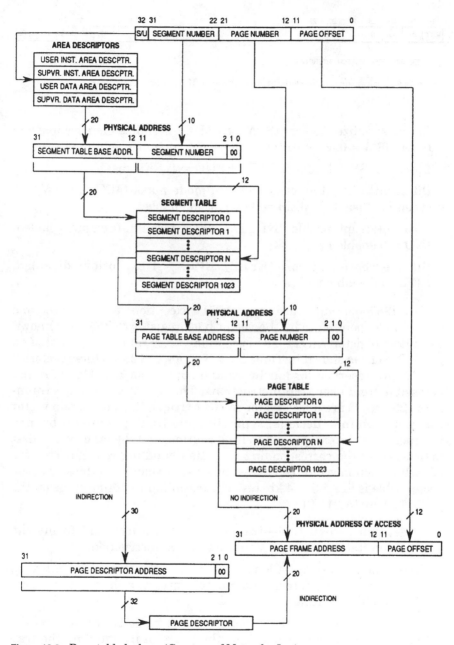

Figure 19.8   Page table lookup. (*Courtesy of Motorola, Inc.*)

UNDEFINED-RESERVED FOR FUTURE USE

Figure 19.9 Processor status register. (*Courtesy of Motorola, Inc.*)

Bit 25, serialize memory (SRM), SRM = 1 serialize memory instructions, SRM = 0, concurrent memory instruction execution

Bits 9—3, SFU disable (SFUD), SFUD = 1, SFU disabled

Bit 2, misaligned access exception mode mask (MXM), MXM = 1 exception disabled, MXM = 0 exception enabled

Bit 1, interrupt disable (IND), IND = 1 external interrupts disabled, IND = 0 enabled

Bit 0, exceptions freeze (EFRZ), EFRZ = 1 exceptions disabled, EFRZ = 0 enabled

The M88000 architecture supports 512 exception vectors, stored in a vector table, pointed to by the vector base register (VBR), also known as control register 7 (cr7). The exception vector table implemented on the MC88110 is shown in Table 19.2. The lower 128 vectors (vectors 0 to 127) are reserved for hardware and supervisor use. They are not accessible from user trap instructions. The upper 384 vectors (numbers 128 to 511) are allocated for software traps. Each exception vector contains two instructions: typically, one instruction is a branch instruction to the exception handling routine, and the other is the first instruction of the corresponding exception handling routine. The size of each exception vector is 8 bytes (two instructions) and the size of the vector table is 512 * 8 = 4 kbytes. The exception handling steps on the MC88110 are [M811 91]:

1. Exception recognition—the processor restores and forms the machine state associated with the faulting instruction.

2. Exception processing—the processor saves the execution context in exception-time registers, and changes program flow to the exception handling routine.

3. Exception handling—the exception handling software corrects the exception condition or performs the function initiated by the trap instruction.

4. Return from exception—the processor restores the execution context which was in effect before the exception occurred and resumes normal execution of program instructions.

TABLE 19.2   Exception Vectors

Number	Vector base address offset	Exception
0	$00	Reset
1	$08	Maskable Interrupt
2	$10	Instruction Access
3	$18	Data Access
4	$20	Misaligned Address
5	$28	Unimplemented Opcode
6	$30	Privilege Violation
7	$38	Bounds Check Violation
8	$40	Integer Divide-by-Zero
9	$48	Integer Overflow
10	$50	Unrecoverable Error
11	$58	Nonmaskable Interrupt
12	$60	Data MMU Read Miss
13	$68	Data MMU Write Miss
14	$70	Instruction MMU ATC Miss
15–113	—	Reserved
114	$390	SFU1—Floating-Point Exception
115	$398	Reserved
116	$3A0	SFU2—Graphics Exception
117	$3A8	Reserved
118	$3B0	SFU3—Unimplemented Opcode
119	$3B8	Reserved
120	$3C0	SFU4—Unimplemented Opcode
121	$3C8	Reserved
122	$3D0	SFU5—Unimplemented Opcode
123	$3D8	Reserved
124	$3E0	SFU6—Unimplemented Opcode
125	$3E8	Reserved
126	$3F0	SFU7—Unimplemented Opcode
127	$3F8	Reserved
128–511	—	Reserved—User Trap Vectors

SOURCE: Courtesy of Motorola, Inc.

The MC88110 has a history buffer, which records the relevant, user-visible machine state as instructions issue. The processor uses information stored in the history buffer to quickly restore the machine state back to the point of the exception. This is the same mechanism the MC88110 uses to recover from mispredicted branches [DiAl 92].

The pinout diagram of the MC88110 is shown in Fig. 19.10, and the signal summary is listed in Table 19.3. A great part of the MC88110 signals is implemented in PowerPC 601 (see Chap. 15).

19.4   The MC88100 and MC88200

The MC88100 is a 1.2-micron HCMOS, 180-pin CPU chip with an on-chip FPU. The MC88200 is a CMMU chip with a 16-kbyte cache. The

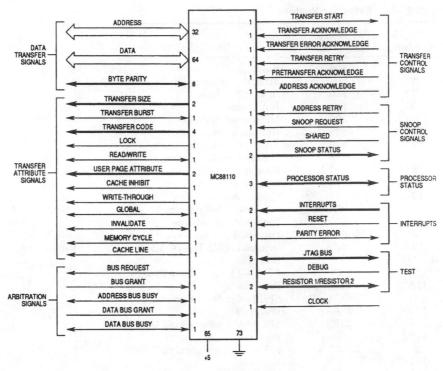

Figure 19.10   MC88110 pinout. (*Courtesy of Motorola, Inc.*)

MC88100 and the MC88200 interconnection is illustrated in Fig. 19.11. The minimum configuration is to connect two CMMU to one CPU chip (two separate buses); one CMMU acts as a Dcache of 16 kbytes, and the other as an Icache of 16 kbytes, for a total of 32 kbytes. However, up to 4 CMMUs can be connected on each side; 64 kbytes data, 64 kbytes instruction cache, for a total of 128 kbytes of cache.

19.5   Concluding Comment

The M88000 family microprocessors were implemented in numerous commercial systems. Of particular note is their implementation in multiprocessors such as the BBN Butterfly and the Encore [Tabk 90a, b]. The MC88110 is an outstanding specimen of a two-issue superscalar system. It has 10 execution units, permitting very high flexibility in dispatching two instructions simultaneously for parallel execution. Barring data dependencies, such parallel execution should almost always be possible (almost, because two multiply instructions,

TABLE 19.3   MC88110 Signal Summary

Function	Mnemonic	Count	Type	Active	Reset
		Data Transfer			
Data Bus	D63–D0	64	I/O	High	Three-state
Address Bus	A31–A0	32	I/O	High	Three-state
Byte Parity	BP7–BP0	8	I/O	High	Three-state
		Transfer Attributes			
Read/Write	$\overline{\text{R/W}}$	1	I/O	High	Three-state
Lock	$\overline{\text{LK}}$	1	Output	Low	Three-state
Cache Inhibit	$\overline{\text{CI}}$	1	Output	Low	Three-state
Write-Through	$\overline{\text{WT}}$	1	Output	Low	Three-state
User Page Attributes	$\overline{\text{UPA1–UPA0}}$	2	Output	Low	Three-state
Transfer Burst	$\overline{\text{TBST}}$	1	I/O	Low	Three-state
Transfer Size	TSIZ1–TSIZ0	2	Output	High	Three-state
Transfer Code	TC3–TC0	4	Output	High	Three-state
Invalidate	$\overline{\text{INV}}$	1	I/O	Low	Three-state
Memory Cycle	$\overline{\text{MC}}$	1	Output	Low	Three-state
Global	$\overline{\text{GBL}}$	1	I/O	Low	Three-state
Cache Line	CLINE	1	Output	High	Three-state
		Transfer Control			
Transfer Start	$\overline{\text{TS}}$	1	Output	Low	Three-state
Transfer Acknowledge	$\overline{\text{TA}}$	1	Input	Low	—
Pretransfer Ack	$\overline{\text{PTA}}$	1	Input	Low	—
Transfer Error Ack	$\overline{\text{TEA}}$	1	Input	Low	—
Transfer Retry	$\overline{\text{TRTRY}}$	1	Input	Low	—
Address Acknowledge	$\overline{\text{AACK}}$	1	Input	Low	—
		Snoop Control			
Snoop Request	$\overline{\text{SR}}$	2	Input	Low	—
Address Retry	$\overline{\text{ARTRY}}$	1	Input	Low	—
Snoop Status	$\overline{\text{SSTAT1–}}$ $\overline{\text{SSTAT0}}$	2	Output	Low	Three-state
Shared	$\overline{\text{SHD}}$	1	Input	Low	—
		Arbitration			
Bus Request	$\overline{\text{BR}}$	1	Output	Low	Negated
Bus Grant	$\overline{\text{BG}}$	1	Input	Low	—
Address Bus Busy	$\overline{\text{ABB}}$	1	I/O	Low	Three-state
Data Bus Grant	$\overline{\text{DBG}}$	1	Input	Low	—
Data Bus Busy	$\overline{\text{DBB}}$	1	I/O	Low	Three-state
		Processor Status			
Processor Status	PSTAT2–PSTST0	3	Output	High	Input
		Interrupt			
Nonmaskable Interrupt	$\overline{\text{NMI}}$	1	Input	Low	Three-state
Interrupt	$\overline{\text{INT}}$	1	Input	Low	Three-state
Reset	$\overline{\text{RST}}$	1	Input	Low	Three-state
Byte Parity Error	$\overline{\text{BPE}}$	1	Output	Low	Three-state

TABLE 19.3   MC88110 Signal Summary *(Continued)*

Function	Mnemonic	Count	Type	Active	Reset
		Clock			
Clock	CLK	1	Input	Rising Clock Edge	—
		Test Pins			
Debug	DBUG	1	Input	Low	—
Resistor 1	RES1	1	Input	N/A	—
Resistor 2	RES2	1	Output	N/A	—
JTAG Test Reset	TRST	1	Input	Low	—
JTAG Test Mode Select	TMS	1	Input	High	—
JTAG Test Clock	TCK	1	Input	Clock Edge	—
JTAG Test Data Input	TDI	1	Input	High	—
JTAG Test Data Output	TDO	1	Output	High	—

SOURCE: Courtesy of Motorola, Inc.

cannot be executed simultaneously, for instance). In any case, the MC88110 has less restrictions for parallel issue of two instructions than other systems. Should Motorola succeed in producing a sample running above 100 MHz, the MC88110 may outperform other RISC superscalar frontrunners.

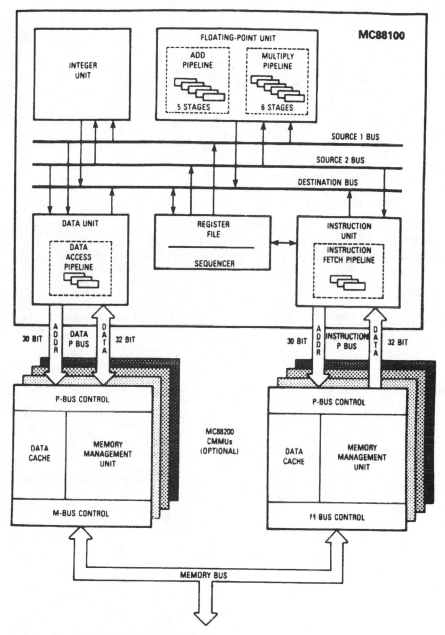

Figure 19.11   MC88100/MC88200 block diagram. *(Courtesy of Motorola, Inc.)*

20

The HP Precision Architecture Family

20.1 Introduction

Hewlett-Packard (HP) was among the first industrial manufacturers to come out with a commercial RISC-type system in the mid-eighties [BiWo 86, Lee 89, Maho 86, PARI 90, Tabk 87, Tabk 90b]. The architecture of this system was called precision architecture (PA) or PA-RISC. A number of microprocessors, implementing the PA, were produced by HP in recent years. One of the latest implementations, the PA7100 [Aspr 93, DWYF 92, Kneb 93], exceeded a number of notable RISC systems in its performance. The PA architecture will be presented in the next section, followed by a description of its microprocessor implementation.

20.2 The PA-RISC Architecture

When the HP team started to work on the PA-RISC, some of the design goals were to create a general-purpose architecture for use in commercial and technical applications [Lee 89]. The architecture was designed to be scalable across technologies, cost ranges, performance ranges, and to provide price-performance advantages. The architecture was designed having in mind architectural longevity, allowing growth and extendability, support multiple operating environments, secure systems, and real-time environments. The PA-RISC is described in the next paragraphs of this section.

CPU registers

The PA-RISC features the following CPU registers [PARI 90]:

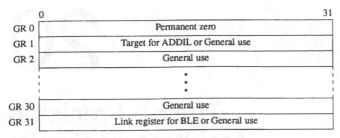

Figure 20.1   General registers. (*Courtesy of Hewlett-Packard, Inc.*)

1. General registers, thirty-two 32-bit, GR0 to GR31, shown in Fig. 20.1. Three of them have special functions:

GR0 is permanently wired to a zero value, as in most RISC systems.

GR1 is the target of the add immediate left (ADDIL) instruction.

GR31 is the instruction address offset link register for the base-relative interspace procedure call instruction branch and link external (BLE).

Registers GR1 and GR31 can also be used as general registers; however, software conventions may at times restrict their use. In the PA bit count starts at the MSB (bit 0) and ends at the LSB (bit 31; big-endian bit ordering), as shown in Fig. 20.1 (same as in PowerPC and IBM architecture; see Chap. 15).

2. Shadow registers, 7 32-bit. Implementations may optionally choose to provide these registers into which the contents of GR1, GR8, GR9, GR16, GR17, GR24, and GR25 are copied upon interrupts (called "interruptions" in PA). The contents of these general registers is restored from their shadow registers upon a return from interrupt. Although shadow registers were initially declared to be optional by the PA-RISC architecture, every PA-RISC 1.1 processor implementation has provided them and they are now required by the architecture.

3. Space registers, 8 registers SR0 to SR7, shown in Fig. 20.2, contain space identifiers for virtual addressing. Instructions specify SRs either directly in the instruction or indirectly through the most significant bits of the GR. Instruction addresses, computed by branch instructions, may use any of the SRs. SR0 is the instruction address space link register for the base-relative interspace procedure call instruction (BLE). Data operands can specify SR1 through SR3 explicitly, and SR4 through SR7 indirectly, via GRs. SR1 through SR7 have no special functions; however, their use will normally be constrained by software conventions. For instance, SR1 through SR3 provide general use virtual pointers. SR4 tracks the instruction address (IA) space

SR 0	Link code space ID
SR 1	General use
SR 2	General use
SR 3	General use
SR 4	Tracks IA space
SR 5	Process private data
SR 6	Shared data
SR 7	Operating system's public code, literals and data

Figure 20.2 Space registers. (*Courtesy of Hewlett-Packard, Inc.*)

and provides access to literal data contained in the current code segment. SR5 points to a space containing process private data, SR6 to a space containing data shared by a group of processes, and SR7 to a space containing the OS public code, literals, and data. SR5 through SR7 can be modified only by code executing at the most privileged level. In different PA-RISC implementations, SRs may be nonexistent (level 0), 16 (level 1), 24 (level 1.5), or 32-bit wide (level 2). The software conventions shown in Fig. 20.2 are an example and do not necessarily correspond to any particular software implementation.

4. Control registers, 25 32-bit CR0, CR8 through CR31, which contain system state information. Registers CR1 through CR7 are reserved for future use.

5. Floating-point registers, 32 64-bit, which can also be used as 64 32-bit locations. Registers 0 through 3 contain the status (32-bit) and exception registers (seven 32-bit). Registers 4 through 31 are data registers. A 32-bit floating-point register is identified by appending a suffix in the instruction to the identifier of the 64-bit register within which it is contained. The suffix for the left side 32-bit register is "L"; the use of this suffix is optional. The suffix for the right-side 32-bit register is "R"; its use is not optional.

Example Left half of the 64-bit register 12 (bits 0 to 31) can be referred to as either 12 or 12L; the right half (bits 32 to 63) is referred to as 12R.

6. Processor status word (PSW), 32-bit, shown in Fig. 20.3. The bit description of the PSW is as follows.

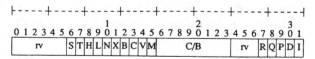

Figure 20.3 Processor status word. (*Courtesy of Hewlett-Packard, Inc.*)

Bit 6—secure interval timer (S). When S = 1, the interval timer is readable only by code executing at the most privileged level. When S = 0, it is readable at any privilege level.

Bit 7—taken branch trap enable (T). When T = 1, any taken branch is terminated with a taken branch trap.

Bit 8—higher-privilege transfer trap enable (H). When H = 1, a higher privilege transfer trap occurs whenever the following instruction is of a higher privilege.

Bit 9—lower-privilege transfer trap enable (L). When L = 1, a lower privilege transfer trap occurs whenever the following instruction is of a lower privilege.

Bit 10—nullify (N). When N = 1, the current instruction is nullified.

Bit 11—data memory break disable (X). When X = 1, data memory break traps are disabled. X may be set by a return from interrupt instruction; it is set to 0 by other instructions.

Bit 12—taken branch (B). B is set to 1 by any taken branch and set to 0 otherwise.

Bit 13—code address translation enable (C). When C = 1, instruction addresses are translated and access rights checked.

Bit 14—divide step correction (V). The integer division primitive instruction records immediate status in bit V to provide a nonrestoring divide primitive.

Bit 15—high-priority machine check mask (M). When M = 1, high-priority machine checks (HPMCs) are not allowed. Normally 0, M is set to 1 after an HPMC and cleared to 0 after all other interrupts.

Bits 16 to 23—carry/borrow (C/B). Set on a carry by a set of arithmetic add instructions. Cleared on a borrow for subtract instructions.

Bit 27—recovery counter enable (R). When R = 1, recovery counter traps occur if bit 0 of the recovery counter is a 1. This bit also enables decrementing of the recovery counter.

Bit 28—interruption state collection enable (Q). When Q = 1, interruption state is collected.

Bit 29—protection identifier validation enable (P). When P = 1 and C = 1, instruction references check for valid protection identifiers (PIDs). When P = 1 and D = 1, data references check for valid PIDs. When P = 1, probe instructions check for valid PIDs.

Bit 30—data address translation enable (D). When D = 1, data addresses are translated and access rights checked.

Bit 31—external interrupt, power failure interrupt, and low-priority machine check interrupt unmask (I). When I = 1, these interrupts are unmasked and can cause an interruption. When I = 0, the interruptions are held pending.

Data types

1. Integer signed and unsigned
   a. Byte: 8-bit
   b. Halfword: 16-bit
   c. Word: 32-bit

2. Integer unsigned
   a. Doubleword 64-bit
   b. Floating-point

3. IEEE 754-1985 standard
   a. Single-precision: 32-bit
   b. Double-precision: 64-bit
   c. Quadruple-word (or: extended double) 128-bit (sign bit, 15-bit exponent, 112-bit mantissa). The IEEE standard does not fully specify a quadruple precision format, but rather a range to which PA-RISC complies.

4. Packed decimal: Consists of 7, 15, 23, or 31 BCD 4-bit digits, aligned on a word boundary, and having a value of 0 to 9, followed by a 4-bit sign. The standard sign for a positive number is X'C (X' denotes hexadecimal), but any value except X'D will be interpreted as positive. A minus sign is denoted by X'D. X'B is not supported as an alternative minus sign [PARI 90].

The alignment of the data operands in memory is illustrated in Fig. 20.4. As we can see, a big-endian byte ordering is used. In the latest version of PA-RISC architecture both big- and little-endian byte ordering is supported. Bit 5 (E-bit) of PSW is cleared to 0 for big-endian, and set to 1 for little-endian byte ordering.

Addressing modes

The PA features the following addressing modes [PARI 90]:

Register

Immediate

Base with offset (displacement)

Base with scaled (shifted) index with offset

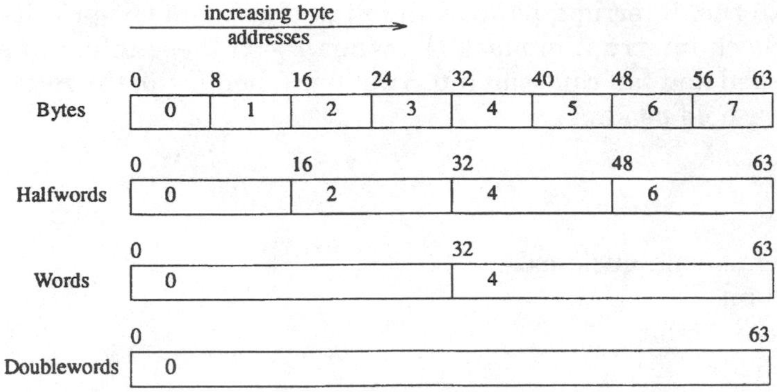

increasing byte
addresses

	0	8	16	24	32	40	48	56	63
Bytes	0	1	2	3	4	5	6	7	

	0	16	32	48	63
Halfwords	0	2	4	6	

	0	32	63
Words	0	4	

	0	63
Doublewords	0	

Figure 20.4 Physical memory addressing and storage units. (*Courtesy of Hewlett-Packard, Inc.*)

Predecrement. If the displacement d is negative, its sign-extended value is added to the base register GRb and the result (the effective offset) is stored in GRb.

Postincrement. If d is positive, the effective offset is the original value in GRb. The sum of the content of GRb and the sign-extended value of d is stored in GRb.

PC-relative (IA-relative). In PA, what is usually called PC, is called IA (instruction address).

Instruction formats

The PA-RISC instruction formats are shown in Fig. 20.5 [Lee 89], along with the abbreviations interpretation. One can easily recognize the three register fields for ALU instructions; with one register field interchangeable with an immediate operand field. In the load or store instruction formats, one register field serves for a destination (load) or a source (store), one register field for the base register, and one for a possible index register (interchangeable for use as an offset). For branch instructions, there are fields for a base register and for an offset.

Instruction set

A summary of the PA-RISC instruction set, subdivided into different categories, is listed in Table 20.1 [Lee 89]. The instructions are listed by their explicit names, without their mnemonics.

Addressing space

The PA-RISC architects felt that the longevity of an architecture lies in the range of its addressing capabilities rather than in the size of its words [Lee 89]. While processing 64-bit integers rather than 32-bit

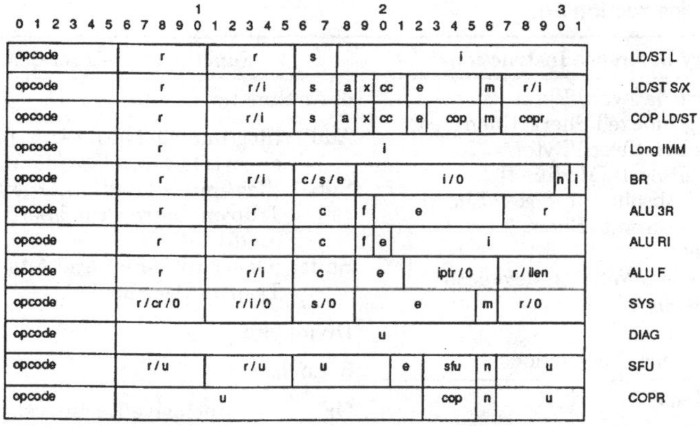

0 1 2 3 4 5 6 7 8 9	1 0 1 2 3 4 5 6	7	8 9	2 0 1 2 3 4 5 6 7 8 9	3 0 1	
opcode	r	r	s	i		LD/ST L
opcode	r	r/i	s	a x cc	e	m r/i	LD/ST S/X
opcode	r	r/i	s	a x cc	e cop m copr		COP LD/ST
opcode	r	i				Long IMM
opcode	r	r/i	c/s/e	i/0	n i	BR
opcode	r	r	c f	e	r	ALU 3R
opcode	r	r	c f e	i		ALU RI
opcode	r	r/i	c	e iptr/0	r/ilen	ALU F
opcode	r/cr/0	r/i/0	s/0	e	m r/0	SYS
opcode	u					DIAG
opcode	r/u	r/u	u	e sfu n	u	SFU
opcode	u		cop n	u		COPR

r — General Register Specifier
s — Space Register Specifier
i — Immediate (or displacement or offest)
a — Premodify, Version Postmodify, or Index Shifted by Data Size
x — Indexed (x = 0)Versus Short Displacement (x = 1)
cc — Cache Hints
e — Subop (opcode extension)
m — Modification Specifier
n — Nullification Specifier
c — Condition Specifier

f — Falsify Condition c
iptr — Immediate Pointer
ilen — Immediate Length
cr — Control Register
0 — Not Used (set to zeros)
u — Undefined (can be defined as instruction extension)
stu — Special Function Unit Identifier
cop — Coprocessor Unit Identifier
copr — Coprocessor Register

Figure 20.5 Instruction formats. (*Courtesy of Hewlett-Packard, Inc.,*© 1989 IEEE.)

integers might increase accuracy, they did not consider the trade-off in the hardware required for 64-bit datapaths throughout the processor to be cost-effective for general-purpose computers. This was a sound decision conforming with the state of the art of the mid-eighties when PA was developed. As can be seen in earlier chapters, the designers of the early nineties Alpha AXP (Chap. 14) and of the R4000 (Chap. 17), did not think that way and have adopted a 64-bit integer datapath. The PA designers noted that computer usage has clearly tended toward the processing of larger programs and more data. For this reason they decided to provide up to a 64-bit virtual address range [Lee 89]. Eventually, PA-RISC will have a full 64-bit architecture extension. It should be noted that HP has been the first to ship full 64-bit virtual address space implementations.

The PA-RISC memory is byte-addressable. The PA-RISC distinguishes between an absolute address and a virtual address. When absolute addresses are used directly, no protection or access rights checks are performed. Virtual addresses are translated to absolute addresses and undergo protection and access rights checking. Memory accesses using absolute addresses are called absolute accesses, and those using virtual addresses are called virtual accesses. PA defines four levels of processor architecture in conjunction with memory access: levels 0, 1, 1.5, and 2. Level 0 systems support only absolute addressing and have no SRs. Level 1, 1.5, and 2 systems provide vir-

TABLE 20.1   Instruction Set

Memory Reference Instructions	
Load	{Word/Halfword/Byte} {Long/Indexed/Short} [Modified]*
Store	{Word/Halfword/Byte} {Long/Short} [Modified]
Load	Word Absolute {Indexed/Short}
Store	Word Absolute Short
Load	Offset
Load	And Clear Word {Indexed/Short}
Store	Bytes Short

Branch Instructions

a. Unconditional

Branch And Link {Displacement/Reg}
Branch Vectored
Branch External [and Link]
Gateway

b. Conditional

Add {Reg/Immed} And Branch if {True/False}
Compare {Reg/Immed} And Branch if {True/False}
Move {Reg/Immed} And Branch if {True/False}
Branch On Bit {Variable/Constant}

System Instructions

a. System Control

System Mask {Set/Reset/Move to}
Move {to/from} Control Register
Move {to/from} Space Register
Load Space ID
Break
Return From Interrupt
Diagnose

b. Memory Management

Insert TLB {Instruction/Data} {Address/Protection}
Purge TLB {Instruction/Data} [Entry]
Probe Access {Read/Write} {Reg/Immed}
Load Physical Address
Load Hash Address

c. Cache Management

Flush {Instruction/Data} Cache [Entry]
Purge Data Cache
Sync

Functional Instructions

a. Arithmetic

Add	{Reg/Immed} [with carry] [and Trap on {overflow/cond/overflow or cond}]
Sub	{Reg/Immed} [with borrow] [and Trap on {borrow/cond/borrow or cond}]
Shift	{One/Two/Three} And Add [and Trap on Overflow]

Divide Step

b. Logical

Or	{Inclusive/Exclusive}
And	{True/Complement}
Compare	{Reg/Immed} And Clear
Add	Logical
Shift	{One/Two/Three} And Add Logical

c. Unit and Decimal

Unit Xor
Unit Add Complement [and Trap on Condition]
Decimal Correct
Intermediate Decimal Correct

d. Bit Manipulation

Extract {Variable Pos/Constant Pos} {Signed/Unsigned}
Deposit {Variable Pos/Constant Pos} {Reg/Immed}
Zero and Deposit {Variable Pos/Constant Pos} {Reg/Immed}
Shift Double {Variable Pos/Constant Pos}

e. Long Immediate

Add Immediate Left
Load Immediate Left

Assist Instructions

a. Special Function Unit Interface

Spop {Zero/One/Two/Three}

b. Coprocessor Interface

Copr Load {Word/Doubleword} {Indexed/Short}
Copr Store {Word/Doubleword} {Indexed/Short}
Copr Operation

*Curly brackets indicate that one alternative within the curly brackets is selected for a given instruction, while square brackets indicate an optional feature that can be specified in the instruction.

Note: cond = condition; Immed = immediate; Pos = position; Reg = register.

SOURCE: Courtesy of Hewlett-Packard, Inc. and IEEE.

tual addressing through the use of 16-, 24-, and 32-bit SRs, respectively [PARI 90].

Virtual memory is structured as a set of virtual spaces, each containing 4 Gbytes (2^{32} bytes). Level 1 processors have 2^{16}, level 1.5 processors have 2^{24}, and level 2 processors have 2^{32} address spaces.

During virtual address translation, a space is selected by a space identifier (space ID) contained in the 32-bit upper portion of the 64-bit virtual address. The byte offset within the space is specified by the lower 32 bits of the virtual address. The memory address space is also subdivided into 4-kbyte pages, as it is in most other systems. Figure 20.6 illustrates the structure of spaces, pages, and offsets. Support is provided for the emulation of larger page sizes (32 kbytes). Eight contiguous pages, with the first of these pages beginning on a 32-kbytes boundary, are referred to as a page group [PARI 90].

Registers SR0 through SR7 are used to compute instruction addresses for Icache flush, TLB instructions, and for some branch target calculations. Addresses for instruction fetch and some branch target calculations are generated from the instruction address (IA)

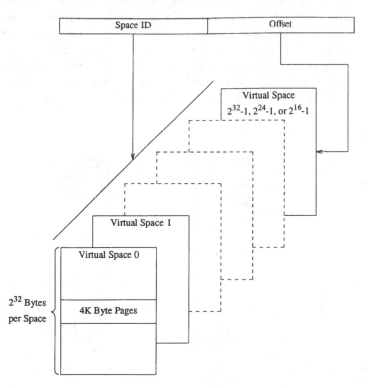

Figure 20.6 Structure of spaces, pages, and offsets. (*Courtesy of Hewlett-Packard, Inc.*)

queues. When an instruction address is computed for an external branch target, the 3-bit s-field in the instruction selects the SR to be used (see Fig. 20.5 for the BR instruction). Instruction addresses are aligned on word (4-byte) boundaries and the least significant 2 bits of the offset are used to hold the privilege level (since these 2 bits are zero in the instruction address offset). The privilege level controls both instruction and data references.

The current IA consists of a space ID and a 32-bit offset (Fig. 20.6). The current IA is maintained in the front elements of the IA space and offset queues. The 32-bit offset is computed by one of the addressing modes, which may use a base register, a scaled index register, and a sign-extended displacement.

The space ID for data references is selected from SR1 through SR7 by the following procedure. The 2-bit s-field of the instruction, when nonzero, selects corresponding SRs 1, 2, or 3. When the s-field is zero, the two MSB (bits 0 and 1) of the base register are used to select one of the SRs (4 through 7). Adding 4 to these 2 bits generates the selected SR. Figure 20.7 illustrates the space ID selection. Data references with the s-field zero permit addressing of four distinct spaces selected by program data. This is called short pointer addressing since a 32-bit value is an offset and selects an SR. Only a fourth of the space is directly addressable by the base register with short pointers and the region corresponds to the quadrant selected by the upper 2 bits. For example, if a base register contains the value X′4000 1000 and the s-field is zero, SR5 is used (01 + 100; or 1 + 4) as the space ID and the second quadrant of the space is directly addressable.

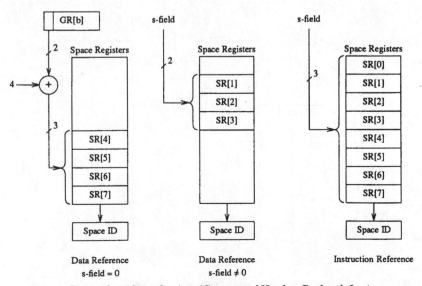

Figure 20.7 Space identifier selection. (*Courtesy of Hewlett-Packard, Inc.*)

20.3   PA-RISC Implementations

The most notable PA-RISC implementation is the PA7100. It is a 0.8-micron CMOS, 850,000-transistor, 504-pin grid array (PGA) microprocessor. It can run at a frequency of up to 100 MHz [Aspr 93, DWYF 92]. The PA7100 is a two-issue superscalar, however, there is a restriction that only an integer and a floating-point instruction can be issued together for execution. It has a 64-bit data bus in and out of the chip. The PA7100 has an on-chip FPU, but its cache is outside, unlike other recent RISC-type microprocessors, described earlier in Chaps. 14 through 19. One of the advantages of the PA7100 outside cache is that it can be a very large one, of several megabytes, as opposed to current on-chip caches of a maximum total of 32 to 36 kbytes. Certainly, a much higher hit ratio can be achieved with a larger cache.

The PA7100 has three bus outlets: one to the system interface (main memory, I/O, graphics), one to the Dcache, and one to the Icache. The external Dcache can be configured from 4 kbytes to 2 Mbytes. The Icache can be configured from 4 kbytes to 1 Mbytes. Both caches are virtually addressed, direct-mapped, with 32 bytes/line. Having no on-chip cache is certainly a disadvantage because of a possible longer access delay. This disadvantage is made up by the fact that the size of the cache can reach a total of 3 Mbytes, tending to significantly increase the hit ratio (see Chap. 4). On the other hand, the cost of a 3-Mbyte primary cache may be quite high. The PA7100 designers made a special effort to tune the processor operation to a relatively lower-cost SRAM cache [Aspr 93, DWYF 92]. The caches are wave-pipelined and hence the ALU needs only to complete the 22 lower address bits before the cache SRAM has the full address. Cache SRAMs that can use a partial address for row address specification before column address specification, improve the access time. Having a virtually addressed cache may pose another disadvantage because of address aliasing problems. On the other hand, such a cache does not need translation to a physical address, covering the tag field (see Chap. 4), which in turn speeds up the cache access.

The PA7100 has a fully-associative TLB with 120 fixed entries for 4-kbyte pages. In addition, there are 16 variable-size entries for spaces ranging from 512 kbytes to 64 Mbytes. Thus, a large page option is implemented in effect, as in other systems.

The PA7100 FPU fully complies with the IEEE 754-1985 standard [IEEE 85]. It is composed of the following subunits [Aspr 93]:

1. Floating-point register file, as described in the previous section.

2. Floating-point ALU, which performs floating-point add, subtract, compare, complement, and convert instructions for both single- and double-precision operands.

3. Floating-point multiplier, performing multiplications of single- and double-precision floating-point operands. In addition, integer multiplications of 32-bit unsigned integers provide a 64-bit product. The carry save method [Hays 88] is used in the multiplication implementation.

4. Floating-point divider, performing division and square root operations. The unit uses a modified radix-4 SRT (Sweeney–Robertson–Tocher) algorithm [Cava 84, Hwan 79, Robe 58].

PA-RISC has a five-operand floating-point instruction that performs:

```
C ← A * B
D ← D + E,
```

in one cycle.

This allows PA-RISC to execute the frequent multiply-add operation at a vector rate of one per clock cycle.

An alternate implementation of PA is the PA7100LC [Kneb 93]. The PA7100LC is a 0.8-micron CMOS, 800,000-transistor, 432-pin ceramic PGA microprocessor. It can operate at frequencies of up to 75 MHz. The PA7100LC implements a 48-bit virtual address. One of the most significant improvements on the PA7100LC is the addition of a second integer ALU, permitting simultaneous execution of two integer instructions, and thus improving the superscalar quality of operation of the PA7100LC. In order to reduce the interfacing cost of the processor, an interface to a unified (as opposed to dual) external (off-chip) cache is provided. The external primary unified cache can be configured from 8 kbytes to 2 Mbytes. As in the PA7100, it is direct-mapped. The purpose of producing the PA7100LC was to provide a lower-cost PA implementation processor, without giving up on performance.

Future PA-RISC implementations include:

PA7150—an upgrade of PA7100 to higher frequencies of 125 to 150 MHz

PA7200 with enhanced cache management, to be featured in workstations and servers including HP's first symmetric multiprocessor desktop

PA8000 which will offer 96-bit functionality via 32-bit segmentation, include a dual FPU, higher bandwidth memory interconnect, and speculative execution

Farther in the future is the PA9000 which will offer a higher level of parallelism

20.4   Concluding Comment

About eight implementation versions of PA microprocessors have been produced by HP. These microprocessors were and still are extensively used in many workstations and other computing systems. A Precision RISC organization (PRO) was established. The designers of HP are working on new and further improved versions of PA-RISC implementation microprocessors. Despite of the fact that PA implementations do not have yet an on-chip cache, as many others do, their performance exceeds that of other notable RISC-type microprocessors. One of the PA implementation advantages that goes for it is their relatively high frequency of operation (up to 100 MHz for the PA7100). This advantage will be further extended in the forthcoming PA-RISC implementations: PA7150 (125 to 150 MHz) and PA8000 starting at 200 MHz.

System Development and Comparison

21

System Development

When one purchases a mainframe or a minicomputer, the system hardware and great part of its software have already been designed, developed, tested, and produced. Of course, the user can then add more software to the system and purchase additional hardware options (such as memory extension boards), as specified and marketed by the manufacturer. Naturally, the same can be said about microcomputers; a large variety of microprocessor-based PCs and workstations exist. However, there is a difference. The user also has the option of designing and building a microcomputer from scratch, using microprocessor and other auxiliary chips (memories, I/O interfaces), as they fit the intended application. Such an option has the great advantage of allowing the user maximal flexibility at a minimal cost.

A manufactured system may have a number of extra costly features not needed for the intended application. A lesser model may not have enough features. A custom-designed microcomputer would contain only the elements needed for the intended application. The user would pay only for what is necessary to perform the tasks at hand. The resulting system will not be burdened with features that will never be used. Moreover, the user has the opportunity of creating software that fits the hardware actually used and that is tailored for the intended application. This would increase the chance of attaining more efficient and reliable software (based on the saying, "If you use out-of-house software, you get out-of-house results").

The creation of an "in-house" microcomputer involves the development of both its hardware and software, usually pursued in parallel [Rafi 84]. A flow diagram of a typical microcomputer development cycle is illustrated in Fig. 21.1. In such a development, there is a very strong interaction and dependence between hardware and software. For instance, each I/O interface chip is controlled, driven by a specific software routine (or routines). Replacing an I/O interface chip with a

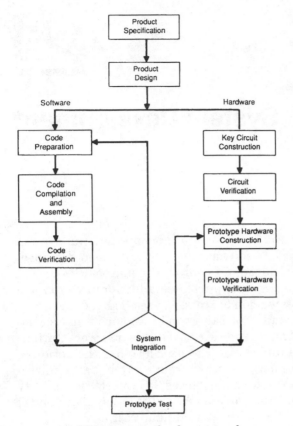

Figure 21.1   Microcomputer development cycle.

different one would usually require changes in the appropriate software. It is therefore advisable to conduct the hardware and software development concurrently. This would also shorten the overall development cycle, directly reducing development cost (the time of the designer is much more costly than the hardware). Moreover, getting the product earlier into the market is crucial to its eventual success.

The initial development stages are common to both hardware and software. First, the product has to be specified in detail: what it is supposed to perform and how fast should it be done. The processor (CPU) should be selected at this stage. At the next stage (see Fig. 21.1), the basic system design is initiated. For the hardware this should result in a drawing of all chip interconnections. Detailed flow diagrams should be prepared for the software. Subsequently, the development of the system hardware and software proceeds concurrently, following the steps shown in Fig. 21.1. At the completion of both stages, the system hardware and software are integrated by loading the software into the appropriate memory chips. Subsequently, the whole system prototype is tested. Due to possible software or hardware errors detected during the integration or testing processes, certain software and/or hardware development steps may have to be corrected and repeated. Such a possibility is illustrated by the feedback paths in Fig. 21.1.

System integration does not necessarily have to occur at the end of the software and hardware development cycles. One can also adopt partial, step-by-step integration of completed hardware with the appropriate completed software. It is easier to test smaller subsystems, module by module, than start testing a large system. It is also easier to locate possible errors in smaller systems.

The development process starts from scratch. This means that in the beginning the designer-developer has no hardware and no software. Let us say that the hardware prototype construction has started by placing step-by-step, chip sockets on the board, along with their appropriate interconnections. The designer will be able to test the software on the prototype system only when it is completed. This precludes the recommended option of partial, intermediate, module-by-module testing. How can it be done? This is where the microcomputer development system (MDS) comes in.

The MDS offers the developer all of the subsystems needed at each stage of the development. The MDS is a microcomputer endowed with hardware and software features needed for efficient and user-friendly development of microprocessor-based systems. A block diagram of a typical MDS is shown in Fig. 21.2 [Rafi 84]. Being a microcomputer, it contains some typical microcomputer components:

CPU

Main memory

Secondary memory (disk)

I/O interface (printer, MODEM, console)

Some MDS may contain more than one CPU to perform different test and interface tasks simultaneously. The other subsystems shown in Fig. 21.2 are specifically intended for hardware development tasks:

PROM programmer

Logic analyzer interface

In-circuit emulator (ICE) interface

Most MDSs have the following features:

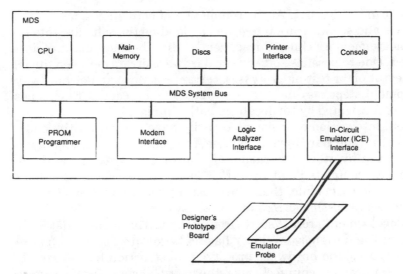

Figure 21.2 Microcomputer development system (MDS).

- A front panel status display to observe the contents of registers and memory locations
- A facility for changing contents of registers and memory locations
- A reset control that start the processor (CPU) in a given state
- A single-step control
- A run control that allows a program to be executed beginning at an instruction in a specified memory location
- A bootstrap loader that enters the initial programs into the memory
- Utility programs that load user programs into the memory from a keyboard
- An assembler
- A system monitor that serves as the operating system (OS) of the MDS
- A facility for setting breakpoints or traces
- Connectors for interfacing external devices
- An editor for minor changes in user programs

Among the interfaces mentioned above, there is an interface to a logic analyzer, which allows the designer to obtain a visual representation of the logic levels occurring in real time in the hardware. The PROM (programmable read-only memory) programmer (see Fig. 21.2) is a device for loading programs into PROM memory chips. When used with an MDS, it may be placed under software control. The front panel of a PROM programmer includes one or more sockets for inserting the PROM chips that are to be programmed or examined.

A crucial part offered by the MDS is the in-circuit emulator (ICE) interface. The ICE is a complete microcomputer system containing its own CPU, timing circuitry, memory, and I/O interface (in addition to similar MDS subsystems shown in Fig. 21.2). The ICE interfaces directly with the designer's prototype via an external cable and emulator probe (see Fig. 21.2), extending the MDS features into the prototype board. There are different ICE subsystems for different microprocessors, such as ICE-86 for 8086-based systems, or ICE-486 for 80486-based systems (see Part 2). During the development period, the ICE is substituted for the prototype's CPU and/or other parts, and executes the system instructions in the environment of the prototype system, under the control of the MDS. A more detailed MDS-ICE-prototype interface is illustrated in Fig. 21.3.

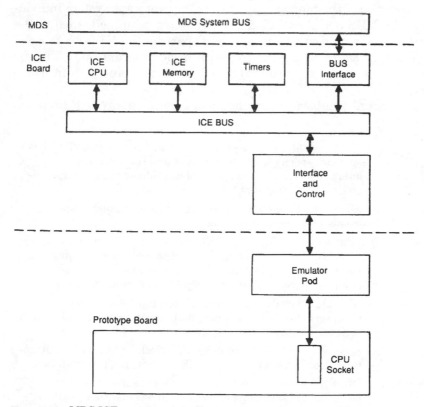

Figure 21.3 MDS-ICE prototype interface.

The ICE intended for the development of a CPU X-based system (ICE-X) consists primarily of an X microprocessor (CPU), memory (ROM, EPROM, RAM), timers, and interface and control logic. The X letter is used as a general notation for some microprocessors. The ICE bus interfaces with the MDS system bus and with the prototype board through an emulator pod that contains all of the necessary interconnection wiring. The emulator pod is connected to the CPU socket on the prototype board (Fig. 21.3). This permits the ICE CPU to replace the prototype CPU (the same X microprocessor) for the duration of the development period. The interface between the MDS, the ICE, and the prototype board allows the system hardware to be shared by the prototype and the developer to control the emulation via the MDS. The ICE can also be operated without being connected to the prototype. This allows preliminary software development even before the developer's prototype hardware is available.

Under the software, usually provided with the ICE, the program to be emulated can be stored either entirely in the ICE memory, entirely in the prototype memory, or divided between the two. This allows the developer to test the memory modules of the prototype system individually. The MDS peripheral interface subsystems can be loaned to the prototype system by substituting the I/O parts of the MDS for those of the prototype board. The ICE can also run the programs stored in the prototype memory using the prototype's memory and I/O interface chips.

Some typical emulator commands available on many ICE products are:

Go	Start emulation until a break condition is satisfied. The developer can specify the starting point and the breakpoints. A break condition can be specified on address values, data values, and/or bus status.
Step	Allows the developer to run a program on a single-step basis from a given starting address.
Trace	Allows the developer to specify the conditions for enabling and disabling trace data collection during real-time and single-step emulation. Trace information can include register or/and memory contents, flags, and signals on certain pins.
Display	Displays trace data and emulation conditions. Various formats of representation (binary, decimal, hexadecimal, ASCII) can be selected for display.
Clock	Specifies whether the internal ICE clock or an external prototype clock is used to run the ICE CPU during an emulation.
Memory map	Displays and declares what developer memory is to be accessed and the range of memory to be borrowed from the MDS (which includes the ICE).

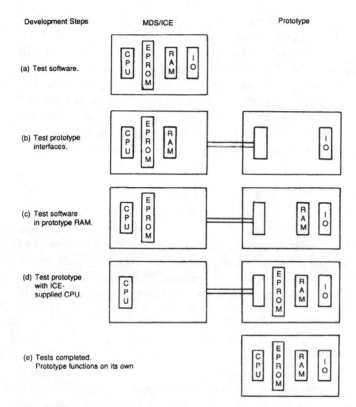

Development Steps   MDS/ICE   Prototype

(a) Test software.

(b) Test prototype interfaces.

(c) Test software in prototype RAM.

(d) Test prototype with ICE-supplied CPU.

(e) Tests completed. Prototype functions on its own

Figure 21.4 MDS-ICE prototype development process.

A possible sequence of development stages using an ICE is illustrated in Fig. 21.4. Initially, at stage (a), the MDS/ICE system is used as a host microcomputer to test the system software. Subsequently, at stage (b), the I/O interface chips are placed on the prototype board, and the ICE is connected to the board's CPU socket (see also Fig. 21.3). The prototype's I/O interfaces are tested at this time. The prototype's memory modules are gradually activated at stages (c) and (d), and the system software is tested using the prototype memory. When all tests are successful, the ICE is disconnected from the prototype board at stage (e), and a CPU chip is placed in the board's socket. The prototype is now ready to function as a microcomputer on its own.

The above development stages are loosely analogous to the development of a fetus into a baby in the mother's womb. The MDS represents the mother, the ICE the womb, and the prototype board the fetus. Initially, there is only the mother in stage (a). The prototype gradually develops and grows in stages (b) to (d). At the completion of the development at stage (e), when the prototype can "stand on its own feet," its connection with the ICE is severed, and a new system is created.

Some examples of modern development systems will be presented in the following paragraphs. The initial discussion will concentrate on the development of Pentium-based systems (see Part 2); however, the approach is in complete analogy with the development of systems based on other microprocessors. In fact, the two development systems to be discussed, were designed to be used in conjunction with a large set of different microprocessors from different manufacturers.

Development of Pentium-based systems

The following discussion will concentrate on two development systems which can be used in conjunction with Pentium-based system development, as well as for other systems. One of the systems was produced by American Arium (Tustin, CA), and the other by Microtek International (Hillsboro, OR). Both companies work in close coordination with Intel in matters connected with the development of Pentium-based systems.

American Arium system

The American Arium development scheme is based on its logic analyzer (LA) ML4400 and on its Pentium processor ICE, also called ICE-I5, or more simply LA/ICE. Figure 21.5 shows the interconnection between the ML4400, the host PC, and the LA/ICE.

The ML4400 LA is a modular system to which various options can be added to create a configuration specific to the user's need. Its features include 400 data channels, 128 kbytes deep trace memory, up to 14 levels of configurable triggering, simultaneous support for up to four micro-

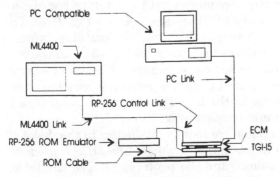

Figure 21.5 ML4400-based development configuration. (*Courtesy of American Arium Co.*)

processors, more than 250 screens of context-sensitive help, calculator for arithmetic calculations, 7-inch CRT display, an MS-DOS-compatible 1.44-Mbyte floppy disk drive, ROM emulator, serial port for data transfer to a PC or a serial printer, and an option card cage for additional features, such as an IEEE-488 control interface. The system can have one to four capture modules for storing and filtering information from different systems such as electronic circuitry. The capture modules are inserted into slots behind the hinged front panel of the system. Connection to the system under test is accomplished by individual probes or by using a target interface adapter (TGI). The TGI connection method, shown in Fig. 21.5, accommodates the high-speed signals encountered in modern CISC and RISC processors, because lead lengths are minimized in this approach. The ML4400 can be used in conjunction with the development of Intel microcontrollers (8031, 8035, 8096), 8-bit 8085, Pentium and other members of the x86 family (Part 2), 80960CA, Motorola 6800, 6809, M68000 family (see Part 3), M88000 family (see Chap. 19), MIPS R3000 (see Chap. 17), and many other systems.

The LA/ICE features 66-MHz operation, complex breakpoints, C level debugging, Intel debug port, ROM emulation memory up to 1 Mbyte, and optional timing analysis up to 1 gigahertz (gHz). The system is upgradable for multiprocessor applications.

The TGI adapter (see Fig. 21.5) allows access to all the processor pins without disconnecting address, data, or bus control signals from the user's circuit. No access time penalty is incurred. The combination of the TGI and an emulator control module (ECM), shown in Fig. 21.5, is controlled by the host PC (any DOS-based, with an i386 and up) to direct data traffic between the target, the capture module in the ML4400, and the ROM emulator. In this fashion the PC host software can automatically tell the ML4400 to set up a highly complex trace structure, load code or modifications into the ROM emulator, and control and trace pins of the CPU. All traditional emulator features are available. In addition, complex hardware breakpoints are available with many event recognizers, timers, counters, pre- and post-trace filtering, and 12 levels of if-then-else breakpoints. Because of the modular nature of the system, it is reasonable to assume that it will be able to support future microprocessors at a low additional cost.

A reduced configuration, not using the LA, which interfaces to the target circuit via a debug port is also possible. Intel defines a 20-pin debug port [Pnt1 93]. Some of its most important signals are (O denotes output from the chip, I denotes input to the chip):

TDO (O)—test data output, a serial output to the test logic. TAP (test access port) instructions and data are shifted from the Pentium through the TDO pin.

TDI (I)—test data input, a serial input for the test logic. TAP instructions and data are shifted into the Pentium through the TDI pin.

TMS (I)—test mode select, controls the sequence of TAP controller state changes.

TCK (I)—testability clock.

TRST# (I)—test reset, when asserted, it allows the TAP controller to be asynchronously initiated.

SMIACT# (O)—system management interrupt active, when activated, it indicates that the processor is operating in system management mode (SMM).

RESET (O)—when it is an output, it indicates to the outside circuitry that the processor has been reset.

INIT (O)—when it is an output, it indicates that the processor has been initialized to begin execution in a known state.

R/S# (I)—a signal interrupting the processor.

PRDY (O)—indicates that the processor stopped normal execution in response to the R/S# signal.

Figure 21.6 shows a minimal debug port implementation. Designs that require the TAP to function while debugging will need to use the American Arium debug port since LA/ICE can emulate simple TAP behavior while still controlling the Pentium CPU. Designs which use the original 20-pin Intel definition for the debug port, can use LA/ICE without this added feature. A special 20- to-30-pin cable (available from American Arium) must be used in that instance.

Microtek in-circuit emulator

The new Microtek Pentium emulator (MPE) system features 66 MHz operation, Windows 3.1 interface, full C source debugging, true hardware execution breakpoints, execution or bus-cycle trace, access to all internal registers, and accurate tracing in cache. The PC host system requirements are: MS-DOS PC i386 at 33 MHz or better, DOS Version 5.0 or later, Microsoft Windows 3.1, at least 4 Mbytes memory (8 Mbytes recommended), video graphics array (VGA) color monitor, 10 Mbytes hard drive space, and a serial port (COM1 or COM2).

More specifically, the MPE features [MPE 93]:

1. Microsoft Windows interface which includes a command line.
2. Source-level debugging, including the capability to set and display breakpoints directly in source code, look through program modules, and examine the procedure call chain.

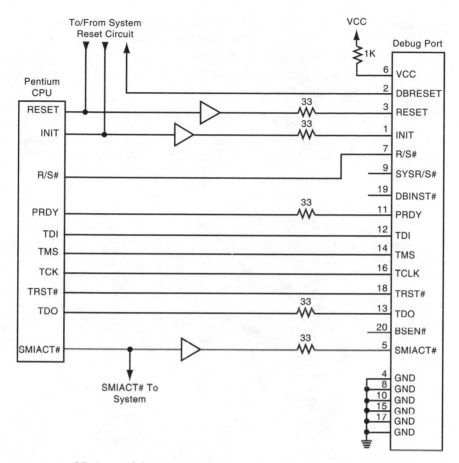

Figure 21.6 Minimum debug port implementation. (*Courtesy of Intel and American Arium Co.*)

3. Watch expressions, selecting specific program variables to display, and watch the values of these variables change while stepping through the program.

4. Capability to define a breakpoint as a source-code line or an assembly instruction, an execution address, an external input, a bus event, or a sequence of events.

5. Single-step capability. A program can be executed as single assembly language steps, HLL statement steps, or HLL procedure calls.

6. Capability to examine and modify processor registers.

7. Capability to display and modify memory and system tables, and display and modify assembler code in memory.

8. Capability to use symbolics to debug programs written in either assembly or C languages. One can also display and modify program memory using program symbolics.

9. Support for loading Microsoft 86 OMF (6.0) and Intel OMF386 simple bootloadable format.

10. Capability to save up to 16 kbytes of trace frames.

An interconnection between the emulator, the host PC, the processor module, and the stand-alone self-test (SAST) board is shown in Fig. 21.7.

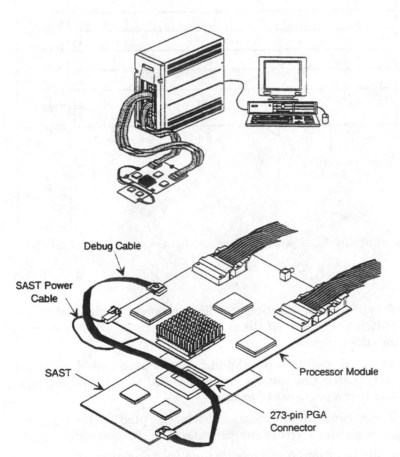

Figure 21.7 MPE system. (*Courtesy of Microtek International, Inc.*)

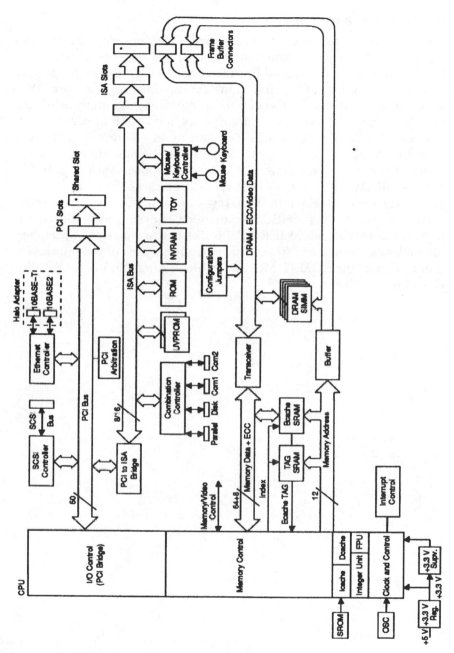

Figure 21.8 EB66 microarchitecture block diagram. (*Courtesy of Digital Equipment Corporation*)

Development of Alpha AXP systems

DEC has announced an evaluation board EB66 intended to serve as a development module for the 21066 and 21068 microprocessors. The EB66 provides a single-board hardware and software peripheral component interconnect (PCI) and industry standard architecture (ISA) development platform for the design, integration, and analysis of supporting logic, subsystems, and software for 21066 and 21068 I/O devices. The EB66 CPU is a 21066 running at 150 or 166 MHz, or a 21068 running at 66 MHz (both chips have the on-chip cache attributes of the 21064 (see Chap. 14). The EB66 also contains 8 to 256 Mbytes DRAM, 256 kbytes or 1 Mbyte backup cache (Bcache), two 100-pin connectors, interface to industry standard PCI local bus, *small computer system interface* (SCSI) controller, Ethernet controller, interface to industry standard IEEE-P996 ISA bus, a slow-speed peripherals controller, mouse and keyboard controller, a time of year clock, and a debug monitor ROM (1 Mbyte). A block diagram of the EB66 is shown in Fig. 21.8.

22

System Comparison

Leading families of CISC and RISC microprocessors were described in Parts 2 to 4 of this text. Some of the above processors will be compared and their relative performance evaluations reported in this chapter. The comparative evaluation will be done primarily between microprocessors belonging to the same generation, to keep the comparison as fair as possible.

Pentium Versus MC68060

Both the Pentium (see Chap. 8) and the MC68060 (see Chap. 12) are 32-bit, dual-issue superscalar with two parallel integer pipelines, 66-MHz, 32-bit separate address bus, dual on-chip cache (8 kbytes code, 8 kbytes data), a dual TLB (called ATC on the MC68060), IEEE standard on-chip FPU microprocessors. Both have a 256-entry branch target address cache (called BTB in the Pentium and BCU in the MC68060). Their different architectural features are summarized in Table 22.1.

While both are 32-bit systems, the Pentium has a 64-bit data bus in and out of the chip, giving it a definite performance advantage over the 32-bit data bus of the MC68060. The Pentium can transfer 8 bytes of instructions or double-precision 64-bit floating-point values in a single bus cycle, as opposed to two bus cycles needed by the MC68060. As in

TABLE 22.1   Comparison of Pentium and MC68060

Feature	Pentium	MC68060
Data bus	64-bit	32-bit
Page sizes	4 kbytes, 4 Mbytes	4 kbytes, 8 kbytes
TLBs (ATCs) entries	$64 + 8 + 32 = 104$	$64 + 64 = 128$
Cache parameters	Two-way, 32 bytes/line	Four-way, 16 bytes/line

the MC68040, the MC68060 features the same sizes of pages: 4 or 8 kbytes, encompassing at most 1 Mbyte of memory. On the other hand, the Pentium features either 4-kbyte or 4-Mbyte pages for large code segments. Although the Pentium has a total of 104 entries in its three TLBs, it can encompass tens and even hundreds of megabytes of memory. The three Pentium TLBs are as follows: 64 entries for Dcache 4-kbyte pages, 8 entries for 4-megabyte Dcache pages, and 32 entries for 4-kbyte or 4-Mbyte Icache pages. As we can see in Table 22.1, Intel changed the cache parameters, while Motorola kept them at the same values. A two-way set associative cache may have a slightly lower hit ratio than the four-way, however, it involves much simpler logic circuitry in its realization.

Since we are promised the same top frequency of 66 MHz for both systems, they can be compared on purely architectural issues. The Pentium has a definite advantage with its double data bus and TLB organization, capable of encompassing a much larger size of memory.

Both systems are completely compatible with the earlier members of the families. Motorola has a double, almost uncommitted CPU register file. As argued in Chap. 6, the Motorola family has an advantage on this point because of its larger register file. This point will be elaborated further later in this chapter. It seems that both systems are quite comparable, and it will be difficult to issue a definite evaluation before some experimental benchmarks are run. Since the MC68060 may become available only a year after the Pentium, it may be a long wait for that.

TABLE 22.2   Comparison of Intel 80486 and Motorola MC68040

Feature	Intel 80486	Motorola MC68040
FPU on chip	Yes (IEEE)	Yes (IEEE)
CPU general-purpose 32-bit registers	8	16; 8 Data
		8 Address
FPU 80-bit registers	8 (stack)	8
MMU on chip	Yes	Yes; dual:
		data, code
Cache on chip	8 kbytes	4-kbyte data
	Mixed	4-kbyte code
Segmentation	Yes	No
Paging	Yes; 4 kbytes/page	Yes; 4 kbytes or
		8 kbytes/page
TLB (or ATC) size	32 entries	64 entries in each:
		data, code ATC
Levels of protection	4	2
Instruction pipeline stages	5	6
Pins	168	179

i486 versus MC68040

A selection of points of comparison between the i486 (Chap. 9) and the MC68040 (Chap. 12) is listed in Table 22.2. Looking carefully at the table, we can perceive only a single line identically marked in both columns: both microprocessors have an on-chip FPU, conforming to the IEEE 754-1985 standard [IEEE 85]. All other data are different, although quite close in some instances. The points of difference between the i486 and the MC68040 will be discussed next in some detail.

CPU general-purpose registers. This discussion is also valid in the comparison of the Pentium with the MC68060, since the Pentium and the i486 have the same general-purpose registers, and the same can be said for MC68060 and MC68040. Both the i486 and MC68040 have 32-bit general-purpose registers; the i486 has 8, while the MC68040 has double that number, 16. The advantages and disadvantages of a large register file were discussed in Chap. 6. The register file of the i486 is definitely too small to avail itself of the above advantages. This is particularly exacerbated by the fact that the ix86 architecture CPU registers (see Chap. 7) are not really as general-purpose as a user might wish. In fact, all of them are dedicated to certain special tasks, such as

EAX, EDX	Dedicated to multiplication/division operations
EDX	Dedicated to some I/O operations
EBX, EBP	Dedicated to serve as base registers for some addressing modes
ECX	Dedicated to serve as a counter in LOOP instructions
ESP	Dedicated to serve as a stack pointer
ESI, EDI	Dedicated to serve as pointers in string instructions and as index registers in some addressing modes.

On the other hand, in the M68000 architecture (see Chap. 11) the eight data registers D0 to D7 are genuinely general-purpose without any restrictions or special tasks imposed on them. Of the eight address registers A0 to A7, only A7 is dedicated as a stack pointer. The user is free to use the other seven registers A0 to A6 in any possible way. From the point of view of the CPU register file, the M68000 architecture has a very clear advantage. It is much better equipped to retain intermediate results during a program run, thus reducing CPU-memory traffic, which in turn enhances the overall performance.

FPU general-purpose registers. Both systems have eight 80-bit registers, providing a large range for floating-point number representation and a high level of precision (see Chap. 3). The only difference between the two is that the i486 FPU registers are organized as a stack, while those of the MC68040 are accessed directly, as its integer CPU registers. Because of the stack organization the x86 architecture might have a slight edge from the standpoint of compiler generation (for that part of the compiler dealing with floating-point operations).

MMU on chip. The i486 has a regular MMU on chip for the control and management of its memory. The MC68040 has a dual MMU: one for code and one for data. This duality, supported by a separate internal (on-chip) instruction data bus and a separate operand data bus, allows the control unit to handle instruction and operand fetching simultaneously in parallel and thus enhances the handling of the instruction pipeline (see also Chap. 5). Of course, the external bus leading to the off-chip main memory is single (32-bit data, 32-bit address), and it is shared by instructions and data operands. With a reasonable on-chip cache hit ratio (see Chap. 4), the off-chip bus would be used less often.

Cache on chip. The total on-chip cache of both systems is 8 kbytes. Interestingly enough, they have the same parameters: both are four-way set-associative with 16 bytes per line. The difference is that while the i486 cache is mixed (unified), the MC68040 cache is dual (4 kbytes code, 4 kbytes data). Each MC68040 cache is controlled within a separate MMU. This results, as explained in Chaps. 4 to 6, in a more efficient handling of the instruction pipeline and enhanced performance.

Segmentation. The Intel x86 family implements segmentation, while the M68000 family does not. The merits and difficulties of segmentation were discussed in Chap. 4. While the earlier Intel systems (8086, 80286) were plagued with the upper 64-kbyte segment size limit, starting with the i386 and on, the segment size can be made as high as 4 Gbytes (maximal size of the physical memory), effectively removing the segmentation feature by the decision of the user. Therefore, as far as segmentation is concerned, the i486 and MC68040 are comparable. The i486 has some edge, since it allows the user to implement segmentation if needed and avail oneself of its advantages. On the other hand, ix86 systems carry extra hardware complexity overhead because of the segmentation feature.

Paging. The MMUs of both systems feature paged virtual memory management (see Chap. 4). The i486 offers a single fixed page size of 4 kbytes. Such a page size is implemented in many other systems. Referring to Chaps. 4 and 7, we can see that for a 4-kbytes page size, we can arrange an address mapping where the page directory and the page tables also have the same page size of 4 kbytes. Thus, the page directory and the page tables can be treated as entire pages and placed within 4-kbyte page frames in the memory. This results in reduced complexity in the MMU hardware and in the OS software, one of whose tasks is to support the management of virtual memory [BiSh 88]. The

MC68040 offers two page sizes, selectable by the user: 4 kbytes and 8 kbytes. This tends to complicate the MMU logic and the OS. On the other hand, with the current trend of offering optional larger size pages (see Chap. 4), the larger 8-kbyte page could be useful to a programmer dealing with large modules of code and data exceeding 4 kbytes.

TLB (or ATC) size.   The i486 MMU has a 32-entry TLB. With a 4-kbyte page it covers 32×4 kbytes = 128 kbytes of memory. The MC68040 offers much more. It has a dual 64-entry ATC, for a total of 128 on-chip entries. For a 4-kbyte page a total of 128×4 kbytes = 512 kbytes of memory is covered (four times that of i486), and for an 8-kbyte page, 1 Mbyte of memory is covered. In this case, a strong advantage of the MC68040 is obvious. Since the ATCs encompass much more memory, the ATC miss probability is considerably smaller. Thus, less time will be wasted in accessing page tables in memory, resulting in faster overall operation.

Levels of protection.   The x86 architecture offers four levels of protection (0, 1, 2, 3; see Chap. 7), while the M68000 architecture has only two: the supervisor and user (see Chap. 11). While the protection mechanism of the Intel x86 is much more sophisticated and, with the segmentation encapsulation of information (see Chap. 4), offers more reliable protection, it also results in more complicated on-chip logic. More time is taken up with protection checks on the x86 microprocessors.

Instruction pipeline stages.   The i486 instruction pipeline is five stages deep, while that of the MC68040 has six stages. This yields a slight edge in favor of the MC68040, however, with both being CISCs, no spectacular differences should be expected due to this point.

Experimental comparison of i486, MC68040 and other systems

A number of comparative experimental benchmark runs were conducted at Motorola in Austin, TX. The systems compared were

Motorola MC68040, at 25 MHz

Intel i486, at 25 MHz

MIPS R3000 (see Chap. 17), at 25 MHz

Sun SPARC (see Chap. 16), at 25 MHz

Motorola MC68030, at 50 MHz

The integer performance was compared using the Dhrystone benchmark [Weic 84]. The results, expressed in kdhrystones/sec, illustrated in Fig. 22.1, were

KDhrystones/s (version 2.0 or 2.1)

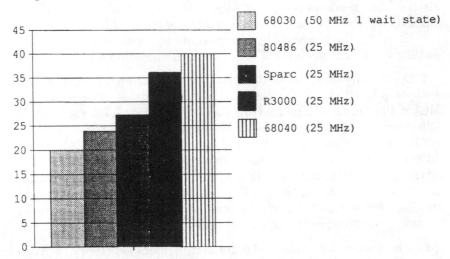

Figure 22.1   Integer performance comparison. (*Courtesy of Motorola, Inc.*)

System	Kdhr/s
MC68040	40
R3000	36
SPARC	27
i486	24
MC68030	20

As we can see, the MC68040 Dhrystone integer performance considerably exceeds that of the i486. It should also be noted that the MC68040 outperforms its predecessor MC68030 by a factor of 2, while the MC68030 operates at a double frequency. The performance of the two RISC-type systems tested (R3000 and SPARC), belonging to the same generation as the MC68040, falls below that of MC68040 and above the i486.

The floating-point performance of the above systems was tested using the Linpack benchmark [Dong 85]. It is summarized in Fig. 22.2, in MFLOPS (millions of floating-point operations per second) units. Since the MC68030 does not have an on-chip FPU, the MC68882 FPU coprocessor was used. The same was done for the R3000 and the SPARC, which do not have an on-chip FPU, using their respective coprocessors. The results were:

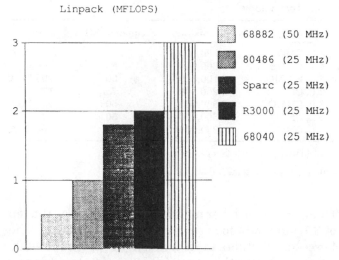

Linpack (MFLOPS)

	68882 (50 MHz)
	80486 (25 MHz)
	Sparc (25 MHz)
	R3000 (25 MHz)
	68040 (25 MHz)

Figure 22.2 Floating-point performance comparison. (*Courtesy of Motorola, Inc.*)

System	MFLOPS
MC68040	3.0
R3000 + coprocessor	2.0
SPARC + coprocessor	1.8
i486	1.0
MC68030 + MC68882	0.5

Interestingly enough, the ordering is the same as in the integer case. Here, however, the MC68040 outperforms the i486 by a factor of 3 and the MC68882 by a factor of 6. This performance ratio is well supported by the discussion given for the Table 22.2 above. The fact that the RISC-type processors, tested above, outperformed the i486 CISC should not escape notice either. This is particularly significant for floating-point performance where the i486 has an on-chip FPU, while the R3000 and the SPARC use off-chip coprocessors.

Comparison of RISC systems

The RISC-type microprocessors, described in Chaps. 14 through 20, will now be compared from several points of view involving their different features.

Technology features. RISC systems technology features are summarized in Table 22.3. They are all designed with 0.8 micron or less CMOS technology, more than one layer metal, with over 150 pins. The latest,

TABLE 22.3   RISC Systems Technology

System	Technology	Transistors	Frequency (MHz)	Pins
DEC 21064	0.7 m, 3 L, CMOS-4	1,680,000	200	431 PGA
PowerPC 601	0.6 m, 4 L, CMOS	2,800,000	80	304
SuperSPARC	0.8 m, 3 L, BiCMOS	3,100,000	50	293 PGA
R4400	0.6 m, 2 L, CMOS	2,200,000	75	179 PGA
i860 XP	0.8 m, 3 L, CHMOS-V	2,550,000	50	262 PGA
MC88110	0.8 m, 3 L, CMOS	1,300,000	50	299
PA7100	0.8 m, 3 L, CMOS	850,000	100	504 PGA

m = micron; PGA = pin grid array; x L = x-layer metal (x = 2, 3, 4).

Note: ALPHA and R4400 are 64-bit systems, all others are 32-bit.

PowerPC 601, is 0.6-micron, four-layer metal. SuperSPARC has the top transistor count of 3.1 million (about the same as the Pentium). The DEC Alpha 21064 exceeds all others in its frequency of operation—200 MHz. The PA7100 exceeds all others in pin count (504) and is the runner-up in frequency of 100 MHz. The R4400 external frequency is 75 MHz, which means that its internal pipeline frequency is double: 150 MHz, catching up with the 21064.

Architecture features.   Some architectural features of the RISC systems (IU and FPU register files, virtual and physical addresses), are listed in Table 22.4. All systems feature separate IU and FPU register files. With the exception of the SuperSPARC with its large Berkeley RISC-style register file, all have 32 IU registers. The 32-register file is still quite large compared to CISC systems (actually double than those of VAX and M68000, and quadruple than ix86), and at the same time conforming to some notable compilers, which explains their choice by most manufacturers. All systems have 32 FPU registers. The PA7100 actually has 32 hardware FPU registers, however, the first 4 are assigned special tasks, so there are only 28 general-purpose FPU reg-

TABLE 22.4   RISC Systems Architecture

System	IU registers	FPU registers	VA bits	PA bits
DEC 21064	32×64	32×64	64	34
PowerPC 601	32×32	32×64	52	32
SuperSPARC	136×32	32×32^1	32	36
R4400	32×64	32×64	64	36
i860 XP	32×32	32×32^1	32	32
MC88110	32×32	32×80	32	32
PA7100	32×32	28×64	64	32

FPU = floating-point unit; IU = integer unit; PA = physical address; VA = virtual address.

[1]Can also be used as 16×64 register file.

isters. The 21064 and R4400 are 64-bit systems and this is why their
IU registers are 64 bits wide. All systems have FPU registers which
are at least 64 bits wide to accommodate IEEE 64-bit double-precision
operands. The SuperSPARC and i860 XP registers can also be used as
16 64-bit registers (16×64). The FPU registers of the MC88110 are 80
bits wide to accommodate the 80-bit extended precision operands.
Most of the systems offer more than 32 bits of virtual address space; 64
bits by 21064, R4400, and PA7100, and 52 bits by PowerPC 601. Two
of the systems (SuperSPARC and R4400) offer a 36-bit physical
address space, and the 21064 has a 34-bit physical address. The
increase of address spaces beyond 32 bits is a trend that is expected to
grow in future systems.

Instruction level parallelism (ILP) features. The ILP features of the
RISC systems are listed in Table 22.5. All practice ILP one way or
another (as do the new CISCs Pentium and MC68060). Only the
R4400 is a two-issue superpipelined; all others are superscalar. Most
are two-issue, the PowerPC 601 and the SuperSPARC are three-issue.
The processors differ in the number of functional units capable of
simultaneous execution, and therefore, there are differences in the
amount of restrictions imposed on simultaneous issue of instructions.
The one with the least restrictions is the MC88110 with its 10 functional units, which include two IUs and two graphics units. The most
restricted is the i860 XP, where only an integer and a floating-point
pair of instructions can be issued simultaneously under a special software arrangement.

RISC memory and cache organization comparison. The RISC systems
cache and memory management features are presented in Table 22.6.
With the exception of PowerPC 601, which has a unified cache, all feature a dual cache. Because of obvious advantages of a dual cache (see

TABLE 22.5   RISC Systems ILP Features

System	ILP issue	IU units	FPU units	Graphics units
DEC 21064	2	1	1	0
PowerPC 601	3	2**	1	0
SuperSPARC	3	2	2	0
R4400	2*	1	3	0
i860 XP	2	1	2	1
MC88110	2	3	3	2
PA7100	2	1	3	0

BPU = branch processing unit; FPU = floating-point unit; ILP = instruction level parallelism; IU = integer unit.

*superpiplined, all others are superscalar.

**The IU and the BPU.

TABLE 22.6   RISC Systems Memory Organization

System	Icache, kbytes	Dcache, kbytes	Ecache, Mbytes	TLB entries	Page size
DEC 21064	8, direct	8, direct	Possible[1]	32 + 12	8, 16, 32, 64 kbytes[3]
PowerPC601	32, eight-way	—	Possible	256	4 kbytes[4]
SuperSPARC	20, five-way	16, four-way	Possible	64	4 kbytes
R4400	16, direct	16, direct	Possible[1]	96	4 kbytes[5]
i860 XP	16, four-way	16, four-way	Possible	64	4 kbytes, 4 Mbytes
MC88110	8, two-way	8, two-way	Possible	40 + 40	4 kbytes[6]
PA7100	0	0	1 I, 2 D[2]	120 + 16	4 kbytes[7]

D = data; Dcache = on-chip data cache; Ecache = off-chip external cache; I = instruction; Icache = on-chip instruction cache.

[1] on-chip secondary cache controller.

[2] primary external caches.

[3] pages can be grouped: 1, 8, 64, 512 to use a single TLB entry.

[4] PowerPC 601 has in addition blocks of 128 kbytes to 8 Mbytes.

[5] R4400 has a page size range of 4 kbytes–16 Mbytes, increasing by multiples of 4.

[6] MC88110 has in addition blocks of 512 kbytes–64 Mbytes.

[7] PA7100 has in addition spaces of 512 kbytes–64 Mbytes.

Chaps. 5 and 6), this is expected to change in future PowerPC systems. The largest total cache on chip is that of the SuperSPARC: 36 kbytes. The only one without an on-chip cache is the PA7100. Although its designers have tuned the external cache access for minimal delay, the access of an on-chip cache is still faster; PA7100's high frequency of operation tends to "cover up" for this obvious disadvantage, which may probably be corrected in future PA implementations. All mappings from direct to eight-way are used. Practically all systems have on-chip logic to interface to an external secondary cache (except PowerPC, but this may also change soon). The largest TLB is featured by the PowerPC 601: 256 entries. With the exception of the SuperSPARC, all systems feature additional page sizes (sometime called "blocks" or "spaces" but offering about the same options as larger pages). This follows some recent research results which point to an advantage of having optional larger page sizes (see Chap. 4).

Performance comparison

A number of benchmark performance experiments on earlier generation RISC-type systems, with the CISC i486 and MC68040, were reported in [Slat 91]. Some of these results are summarized in a different format in Table 22.7. The large disparity between the operating frequencies of the different processors being tested, should be noted. Naturally, the PA-RISC (a PA microprocessor preceding the PA7100), running at 66 MHz, with the next runner-up at 40 MHz, is ahead of any

TABLE 22.7   Geometric Means SPECmark Results

Processor	SPECmark	Integer-only	Floating-point-only
PA-RISC, 66 MHz	72.2	51.0	91.0
RS/6000, 30 MHz	34.7	24.0	44.3
SPARC, 40 MHz	21.2	20.8	21.6
R3000, 33 MHz	26.5	27.1	26.1
MC88100, 33 MHz	17.8	21.4	15.8
MC68040, 25 MHz	11.8	12.9	11.0
i486, 33 MHz	12.1	18.2	9.2
i860 XR, 40 MHz	24.7	19.3	29.2

Note: The above processors were within the following systems. PA-RISC: HP9000 Model 730, IBM RS/6000 Model 540; SPARC: Sun SPARCstation 2; R3000: MIPS RC3360; MC88100: Motorola Delta Model 8612; i486: HP 425s; i860: Alacron AL860 [Slat 91].

other processor both in integer and floating-point performance. It is felt that a fairer experimentation would be to test all processors at the same frequency. At this point, as a purely computational experiment, the results in Table 22.7 were scaled down to 25 MHz, the lowest frequency reported in the table (for the MC68040). If the result in Table 22.7 was X, at a frequency of Y, the scaled-down value $Z = 25X/Y$. The scaled-down results are shown in Table 22.8. Now the PA-RISC is left slightly behind the IBM RS/6000 both in integer and floating-point operation. It is also slightly outperformed by the MIPS R3000 in integer operation. Another interesting thing to observe is that while i486 (at 33 MHz) outperforms MC68040 (at 25 MHZ) in the overall SPECmark figure in Table 22.7, the situation is reversed in Table 22.8, in which the i486 is assumed to be operating at 25 MHz as well.

The last-generation RISC systems are compared with the Pentium in Table 22.9, based on reports of experiments by Intel, also documented in [Stam 93]. There is again the disparity of frequencies for different processors. Even so, it is interesting to note that the CISC Pentium outperforms the RISC PowerPC 601 at the same frequency

TABLE 22.8   SPEC Results at 25 MHz

Processor	SPECmark	Integer-only	Floating-point-only
PA-RISC	27.3	19.3	34.5
RS/6000	28.9	20.0	36.9
SPARC	13.3	13.0	13.5
R3000	20.0	20.5	27.3
MC88100	13.5	16.2	12.0
MC68040	11.8	12.9	11.0
i486	9.2	13.8	7.0
i860 XR	15.4	12.0	18.3

TABLE 22.9 SPEC Performance Comparison

CPU (MHz)	SPECint92	int,66MHz	SPECfp92	fp,66MHz
i486DX (66)	32	32	16	16
Pentium (66)	65	65	57	57
21064 (150)	74	33	126	55
R4000 (100)	62	41	63	42
R4400 (150)	82	36	86	38
PA7100 (99)	80	53	151	101
MPC601 (66)	50	50	80	80
SSPARC (40)	53	87	63	104

Note: MPC601 is the PowerPC 601; SSPARC is the SuperSPARC.

and the R4000 (100 MHz inside, 50 Mhz outside) for integer operations. This should not come as a complete surprise. Intel has considerably improved the design of the Pentium, compared to the earlier members of the x86 family. Although the Pentium is a CISC, it is a two-issue superscalar, with an on-chip dual cache (8 kbytes code, 8 kbytes data), and with a 64-bit data bus in and out of the chip. This and other innovative features should explain why the CISC Pentium can successfully compete with the leading RISCs. In order to equalize the comparison, the results in columns SPECint92 and SPECfp92, were scaled to 66 MHz for all systems, as was done in Table 22.8. While the R4400 outperformed all systems in integer operations, and the PA7100 excelled in floating-point operations, the SuperSPARC comes out on top both in integer and floating-point operations in the equalized frequency model. This should come at no surprise; the SuperSPARC is a three-issue superscalar with multiple execution units (both integer and floating-point), which alleviates restrictions on parallel execution of simultaneously issued instructions (see Chap. 5).

Independent experimental results were also reported by TI, the manufacturer of SuperSPARC and its codesigner with Sun. These results for SPECint92 are listed in Table 22.10. According to this report, the SuperSPARC at 50 MHz outperforms all other systems.

TABLE 22.10 Performance Comparison Reported by TI

System	Freq, MHz	SPECint92	SPECint/MHz
SuperSPARC	50	68	1.33
DEC 21064	133	65	0.54
Pentium	66	64	0.97
R4000	100	58	0.56
RS/6000	50	48	0.95
HP-PA 7100	50	37	0.81

TI = Texas Instruments.

This becomes even clearer when the results are shown per MHz in the rightmost column of Table 22.10.

According to TI projections into the future, in 1994 the next version of the SuperSPARC is expected to reach a frequency of 90 MHz, with a SPECint performance of up to 140, and a SPECfp performance of up to 210. This of course remains to be seen and compared with other systems in the future.

A comparison of the Pentium performance with that of the PowerPC 601, at the same frequency of 60 MHz, was reported in [Half 93]. Using *BYTE Magazine* benchmarks, the results indicated that the 601 outperformed the Pentium by a factor ranging from 1.44 to 4.7. The lower figures are for simple FPU operations, and the higher for bit-field operations. The Pentium actually outperformed the 601 on transcendental FPU operations.

A comparison between DEC 21064 and PowerPC 601-based workstation performance was reported in [Amar 93]. The 21064-based workstation tested was the DEC 3000 Model 500X AXP, and the PowerPC 601-based was the HP 9000 Model 735. The 500X AXP has a 200-MHz 21064 processor with a 512-kbyte secondary off-chip cache. Although the 500X can support up to 256 Mbytes RAM, only a minimal configuration of 96 MB was used. The HP 9000/735 had a 99-MHz PA7100 processor with an external primary cache of 256 kbytes instruction and 256 kbytes data (a total of 512 kbytes) and 32 Mbytes RAM. The experimentation used the DN&R Labs' standard suite of Unix workstation benchmarks. The performance was measured in units of Microvax II processing (MVUPs). Naturally, the 21064-based system, running at a double frequency with an on-chip cache, came out on top with an average 277.86 MVUPs, as opposed to the PA7100-based systems which achieved 185.85 MVUPs.

As impressive as some of the above benchmark results may be, it should be remembered that these results are for a restricted and limited quantity of programs. The results for many other programs may turn out to be quite different. Thus, the above results should be taken as indicative, but not conclusive. The same caution should be exercised with the numbers obtained by frequency scaling in Tables 22.8 and 22.9. After all, they do not represent the realistic state of affairs; if a microprocessor cannot run above a certain frequency, any numbers obtained assuming that it can, are only an indication of what might possibly be, but is not, or not yet. If microprocessor A attains a higher performance compared to B because A runs at a higher frequency, well, that is the reality. If the manufacturer of microprocessor A succeeded in creating a system capable of operating at a very high frequency, it should certainly enjoy the advantages coming out of such an achievement. One should also take into account that filling the chip with extra resources and increasing its density beyond certain limits may also limit the capability of the chip to operate at higher frequencies. The balance is somewhere in the middle; enough resources to run simultaneous operations in a superscalar environment, but not too many to inhibit high-frequency operation.

23

Microprocessors
Then and Now

The microprocessor has come a long way since its inception in the early seventies as an 8-bit primitive and relatively slow processor, operating at 1 to 5 MHz. Over 20 years later, in the mid-nineties, the microprocessor is a full-scale 32- or 64-bit computing system. The performance of some of the most advanced microprocessors surpasses that of some well-known mainframe computers and competes with the performance of some supercomputers. In fact, the new Cray massively parallel system (T3D) uses individual Alpha AXP microprocessors as its CPUs (see Chap. 14). All supercomputers of the future may well be composed of high-performance microprocessors.

While the first microprocessors of the early seventies had only tens of thousands of transistors per chip, the number of transistors on a chip in 1993 exceeded 3 millions on some microprocessors, such as the Pentium and the SuperSPARC.

The first microprocessors could barely contain a simple ALU and a control unit. Initially, a microprocessor ALU could only add and subtract; even multiplication and division were not included in the instruction set. Today's microprocessors handle not only all arithmetic and logical operations; all of the new generation have an on-chip FPU that adheres to the IEEE 754-1985 standard [IEEE 85]. Just a couple of years ago, most microprocessors of the preceding generation had to use an off-chip coprocessor for floating-point computations.

The early microprocessors had no memory management features. Subsequently, MMU coprocessors (such as MC68851) were used. Practically all current advanced microprocessors have an on-chip paged MMU. Some have even a dual MMU, one for instructions and one for data (MC68060, MC68040). Almost all modern microprocessors have an on-chip cache. In almost all cases the cache is dual. The largest on-chip cache so far is that of 36 kbytes (20 kbytes code, 16 kbytes data on the SuperSPARC). The size of on-chip caches will undoubtedly continue to grow in the future. Most modern microprocessors have on-chip logic to interface to an external secondary cache of several megabytes.

In the past, a microprocessor CPU had from two (MC6800) to seven (Intel 8080, 8085) CPU general-purpose registers. Some of the modern microprocessors, particularly those of RISC-type, have 32, and some even more (SPARC: 136, Am29050: 192). All last-generation advanced RISC-type microprocessors have a separate set of 32 FPU general-purpose registers.

The address space of early 8-bit microprocessors was limited to 64 kbytes with a 16-bit address. With the transition to 16-bit microprocessors, the address space grew: 1 mbyte on 8086, 16 Mbytes on i286 and MC68000, and finally the address space developed into a full 32-bit address for a 4-Gbyte address space with the advent of 32-bit microprocessors (i386, MC68020, and all RISCs). Today even the 32-bit address is exceeded. All modern systems offer the virtual memory feature. Some systems already offer a 64-bit virtual address option (21064, R4400, PA7100), and some offer a 36-bit physical address (SuperSPARC, R4400). No doubt the physical address size will increase in future implementations.

With the achievement of over 3 million transistors per chip in 1993, some microprocessor manufacturers promise a tremendous growth in that direction. Some even mention figures of 50 to 100 millions of transistors per chip by the year 2000. How will the extra on-chip space be used? There are a number of obvious directions for expansion:

1. Increase the on-chip cache size. This has already been happening in recent years.

2. Expand to 64-bit systems for both data and address. This is already happening for some systems such as 21064 and R4400. More manufacturers are expected to join this trend. Some systems, while still having a 32-bit integer unit, offer a 64-bit data bus in and out of chip (Pentium, i860).

3. Increase the number of SFUs on chip. Some systems have already started this trend, with two graphics units on the MC88110 and one on i860 XP.

4. Introduce main memory on chip. The INMOS transputer has already started this trend in the mid-eighties with an on-chip 4-kbyte main memory storage [Tabk 90b]. With several millions of transistors per chip, other manufacturers may follow.

5. Introduce I/O interfaces and other support functions (such as timing circuitry) on chip. This was already done for controller chips (such

as the i960 family [HiTa 92]) and on the Intel 80186 [LiGi 86]. With the tremendous increase of transistors per chip, many other micro-processors, not necessarily microcontrollers, will have such on-chip features.

There is a great amount of development effort going on within all microprocessor manufacturing companies. The creators of micro-processors, besides causing a revolution in the computer industry (and every other industry for that matter), have never ceased to fascinate, amaze, and surprise a very large sector of the world's population, interested in this area, with new features which only 25 years ago could be regarded as science fiction. This trend will undoubtedly con-tinue stronger than ever.

Acronyms

A   Accessed

AC   1. Alternating current; 2. Alignment check

AF   Auxiliary carry flag

ALU   Arithmetic logic unit

AM   Alignment mask

AMD   Advanced micro devices

ANSI   American National Standards Institute

AO   Access override

AP   Argument pointer

ARI   Address register indirect

ASC   American Standard Code

ATC   Address translation cache

AU   Address unit

AxCase   Advanced cross-computer-aided software engineering

AxDB   Advanced cross-debug system

AxLS   Advanced cross-language system

B   1. Byte; 2. Busy

BATC   Block ATC

BBA   Basis branch analyzer

BBN   Bolt, Beranek, and Newman

BCD   Binary coded decimal

BE   Byte enable

BiCMOS   Bipolar CMOS

BIST   Built-in self-test

BIT   Bipolar integrated technology

BIU   Bus interface unit

BMIC   Bus master interface controller

BPC   Breakpoint program counter

BPU   Branch processing unit

BU   Bus unit

C   1. Carry; 2. Conforming

CA   Core architecture

CACP   Centralized arbitration control point

CACR   Cache control register

CAD   Computer-aided design

CAR   Compare address register

CCR   Condition code register

CD   Cache disable

CDC   Control Data Corporation

CF   Carry flag

CFG   Configuration

CIS   Central instruction sequencer

CISC   Complex instruction set computer

CLA   Carry look ahead

CMMU   Cache/MMU

CMOS   Complementary MOS

CPL   Current PL

CPU   Central processing unit

CR   1. Control register; 2. Condition register

CROM   Control ROM

CSA   Carry save adder

CTR   Count register

CU   Control unit

CWP   Current window pointer

DAR   Data address register

DARPA   Defense Advanced Research Project Agency

DC   1. Direct current; 2. Data cache

DCU   Data cache unit

DE   Data execution

DEC   Digital Equipment Corporation

DF   Direction flag

DFC   Destination function code

DIP   Dual in-line packages

DM   Data memory

DMA   1. Direct memory access; 2. Data memory address

DMAC   DMA controller

DMD   Data memory data

DMMU   Data MMU

DMT   Data memory transaction

DMU   Data memory unit

DOS   Disk OS

DP   Data parity

DPL   Descriptor PL

DR   Debug status register

DSL   Description language

DRAM   Dynamic RAM

DS   Dual space

DSR   Debug status register

DSISR   Data storage interrupt status register

DST   Destination

DTB   Data translation buffer

DTT   Data transparent translation

EA   Effective address

EBB   EISA bus buffer

EBC   EISA bus controller

ECL   Emitter current logic

ECM   Emulator control module

EISA   Extended industry standard architecture

EM   Emulate coprocessor

EMI   Electromagnetic interference

EPSR   Exception-time PSR

ES   Error summary

EU   Execution unit

FCC   Federal Communications Commission

FCOP   Floating-point coprocessor

FCR   Floating-point control register

FCU   Floating-point conversion unit

FDC   Floppy disk controller

FEU   Floating-point execution unit

FFT   Fast Fourier transform

FGR   Floating-point general-purpose register

FIFO   First-in first-out

FNU   Floating-point normalization unit

FP   Frame pointer

FPA   Floating-point accelerator

FPC   Floating-point controller

FPCR   Floating-point control register

FPECR   Floating-point exception cause register

FPIAR   Floating-point instruction address register

FPP   Floating-point processor

FPR   Floating-point control register

FPSR   Floating-point status register

FPU   Floating-point unit

FQ   Floating-point queue

FRF   Floating-point register file

FSR   Floating-point status register

FWB   Floating-point write-back

FXU   Fixed-point unit

G   Giga (times 10^9)

GaAs   Galium arsenide

GDT   Global descriptor table

GDTR   Global descriptor table register

GMU   George Mason University

GPR   General-purpose register

GR   General register

GRF   General register file

HLL   High-level language

HP   Hewlett-Packard

HPMC   High-priority machine check

Hz   Hertz

IAR   Instruction address register

IBM   International Business Machines

IC   1. Integrated circuit; 2. Instruction cache

ICD   In-circuit debugger

ICE   In-circuit emulator

ICR   Interrupt control register

ICU   1. Interrupt control unit; 2. Instruction cache unit

ID   Instruction decode

IDT   1. Integrated device technology; 2. Interrupt descriptor table

IDTR   IDT register

IEEE   Institute of Electrical and Electronics Engineers

IF   1. Instruction fetch; 2. Interrupt enable flag

IFU   Instruction fetch unit

ILP   Instruction level parallelism

IMMU   Instruction MMU

IMU   Instruction memory unit

IO   Input output

IOC   IO controller

IOCC   I/O channel controller

IOP   Input output processor

IOPL   I/O privilege level

IP   1. Instruction pointer; 2. Instruction prefetch

IPL   Initial program load

IPR   Internal processor register

IR   Instruction register

ISA   Instruction set architecture

ISP   1. Integrated system peripheral; 2. Interrupt stack pointer

ISR   Interrupt service routine

ITB   Instruction translation buffer

ITT   Instruction transparent translation

IU   1. Integer unit; 2. Instruction unit

k   Kilo (times 10^3)

kHz   Kilohertz (10^3 hertz)

L   Longword

LA   Logic analyzer

LAN   Local area network

LBA   Logical block address

LDC   Lock data cache

LDT   Local descriptor table

LDTR   LDT register

LIC   Lock instruction cache

LIFO   Last-in first-out

LIO   Local I/O

LPA   Logical page address

LR   1. Logical register; 2. Link register

LRU   Least recently used

LSB   Least-significant bit

LSI   Large-scale integration

M   1. Memory; 2. Mega (times 10^6)

MAG   Memory address generator

MAR   Memory address register

MBUS   Memory bus

MC   Motorola Company

McD   McDonnell-Douglas

MCR   Memory management control

MDS   Microcomputer development system

MEMC   Memory controller

MFLOPS   Millions of floating-point operations per second

MHz   Megahertz (10^6 hertz)

MIPS   1. Microprocessor without interlocked pipeline stages; 2. Millions of instructions per second

MM   Main memory

MMU   Memory management unit

MMUSR   MMU status register

MODEM Modulator/demodulator

MOS Metal oxide semiconductor

MP Monitor coprocessor

MPE Multiprogramming executive

MQ Multiplier and quotient

MSB Most-significant bit

MSI Medium-scale integration

MSP Master stack pointer

MSR 1. Memory management status register; 2. Machine state register

MSW Machine status word

MVUP MicroVAX II units of processing

N Negative

NaN Not a number

NE Numerics exception

NEC Nippon Electric Company

NIP Next instruction pointer

NMI Nonmaskable interrupt

NMOS N-channel MOS

NS National Semiconductor

NT Nested task

NWINDOWS Number of windows

OB Overflow bit

OD Operand decode

OF 1. Operand fetch; 2. Overflow flag

OPR Overflow pointer register

OS 1. Operating system; 2. Operand store

OT Overflow type

PA Precision architecture

PAL Privileged architecture library

PATC Page ATC

PBA Physical block address

PBUS Processor bus

PC 1. Program counter; 2. Personal computer

PCB Process control block

PCC   Process cycle counter

PCD   Page cache disable

PCI   Peripheral component interconnect

PCS   Program control section

PCU   Program control unit

PDE   Page directory entry

PE   1. Protection enable; 2. Processing element

PF   Parity flag

PFA   Page frame address

PFN   Page frame number

PFP   Previous frame pointer

PFT   Page frame table

PG   Paging enable

PGA   Pin grid array

PL   Privilege level

PM   Program memory

PMMU   Paged MMU

PMOS   P-channel MOS

POWER   Performance optimized with enhanced RISC

PR   Parameter register

PRO   Precision RISC organization

PROM   Programmable ROM

PSCU   Program status control unit

PSR   Processor status register

PSW   Processor status word

PT   Program time

PTE   Page table entry

PVAM   Protected virtual address mode

PWT   Page write-through

QNaN   Quiet NaN

R   Referenced

RA   Real address

RAM   Random-access memory

RCA   Radio Corporation of America

RF   1. Register file; 2. Resume flag

RIP   Return instruction pointer

RISC   Reduced instruction set computer

RMW   Read-modify-write

ROM   Read-only memory

ROMP   Research office product division microprocessor

ROS   Read-only storage

RPL   Requestor PL

RPN   Real page number

RWM   Read-write memory

S   Sign

SAPR   Supervisor area pointer register

SAST   Stand-alone self-test

SB   1. Scoreboard; 2. Static base

SCSI   Small computer systems interface

SCU   Storage control unit

SF   1. Sign flag; 2. Stack flag

SFC   Source function code

SFIP   Shadow FIP

SFU   Special function unit

SIO   System I/O

SLA   Serial link adapter

SMD   Storage module device

SNaN   Signaling NaN

SNIP   Shadow NIP

SP   Stack pointer

SPARC   Scalable processor architecture

SPR   Special-purpose register

SR   1. Status register; 2. Segment registe

SRAM   Static RAM

SRP   Supervisor root pointer

STDIO   Standard I/O

SWP   Saved window pointer

SX   Storage and execution

SXIP Shadow XIP

TB Translation buffer

TBR Trap base register

TC Translation control register

TCW Tag and translation control word

TEAR Translation exception address register

TEX Translation exception

TF Trap enable flag

TGI Target interface adapter

TI 1. Table index; 2. Texas Instruments

TIC Target instruction cache

TID Transaction identifier

TLB Translation lookaside buffer

TR 1. TSS register; 2. Test register

TS 1. Task switched; 2. Translate supervisor

TSS Task state segment

TSSR TSS register

TTL Transistor-transistor logic

TU Translate user

UAPR User area pointer register

UCLA University of California at Los Angeles

URP User root pointer

USP User stack pointer

UWS Ultrix worksystem software

V 1. Overflow; 2. Valid

VA Virtual address

VBR Vector base register

VGA Video graphics array

VIDC Video controller

VLSI Very large-scale integration

VM 1. Virtual memory; 2. Virtual mode

VRAM Video RAM

W Word

WB Write-back

WIM   Window invalid mask

WP   Write protect

WT   1. Windows transferred; 2. Writes transparent

X   Extend flag

XER   Exception register

XIP   Execution instruction pointer

XRF   Extended register file

Z   1. Zero; 2. Zilog

ZF   Zero flag

References

[AlAv 93] D. Alpert, D. Avnon, Architecture of the Pentium Microprocessor, IEEE MICRO, Vol. 13, No. 3, pp. 11–21, June 1993.

[AlBe 93] M. S. Allen, M. C. Becker, Multiprocessing Aspects of the PowerPC 601, Proc. COMPCON 93, pp. 117–126, San Francisco, CA, Feb. 22–26, 1993.

[Alex 93] N. Alexandridis, *Design of Microprocessor-Based Systems*, Prentice-Hall, Englewood Cliffs, NJ, 1993.

[AlGo 89] G. S. Almasi, A. Gottlieb, *Highly Parallel Computing*, Benjamin/Cummings, Redwood City, CA, 1989.

[Alsp 90] M. Alsup, Motorola's 88000 Family Architecture, IEEE MICRO, Vol. 10, No. 3, pp. 48–66, June 1990.

[Amar 93] C. Amaru, DEC, HP Seek High-end Workstation Supremacy, Digital News & Review, Aug. 23, 1993.

[ArCu 86] Arvind, D. E. Culler, Dataflow Architectures, Annual Reviews in Computer Science, Vol. 1, pp. 225–253, Annual Reviews Inc., Palo Alto, CA, 1986.

[Aspr 93] T. Asprey et al., Performance Features of the PA7100 Microprocessor, IEEE MICRO, Vol. 13, No. 3, pp. 22–35, June 1993.

[Atki 91] M. Atkins, Performance and the i860 Microprocessor, IEEE MICRO, Vol. 11, No. 5, pp. 24–27, 72–78, Oct. 1991.

[BAMo 93] M. C. Becker, M. S. Allen, C. R. Moore, J. S. Muhich, D. P. Tuttle, The PowerPC 601 Microprocessor, IEEE MICRO, Vol. 13, No. 5, pp. 54–68, Oct. 1993.

[BaWh 90] H. B. Bakoglu, T. Whiteside, RISC System/6000 Hardware Overview, *IBM RISC System/6000 Technology*, pp. 8–15, SA 23-2619, IBM Corporation, Austin, TX, 1990.

[BiSh 88] L. Bic, A. C. Shaw, *The Logical Design of OS*, Prentice-Hall, Englewood Cliffs, NJ, 1988.

[BiWo 86] J. S. Birnbaum, W. S. Worley, Beyond RISC: High Precision Architecture, HP Journal, Vol. 36, No. 8, pp. 4–10, Aug. 1985 (also in Proc. COMPCON 86, pp. 40–47, San Francisco, CA, March 1986).

[BlKr 92] G. Blanck, S. Kreuger, The SuperSPARC Microprocessor, Proc. COMPCON 92, San Francisco, CA, Feb. 1992.

[BuPa 93] M. Butler, Y. Patt, A Comparative Performance Evaluation of Various State Maintenance Mechanisms, Proc. 26th Annual Symposium on Microarchitecture, MICRO-26, pp. 70–79, Austin, TX, Dec. 1–3, 1993.

[Cava 84] J. J. F. Cavanagh, *Digital Computer Arithmetic*, McGraw-Hill, NY, 1984.

[ChBJ 92] J. B. Chen, A. Borg, N. P. Jouppi, A Simulation Based Study of TLB Performance, Proc. 19th Annual Int. Symp. on Computer Architecture (ISCA 92), pp. 114–123, Gold Coast, Queensland, Australia, May 19–21, 1992.

[Clem 92] A. Clements, *Microprocessor System Design*, PWS-Kent, Boston, MA, 1992.

[Clem 94] A. Clements, *68000 Family Assembly Language*, PWS Publ. Co., Boston, MA, 1994.

[ClSt 80] D. W. Clark, W. D. Strecker, Comments on "The Case for the RISC," Computer Architecture News, Vol. 8, No. 6, pp. 34–38, October 15, 1980.

[Colw 85] R. P. Colwell et al., Computers, Complexity and Controversy, IEEE Computer, Vol. 18, No. 9, pp. 8–19, Sept. 1985.

[Colw 88] R. P. Colwell et al., A VLIW Architecture for a Trace Scheduling Compiler, IEEE Trans. on Computers, Vol. 37, No. 8, pp. 967–979, Aug. 1988.

[Craw 90] J. H. Crawford, The i486 CPU: Executing Instructions in One Clock Cycle, IEEE MICRO, Vol. 10, No. 1, pp. 27–36, Feb. 1990.

[Cray 93] *Cray T3D System Architecture Overview,* Revision 1. C, Cray Research, Inc., September 23, 1993.

[DECM 92] DECchip 21064-AA Microprocessor Hardware Reference Manual, DEC, Order No. EC-N0079-72, Maynard, MA, Oct. 1992.

[Denn 72] P. J. Denning, on Modeling Program Behavior, Proc. Spring Joint Computer Conference, Vol. 40, pp. 937–944, AFIPS Press, Arlington, VA, 1972.

[DiAl 92] K. Diefendorff, M. Allen, Organization of the Motorola 88110 Superscalar RISC Microprocessor, IEEE MICRO, Vol. 12, No. 2, pp. 40–63, April 1992.

[DiOH 94] K. Diefendorff, R. Oehler, R. Hochsprung, Evolution of the PowerPC Architecture, IEEE MICRO, Vol. 14, No. 2, pp. 34–49, April 1994.

[Dobb 92] D. W. Dobberpuhl et al., A 200 MHZ 64-bit Dual-issue CMOS Microprocessor, Digital Technical Journal, Vol. 4, No. 4, pp. 35–50, special issue 1992.

[Dong 85] J. J. Dongarra, Performance of Various Computers using Standard Linear Equations Software in a Fortran Environment, Comp. Arch. News, Vol. 13, No. 1, pp. 3–11, March 1985.

[DWYF 92] E. DeLano, W. Walker, J. Yetter, M. Forsyth, A High Speed Superscalar PA-RISC Processor, Proc. COMPCON 92, pp. 116–121, San Francisco, CA, Feb. 24–28, 1992.

[Eden 90] R. W. Edenfield et al., The 68040 Processor, IEEE MICRO, Vol. 10, No. 1, pp. 66–78, Feb. 1990, Part II, No. 3, pp. 22–35, June 1990.

[Fair 82] D. A. Fairclough, A Unique Microprocessor Instruction Set, IEEE MICRO, Vol. 2, No. 2, pp. 8–18, May 1982.

[FlMM 87] M. J. Flynn, C. L. Mitchell, J. M. Mulder, And Now a Case for More Complex Instruction Sets, IEEE Computer, Vol. 20, No. 9, pp. 71–83, Sept. 1987.

[Garn 88] R. B. Garner, SPARC—Scalable RISC Architecture, Sun Technology, Vol. 1, No. 3, pp. 42–55, Summer 1988.

[GrOe 90] R. D. Groves, R. Oehler, IBM's RISC System/6000 Processor Architecture, Microprocessors and Microsystems, Vol. 14, No. 6, pp. 357–366, July/Aug. 1990.

[GrSh 89] E. Grochowski, K. Shoemaker, Issues in the Implementation of the i486 Cache and Bus, Proc. International Conference on Computer Design (ICCD 89), pp. 193–198, Boston, MA, Oct. 1989.

[Half 93] T. R. Halfhill, PowerOpen Gives Users Freedom of Choice, BYTE, August 1993.

[HaVZ 90] V. C. Hamacher, Z. G. Vranesic, S. G. Zaky, *Computer Organization,* 3rd ed., McGraw-Hill, New York, 1990.

[Hays 88] J. P. Hayes, *Computer Architecture and Organization,* 2nd ed., McGraw-Hill, New York, 1988.

[Hein 93] J. Heinrich, *MIPS R4000 Microprocessor User's Manual,* PTR Prentice-Hall, Englewood Cliffs, NJ, 1993.

[Henn 82] J. L. Hennessy et al., The MIPS Machine, Proc. COMPCON 82, pp. 2–7, San Francisco, CA, Feb. 1982.

[Henn 84] J. L. Hennessy, VLSI Processor Architecture, IEEE Trans. on Computers, Vol. C-33, No. 12, pp. 1221–1246, Dec. 1984.

[HePa 90] J. L. Hennessy, D. A. Patterson, *Computer Architecture: A Quantitative Approach,* M. Kaufmann, San Mateo, CA, 1990.

[HePa 94] J. L. Hennessy, D. A. Patterson, *Computer Organization and Design, The Hardware / Software Approach,* M. Kaufmann, San Mateo, CA, 1994.

[Hill 88] M. D. Hill, A Case for Direct-Mapped Caches, IEEE Computer, Vol. 21, No. 12, pp. 25–40, Dec. 1988.

[HiSm 89] M. D. Hill, A. J. Smith, Evaluating Associativity in CPU Caches, IEEE Trans. on Computers, Vol. 38, No. 12, pp. 1612–1630, Dec. 1989.

[HiTa 92] K. J. Hintz, D. Tabak, *Microcontrollers: Architecture, Implementation, and Programming,* McGraw-Hill, New York, 1992.

[Hwan 79] K. Hwang, *Computer Arithmetic,* Wiley, New York, 1979.

[Hwan 93] K. Hwang, *Advanced Computer Architecture,* McGraw-Hill, New York, 1993.

[HwPa 87] W. W. Hwu, Y. N. Patt, Checkpoint Repair of Out-of-Order Execution Machines, Proc. 14th Annual International Symposium on Computer Architecture, ISCA 87, pp. 18–26, Pittsburgh, PA, June 1987.

[i486 90] *i486 Microprocessor Hardware Reference Manual,* Intel Corporation Order No. 240552-001, 1990.

[i860 91] *i860 XP Microprocessor Data Book,* Intel Corporation Order No. 240874-002, Nov. 1991.

[IEEE 81] A Proposed Standard for Floating-Point Arithmetic, IEEE Computer, Vol. 14, No. 3, pp. 51–62, March 1981.

[IEEE 85] IEEE *Standard 754-1985 for Binary Floating-Point Arithmetic,* IEEE Computer Society Press, Los Alamos, CA, 1985.

[Inte 87] *80387 Programmers' Reference Manual,* Intel Corporation, order number 231917-001, 1987.

[Joup 89] N. P. Jouppi, The Nonuniform Distribution of Instruction- Level and Machine Parallelism and Its Effect on Performance, IEEE Trans. on Computers, Vol. 38, No. 12, pp. 1645–1658, Dec. 1989.

[JoWa 89] N. P. Jouppi, D. Wall, Available Instruction Level Parallelism for Superscalar and Superpipelined Machines, Proc. Third Int. Conf. on Architectural Support for Programming Languages and OS (ASPLOS), pp. 272–282, Boston, MA, April 1989.

[KaHe 92] G. Kane, J. Heinrich, *MIPS RISC Architecture,* Prentice-Hall, Englewood Cliffs, NJ, 1992.

[Kate 85] M. G. H. Katevenis, *Reduced Instruction Set Computer Architectures for VLSI,* MIT Press, Cambridge, MA, 1985.

[KlWi 88] S. R. Kleinman, D. Williams, Sun OS on SPARC, Sun Technology, Vol. 1, No. 3, pp. 56–63, Summer 1988.

[Kneb 93] P. Knebel et al., HP's PA7100LC: A Low-Cost Superscalar PA- RISC Processor, Proc. COMPCON 93, pp. 441–447, San Francisco, CA, Feb. 22–26, 1993.

[KoMa 89] L. Kohn, N. Margulis, Introducing the Intel i860 64-bit Microprocessor, IEEE MICRO, Vol. 9, No. 4, pp. 15–30, Aug. 1989.

[Latt 81] W. W. Lattin et al., A Methodology for VLSI Chip Design, Lambda, second quarter 1981, pp. 34–44.

[Lee 89] R. B. Lee, Precision Architecture, IEEE Computer, Vol. 22, No. 1, pp. 78–91, Aug. 1989.

[LeEc 84] H. M. Levy, R. H. Eckhouse, Jr., *Computer Programming and Architecture: The VAX-11,* Digital Press, Bedford, MA, 1984.

[Leon 87] T. E. Leonard, ed., *VAX Architecture Reference Manual,* Digital Press, Bedford, MA, 1987.

[LiGi 86] Y. C. Liu, G. A. Gibson, *Microcomputer Systems: The 8086/8088 Family,* 2nd ed., Prentice-Hall, Englewood Cliffs, NJ, 1986.

[M680 89] *M68000 Family Programmer's Reference Manual,* Motorola, Inc., Austin, TX, 1989.

[M811 91] *MC88110 User's Manual,* Motorola, UM88110/AD, 1991.

[Maho 86] M. J. Mahon et al., HP PA: The Processor, HP Journal, Vol. 37, No. 8, pp. 4–21, Aug. 1986.

[MaTA 91] S. Makhdoom, D. Tabak, R. Auletta, Register File/Cache Microarchitecture Study Using VHDL, Proc. 24th Annual Int. Symp. on Microarchitecture (MICRO-24), pp. 217–222, Albuquerque, NM, Nov. 18–20, 1991.

[MC40 89] *MC68040 32-bit Microprocessor User's Manual,* Motorola, Inc., Austin, TX, 1989.

[McKi 94] D. L. McKinney et al., Digital's DEC chip 21066: The First Cost-focused Alpha AXP Chip, Digital Technical Journal, Vol. 6, No. 1, pp. 66–77, winter 1994.

[McLe 93] E. McLellan, The Alpha AXP Architecture and 21064 Processor, IEEE MICRO, Vol. 13, No. 3, pp. 36–47, June 1993.

[Mele 89] C. Melear, The Design of the 88000 RISC Family, IEEE MICRO, Vol. 9, No. 2, pp. 26–38, April 1989.

[Milu 86] V. M. Milutinovic, ed., special issue of IEEE Computer, Vol. 19, No. 10, Oct. 1986.

[MiWV 92] S. Mirapuri, M. Woodacre, N. Vasseghi, The MIPS R4000 Processor, IEEE MICRO, Vol. 12, No. 2, pp. 10–22, April 1992.

[Moor 93] C. R. Moore, The PowerPC 601 Microprocessor, Proc. COMPCON 93, pp. 109–116, San Francisco, CA, Feb. 22–26, 1993.

[MPE 93] MPE Emulator Configuration Guide, Microtek International Doc. No. I49-000794, Hillsboro, OR, Oct. 1993.

[MPVa 93] M. Moudgill, K. Pingali, S. Vassiliadis, Register Renaming and Dynamic Speculation: An Alternative Approach, Proc. 26th Annual Symposium on Microarchitecture, MICRO-26, pp. 202–218, Austin, TX, Dec. 1–3, 1993.

[MR40 91] MIPS R4000 Microprocessor User's Manual, MIPS Computer Systems, Inc., Mountain View, CA, 1991.

[Myrs 82] G. J. Myers, Advances in Computer Architecture, 2nd ed., Wiley, New York, 1982.

[OeBl 91] R. E. Oehler, M. W. Blasgen, IBM RS/6000: Architecture and Performance, IEEE MICRO, Vol. 11, No. 3, pp. 14–17, 56–62, June 1991.

[PaDi 80] D. A. Patterson, D. R. Ditzel, The Case for the RISC, Computer Architecture News, Vol. 8, No. 6, pp. 25–33, Oct. 15, 1980.

[Papd 91] G. M. Papadopoulos, Implementation of a General-Purpose Dataflow Multiprocessor, MIT Press, Cambridge, MA, 1991.

[PARI 90] PA-RISC 1.1 Architecture and Instruction Set Reference Manual, HP No. 09740-90039, Cupertino, CA, Nov. 1990.

[PaSe 82] D. A. Patterson, C. H. Sequin, A VLSI RISC, IEEE Computer, Vol. 15, No. 9, pp. 8–21, Sept. 1982.

[PaSi 93] G. Paap, E. Silha, PowerPC: A Performance Architecture, Proc. COMPCON 93, pp. 104–108, San Francisco, CA, Feb. 22–26, 1993.

[Patt 82] D. A. Patterson, A RISCy Approach to Computer Design, Proc. COMPCON 1982, pp. 8–14, San Francisco, CA, 1982.

[Patt 85] D. A. Patterson, Reduced Instruction Set Computers, Comm. ACM, Vol. 28, No. 1, pp. 8–21, Jan. 1985.

[Paul 94] R. P. Paul, SPARC Architecture, Assembly Language Programming, and C, Prentice-Hall, Englewood Cliffs, NJ, 1994.

[PeSW 91] C. Peterson, J. Sutton, P. Wiley, iWarp: A 100-MOPS LIW Microprocessor for Multicomputers, IEEE MICRO, Vol. 11, No. 3, pp. 26–29, 81–87, June 1991.

[Pnt1 93] Pentium Processor User's Manual, Vol. 1: Pentium Processor Data Book, Intel order No. 241428-001, Santa Clara, CA, 1993.

[Pnt2 93] Pentium Processor User's Manual, Vol. 2: 82496 Cache Controller and 82491 Cache SRAM Data Book, Intel order No. 241429-001, Santa Clara, CA, 1993.

[Pnt3 93] Pentium Processor User's Manual, Vol. 3: Architecture and Programming Manual, Intel order No. 241430-001, Santa Clara, CA, 1993.

[PoPC 93] PowerPC 601 User's Manual, Motorola, Inc., MPC601UM/AD, REV 1, Austin, TX, June 1993.

[Prot 88] D. A. Protopapas, Microcomputer Hardware Design, Prentice-Hall, Englewood Cliffs, NJ, 1988.

[Przy 84] S. A. Przybylski et al., Organization and VLSI Implementation of MIPS, J. of VLSI and Computer Systems, Vol. 1, No. 2, pp. 170–208, Spring 1984.

[Przy 90] S. A. Przybylski, Cache and Memory Hierarchy Design, Morgan Kaufmann, San Mateo, CA, 1990.

[Radi 83] G. Radin, The 801 Minicomputer, IBM J. of R&D, Vol. 27, No. 3, pp. 237–246, May 1983.

[Rafi 84] M. Rafiquzzaman, Microprocessors and Microcomputer Development Systems, Harper & Row, New York, 1984.

[RauF 93] B. R. Rau, J. A. Fisher, Instruction Level Parallel Processing: History, Overview, and Perspective, The Journal of Supercomputing, Vol. 7, No. 1/2, pp. 9–50, May 1993.

[Robe 58] J. E. Robertson, A New Class of Digital Division Methods, IEEE Trans. on Computers, Vol. C-7, No. 3, pp. 218–222, Sept. 1958.

[SBNe 82] D. P. Siewiorek, C. G. Bell, A. Newell, *Computer Structures: Principles and Examples,* McGraw-Hill, New York, 1982.

[Sits 92a] R. L. Sites, Alpha AXP Architecture, Digital Technical Journal, Vol. 4, No. 4, pp. 19–34, special issue 1992.

[Sits 92b] R. L. Sites, ed., *Alpha Architecture Reference Manual,* Digital Press, Burlington, MA, 1992.

[Sits 93] R. L. Sites, Alpha AXP Architecture, Comm. ACM, Vol. 36, No. 2, pp. 33–44, Feb. 1993.

[Skin 88] T. P. Skinner, *Assembly Language Programming for the 68000 Family,* Wiley, New York, 1988.

[Slat 91] M. Slater, PA Workstations Set Price/Performance Records, Microprocessor Report, Vol. 5, No. 6, April 3, 1991.

[Smit 78] A. J. Smith, A Comparative Study of Set Associative Memory Mapping Algorithms and Their Use for Cache and Main Memory, IEEE Trans. on Software Eng., Vol. SE-4, No. 2, pp. 121–130, March 1978.

[Smit 82] A. J. Smith, Cache Memories, Computing Surveys, Vol. 14, No. 3, pp. 473–530, Sept. 1982.

[Smit 85] A. J. Smith, Cache Evaluation and the Impact of Workload Choice, Proc. 12th Annual Int. Symp. on Computer Architecture (ISCA 85), pp. 64–73, Boston, MA, June 17–19, 1985.

[Smit 87] A. J. Smith, Line (Block) Size Choice for CPU Cache Memories, IEEE Trans. on Computers, Vol. C-36, No. 9, pp. 1063–1075, Sept. 1987.

[SmPl 85] J. E. Smith, A. R. Pleszkun, Implementation of Precise Interrupts in Pipelined Processors, Proc. 12th Annual International Symposium on Computer Architecture, ISCA 85, pp. 36–44, Boston, MA, June 1985.

[Sprc 90] *SPARC RISC User's Guide,* 2nd ed., Ross Technology, Inc., Austin, TX, Feb. 1990.

[SSPC 92] *SuperSPARC User's Guide,* TI document 2647726-9721, Houston, TX, Oct. 1992.

[Stam 93] N. Stam, Inside Pentium, PC Magazine, April 27, 1993, pp. 123–144.

[Ston 93] H. S. Stone, *High-Performance Computer Architecture,* 3rd ed., Addison-Wesley, Reading, MA, 1993.

[Strs 86] E. Strauss, *Inside the 80286,* Brady/Prentice-Hall, New York, 1986.

[Tabk 87] D. Tabak, *RISC Architecture,* Research Studies Press, UK, and Wiley, New York, 1987.

[Tabk 90a] D. Tabak, *Multiprocessors,* Prentice-Hall, Englewood Cliffs, NJ, 1990.

[Tabk 90b] D. Tabak, *RISC Systems,* Research Studies Press, UK, and Wiley, New York, 1990.

[Tall 92] M. Talluri, S. Kong, M. D. Hill, D. A. Patterson, Tradeoffs in Supporting Two Page Sizes, Proc. 19th Annual Int. Symp. on Computer Architecture (ISCA 92), pp. 415–424, Gold Coast, Queensland, Australia, May 19–21, 1992.

[Turl 88] J. L. Turley, *Advanced 80386 Programming Techniques,* Osborne/McGraw-Hill, Berkeley, CA, 1988.

[Uffe 91] J. Uffenbeck, *Microcomputers and Microprocessors,* 2nd ed., Prentice-Hall, Englewood Cliffs, NJ, 1991.

[Wake 89] J. F. Wakerly, *Microcomputer Architecture and Programming: The 68000 Family,* Wiley, New York, 1989.

[Wall 88] D. W. Wall, Register Windows vs. Register Allocation, Proc. Conf. on Programming Language Design and Implementation (SIGPLAN '88), pp. 67–78, Atlanta, GA, June 22–24, 1988.

[Weic 84] R. P. Weicker, Dhrystone: A Synthetic Systems Programming Benchmark, Comm. ACM, Vol. 27, No. 10, pp. 1013–1030, Oct. 1984.

Index

ABOUT THE AUTHOR

Daniel Tabak is a professor of electrical and computer engineering at George Mason University, Fairfax, Virginia. He was born in 1934 in Wilno, Poland (currently called Vilnius, Lithuania). He attained his B.S. EE (1959) and M.S. NS (1963) degrees at the Technion in Israel, and his Ph.D. EE (1968) at the University of Illinois, Urbana. He was associated with General Electric, Wolf R&D (EG&G), Rensselaer Polytechnic Institute Hartford Graduate Center, Ben Gurion University (Israel), and the Boston University before joining GMU in 1985. Throughout his career he has conducted research and published extensively in the areas of computer-based applications, process control, parallel processing, computer architecture, and microprocessors. He is the author of *RISC Architecture* (translated into Italian), *RISC Systems* (translated into Japanese), *Multiprocessors, Advanced Microprocessors,* and coauthor of *Optimal Control by Mathematical Programming* (translated into Russian), *Engineering Applications of Stochastic Processes,* and *Microcontrollers.* Dr. Tabak is on the editorial board of the international journals *Engineering Applications of AI, Journal of Microcomputer Applications, Microprocessors* and *Microsystems,* and *Control Engineering Practice.* Dr. Tabak is a senior member of IEEE and a member of ACM and Euromicro. He is also a director of Euromicro, representing the United States.